Introduction to Industrial Automation

Stamatios Manesis
George Nikolakopoulos

CRC Press
Taylor & Francis Group
Boca Raton London New York

CRC Press is an imprint of the
Taylor & Francis Group, an **informa** business

CRC Press
Taylor & Francis Group
6000 Broken Sound Parkway NW, Suite 300
Boca Raton, FL 33487-2742

First issued in paperback 2020

© 2018 by Taylor & Francis Group, LLC
CRC Press is an imprint of Taylor & Francis Group, an Informa business

No claim to original U.S. Government works

ISBN-13: 978-0-367-57183-2 (pbk)
ISBN-13: 978-1-4987-0540-0 (hbk)

Visit the Taylor & Francis Web site at
http://www.taylorandfrancis.com

and the CRC Press Web site at
http://www.crcpress.com

To my wife, Lena.

Stamatios Manesis

To my father, Nestoras.

George Nikolakopoulos

Contents

Preface

The book that you are holding is the result of many efforts in teaching the concept of industrial automation to young and promising engineers during the last few decades in academia. The authors aimed to create a system so that engineers will have a full and in-depth overview of the industrial automation field, with a strong connection to real-life applications and that will provide a constant inspiration for the problems that are commonly found in an industrial environment. In the adopted teaching approach, one of the fundamental learning outcomes was to enable independent and outside-the-box creative thinking for automation engineers, in order to produce functional solutions to difficult problems. In all these years, the authors have identified that there has not been a book in the existing literature that provides all the necessary learning directions for students and with a full focus on real-life demands, a book that could prepare them immediately and with the fundamental deep knowledge to deal with the field of industrial automation. Thus, the book that you are holding now is the outcome of a writing and integration process that lasted more than three years, a book that we hope to be a constant reference in the field of industrial automation.

Industrial automation is a multidisciplinary subject that requires knowledge and expertise from various engineering sectors, such as electrical, electronics, chemical, mechanical, communications, process, and software engineering. Nowadays, the application of industrial automation has been transformed into an ubiquitous infrastructure that automates and improves everyday life. Characteristic examples of industrial automation systems can be found in the car industry, the aviation sector, the marine industry, healthcare industry, rail transportation, electrical power production and distribution, the pulp and paper industry, and numerous other applications. Our society has become so dependent on automation, that it is difficult to imagine what life would be like without automation engineering. With the current developments in the field of Industry 4.0, industrial automation is unified with the concept of the Internet of things, embedded systems, and Cyber-Physical systems, in order to create an integrated ecosystem with the vision to enable pure automation for all aspects of life, a future in which everything will be connected, integrated, and automated.

Nowadays, the concept of automation in an industrial production process is a very attractive subject for electrical engineers, because it perfectly combines all the principles and methods of classic automatic control with microcomputer or microprocessor technology. The introduction of microcomputer technology in the field of industrial production, coupled with development in the area of robotics, resulted in the creation of a special scientific field known as "automation and robotics". Although this field appears to be separated from "automatic control", this is mainly due to its great extent and not to the differentiation of its theoretical principles. The rapid development of automation and robotics in today's technological world has made it necessary for some time to introduce industrial automation and robotics courses in the curriculum of various electrical and computer engineering departments.

Each industrial production process consists of a series of simple or complex machines through which the raw material undergoes a sequential treatment in order to achieve the production of an end product, while satisfying the goal of increased production, improving product quality, lowering costs, and increasing production flexibility. Initially, industrial automation systems were implemented in a conventional way, i.e., with independent specific devices (timers, counters, auxiliary relays, etc.), and wiring them according to the desired operating mode. Today, the implementation of an industrial automation system takes place in specific digital devices called programmable logic controllers (PLCs). The key feature of PLC technology is the need for programming (versus wiring) the control logic of the industrial system. The multiple advantages of PLCs have made them the leading tool for controlling an industrial system, but also for other non-industrial systems that we encounter in our everyday lives, such as traffic control for a street intersection, a tower elevator, an automated car wash, etc.

The first part of the book, including Chapters 1, 2, 3, 4, 5 and 6, presents the basic devices, sensors, and actuators, in which the aid of an automation system is implemented, as well as the heuristic composition of automation arrangements and the methodical design of automation circuits. Also, the basic elements of electro-pneumatic technology are introduced. The second part of the book, including Chapters 6, 7, 8, 9 and 10, presents the extended subject of PLCs, whose hardware and software are both described in detail. The basic concepts of industrial networks for networking PLCs or other controllers and the applied use of the Proportional-Integral-Derivative (PID) control law in the industry are also presented. Appendix A presents briefly the arithmetic systems required for PLC programming, while Appendix B introduces the subject of analog input and output values scaling. This book was written based on the assumption that the reader has no prior experience in industrial automation systems, but should be easy to understand if they have any background knowledge related to motor control, digital logic, and digital electronics.

The projected learning outcomes after reading this book will be to understand the functionality of the basic elements of automated systems and the fundamental principles of operation; analyze real-life problems from an industrial automation perspective and understand what is effective and what is not, based on engineering and cost-oriented thinking; identify and select proper sensory and actuation equipment for synthesizing and integrating industrial automation tasks; integrate and synthesize a classical relay-based industrial automation; integrate, synthesize, and program a PLC based on industrial automation; and gain fundamental knowledge in the field of electro-pneumatic automation, industrial networks, and PID control in the industry.

Authors

Stamatios Manesis earned his PhD from the School of Engineering, University of Patras, Rio, Greece, in 1986. He is a professor of industrial automation with the Division of Systems and Control, Electrical and Computer Engineering Department, University of Patras. In 1998–1999, he was with the Industrial Control Centre, Strathclyde University, Glasgow, U.K. In 2008, he was an academic visitor with Eidgenössische Technische Hochschule (ETH) Zurich, Zurich, Switzerland. He has designed various industrial automation systems for Hellenic industries. He has published more than 100 conference and journal papers and has written 5 textbooks. His main research interests include industrial control, industrial automation, industrial networks, expert fuzzy control systems, intelligent controllers, and supervisory control and data acquisition (SCADA) systems. His research has been funded by several national projects (Program for Development of Industrial Enterprises [PDIE], National Program for Energy [NPE], and Karatheodori program). He has participated in various European Union projects such as Science and Technology for Regional Innovation and Development in Europe (STRIDE), European Strategic Program on Research in Information Technology (ESPRIT), and European Social Fund (ESF).

George Nikolakopoulos is a professor of robotics and automation in the Department of Computer Science, Electrical and Space Engineering at Luleå University of Technology, Luleå, Sweden. His work focuses in the area of robotics, control applications, and cyberphysical systems, while he has extensive experience in creating and managing European and national research projects. Previously, he has worked as a project manager and principal investigator in several R&D&I projects funded by the European Union (EU), European Space Agency (ESA), and Swedish and the Greek National Ministry of Research, on projects such as: (a) EU-funded projects: AEROWORKS, DISIRE, Compinnova, FLEXA (IP), C@R (IP), NANOMA (STREP), SYMBIOSIS-EU (STREP), CONFIDENCE (STREP), PROMOVEO (STREP), and CommRob (STREP); (b) ARTEMIS-funded projects: R5-COP; and (c) Swedish-funded projects: Mine Patrolling Rovers, Pneumatic Muscle Dancer, etc. In 2014, he was the coordinator of the H2020-ICT AEROWORKS project in the field of collaborative unmanned aerial vehicles (UAVs) and the H2020-SPIRE project DISIRE in the field of integrated process control with a total budget of 12M Euros. In 2013, he has received a grant of 220K Euros for establishing the largest outdoor motion capture system in Sweden and most probably in Europe as part of the FROST Field Robotics Lab at Luleå University of Technology. In 2014, Prof. Nikolakopoulos was nominated as Luleå University of Technology (LTU's) Wallenberg candidate, one of three nominees from the university and 16 in total engineering nominees in Sweden. In 2003, he received the Information Societies Technologies (IST) Prize Award for best paper that promotes the scope of the European IST (currently known as Information and Communication Technologies [ICT]) sector. Prof. Nikolakopoulos' publications in the field of UAVs have received top recognition from the related scientific community, and these publications have been listed several times in the TOP 25 most popular publications in control engineering practice from Elsevier. In 2014, he has received the 2014 Premium Award for Best Paper in *IET Control Theory and Applications*, Elsevier for research work in the area of UAVs. This premium award recognizes the best research papers published during the past two years in this journal. Finally, his published scientific work includes more than 150 published international journals and conferences in the fields of his interest.

Chapter 1

Industrial Automation

1.1 The Industrial Control System

Every industrial production process consists of a series of simple or complicated machines that, through the combination of raw materials, undergo a sequential transformation and integration in order to produce a final product. The term "machine" denotes every kind of electromechanical device on the industrial floor, e.g., from a simple motor (such as a drilling or a cutting machine) up to a complicated chemical machine (e.g., a chemical combustion machine). The whole set of machines (namely non-homogeneous machines), which are being integrated and combined in an industrial production process, will be referred to as an "integrated machine".

As an example of an integrated machine, Figure 1.1 depicts the typical production line of an integrated paper machine, where the initial raw pulp is undergoing the sequential processes of pretreatment and grinding, refining, pulp bleaching, and pulp pressing and drying, until it is transformed into the final paper of predefined quality. Figure 1.2 shows the various stages of the papermaking process. During the pretreatment and grinding, in the first stage of the papermaking process, debarked and washed wood logs are preheated in order to become easier to grind and are inserted into large wood log grinders, which produce wood chips. Refining is the second stage of the paper manufacturing process, when the quality of the final product is highly dependent on that specific subprocess. During that stage, the wood chips are being received and transformed into pulp via high energy consumption, water infusion, and addition of chemical compounds. During the next stage of pulp bleaching, the pulp produced by the refining system is fed to the machine that is responsible for the discoloration of the mixture. Bleaching is a chemical process applied to cellulosic materials in order to increase their brightness. The last stage of the paper manufacturing process is the drying and pressing process. During this stage, bleached pulp is dried and pressed in order to form the desired production paper.

In the case of an integrated machine, the whole sequence of operations for all the involved machines, the exact transformations and integrations of the raw materials, as well as the overall operational requirements, are *a priori* detailed and clearly defined for the industrial automation engineer, who is in charge of designing and implementing the desired process automation. For a specific production line, the sequence of operations and transformations,

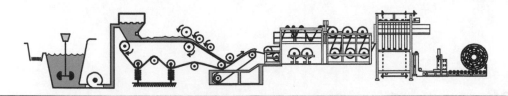

Figure 1.1   Schematic of a pulp and paper industrial process.

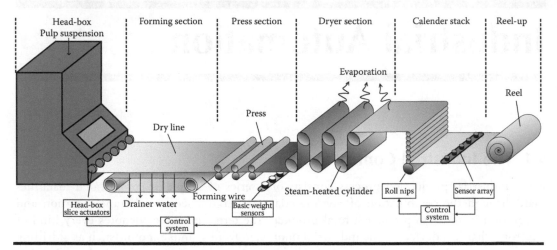

Figure 1.2   Simplified visualization of the pulp processing stages.

applied to the products, are generated from the production process itself and it is not possible, due to simplifications, these process stages to be altered. For example, in the case of an integrated machine of producing biscuits, it has been already defined from the process of production (the total manual and human-based processes) that in the mixture chamber, first the milk should be inserted and in a certain quantity, while in the sequence, the flour should be inserted at specific feeding rates and quantities. In this example, it is not possible, in order to simplify the overall automation process, to override this procedure by either designing an automation system that will either inverse the previous sequence of operations (e.g., first the flour will be inserted and then the milk) or completely ignore the predescribed sequence by allowing both materials to be inserted at the same time in the mixture chamber. Overall, and for all the produced industrial automations, the automated procedure should always satisfy the rules and sequences of the manual produced product, independently of the related complexity in the automation solution.

From the beginning of the industrial era, the main aim of every production process was the achievement of a higher possible level of automation. Reducing the number of personnel involved was also a common aim from the factory owners, as well as the machine suppliers, in order to follow the significantly increasing trend in automation, while the overall development of the industrial processes has taken place in parallel with the technological breakthroughs at the beginning of the nineteenth century and especially when human labor began to be replaced by machines. The transition of the production model, from the initially multi-interrupted form to the continuous form, required the development of specific methodologies and tools that would allow the central coordination of all the various procedures with minimal human intervention in the overall process.

Before analyzing the procedures needed for automating an industrial production line, it is of paramount importance to initially define in detail the various components that the automation and their specific functionalities and properties consist of. In an industrial production line, the "movement" is the fundamental and generalized characteristic of the overall process, since it is impossible to consider an industrial process without the existence of a linear, circular, or any other form of movement. Even in the case of a chemical reaction, where the existence of motion is not obvious, the movement also exists in this case and more specifically in the form of an electrovalve control, which opens in order to supply the reactor with the necessary amount of the reacted components. Furthermore, the existence of the need for movement is significantly evident, either in the cases where the product should be transferred to the various process points of the production line, or in the cases of integrated machines, where parts of the machines should be moved in order to produce the desired processing of the developing product. The machines that can be utilized for the creation of the movement can be categorized into two large categories, as displayed in Figure 1.3.

The first category includes the different types of motors, independently of the operating principle (e.g., one-phase motors, three-phase motors, motors with short-circuit rotors, motors with direct start, motors that start in a Y/Δ mode, etc.) that creates a primary rotational movement, which can be further transformed by the utilization of appropriate mechanisms in a linear or other type of movements. The second category includes all the actuators, where a linear movement is created as the result of the attraction generated by an electromagnet (coil) on a ferromagnetic core, such as the various forms of electro-vanes, electrovalves, etc. The common characteristic of motors and actuators is the fact that they have only two possible states of operation. For expressing these states,

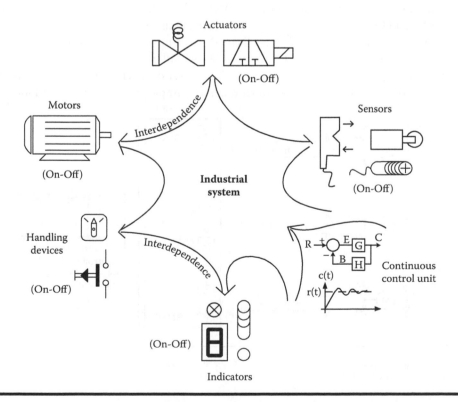

Figure 1.3   Basic kinds of industrial-type equipment composing an "industrial system".

usually we refer to them as "the motor is in operation", "the motor is not operating", "the electrovalve is energized", "the electrovalve is not energized", "the coil is under voltage", and "the coil is not under voltage". In general, there are two states of operation that can be defined as the ON and the OFF operation, which can be further associated directly with the digital logic symbols of 1 and 0.

If one motor has, for example, two rotation directions or two rotational speeds and thus two states of operation, ON1 and ON2, then this consideration is not in conflict with the previous association. Actually, it can be considered as the case of having two motors, where one motor has the two states OFF-ON1 and the other one has the states OFF-ON2. The operation of the two motors, and more specifically the supply of the motors with the required electrical power, is achieved by the power relays that also have two states of operation, the ON and the OFF state. The control of the motors is achieved through the proper control of the relays, and thus the desired control system is applied on the corresponding relays controlling the electrical supply to the motors and it is not applied directly on the motors.

After the definition of the control action being applied directly on the relays, the previous situation with the existence of multiple ON states for a machine will be explained more, through the following example. A three-phase motor is being considered with two directions of rotation. For the operation of the motor, two power relays are needed, which will be denoted by C_1 and C_2, as shown in Figure 1.4. When the C_1 relay is energized (relay C_2 is not energized), the motor's coil ends are connected to the phases R, S, and T of the power network and thus the motor has a certain direction of rotation. When the power relay C_2 is energized (relay C_1 is not energized), the same motor's coil ends are connected to the phases T, S, and R of the power network and thus the motor has the opposite direction of rotation. As has been explained before, due to the fact that the control system is being applied on the power relays, the two states ON1 and ON2, of the same motor, correspond to the states of ON and OFF of two different devices, which are the power relays C_1 and C_2. As a result, the control system, instead of the states OFF-ON1-ON2 of a motor, with two directions of rotation, is being equivalently applied on the OFF-ON states of two different power relays.

To control the operation of an integrated machine, a set of specific operation control devices needs to be incorporated in the overall automation, like a simple push button, a rotational selector

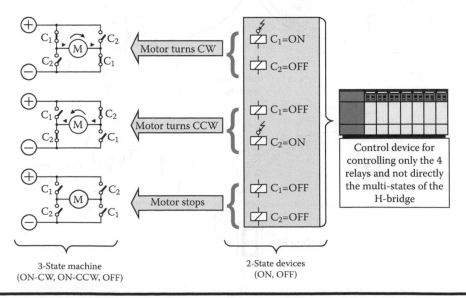

Figure 1.4 Multi-state electric motors are controlled by two state power relays.

switch (knob), etc. In the case that the operation of the integrated machine is set in the "manual" mode, the operator is utilizing the operation devices for turning on the desired motors or the actuators and in the proper sequence. In the case that the integrated machine is set in the "auto" mode, the operator is again utilizing the operation devices, either for initiating the operation mode, or for instructing the integrating machine to change the operational state. As an example, in an integrating machine for chocolate production, the operator is capable, by the press of a button, to order the control system to alter the current recipe production for another one. In this case, the control system should allow the integrated machine to complete the current operation and afterwards, ensuring the prerequisite quantities for executing the ordered recipe change, to command the integrated machine in executing it. In most cases, the automation system of an integrated machine provides both the functionalities of an automatic or manual mode of operation, especially for dealing with the emergency fault situations, where direct manual control of all the provided automatic functionalities of the integrated machine is needed.

The operation control devices have also two states of operation, OFF and ON, similar to the cases of the motors and the actuators. As presented in Figure 1.5, a pressed button is energized and thus is in the ON state, while a non-pressed button is not energized and is in the OFF state. The ON state is independent of the time duration that the button is pressed and of the switching contact type (open or closed) in the not-energized state.

The next category of the machines presented in Figure 1.3 are very commonly installed in industrial environments, and are the indication devices that are utilized for transmitting process information from the integrated machine to the operator. In most cases, the industrial process is widely geographically distributed, and thus the operator that is in charge of the whole process has no direct visual or audio feedback from the process and the overall operation of the multiple integrated machines. However, even in the case that a visual or audio feedback is available, for safety reasons human senses are considered unreliable, and these monitoring, displaying, and visualizing devices are still needed to track the performance and state of operation of the industrial process. Especially in the cases of measuring variables without a direct visual or audio effect (like the variables of pressure, temperature, flow of a liquid in a non-transparent tube, etc.), such monitoring devices are of paramount importance. In most cases, this information is transmitted to the operators through the utilization of light or audio indicators, which can be again considered as devices with an ON and OFF state.

In a large set of integrated machines, specialized sensors for performing specific measuring of quantities are utilized extensively. For example, such sensors can be utilized to sense if there is a flow of a liquid in an opaque tube, if the level of a tank has reached a certain height, if the moving part of a machine has reached the desired place, if the temperature of a reactor has been set to the nominal one, etc. In general, these sensors can be categorized in digital and analog sensors. The digital sensors are characterized by two states of operation, namely ON and OFF or 1 and 0, correspondingly. The analog sensors are able to produce an analog (continuous) measurement of the quantity under study and thus more complicated hardware and software is needed to incorporate the industrial automation for utilizing this information.

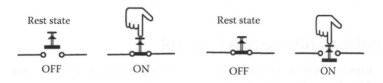

Figure 1.5 Manually operated control devices that have two operational states.

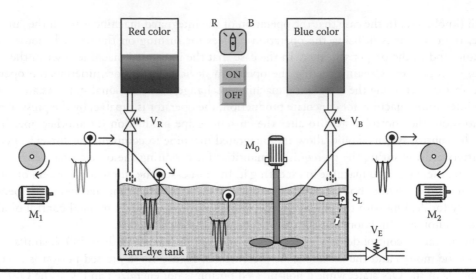

Figure 1.6   Schematic of a simplified batch-dyeing process for textile materials.

The operation of an industrial automation device in an industrial process is not independent of the operation of the rest of the devices in the same process. In all cases, there is a strong dependence among all the utilized devices in the automation, as is also highlighted in the following example. Consider an industrial process of adding color to textile products like fibers or yarns, called "dyeing", where (as shown in Figure 1.6) it is desired to have the following operation:

> … *IF the selector switch is in the "R" position AND by pressing the push button "ON", the electric valve V_R opens AND motors M_1 and M_2 operate AND when the level sensor S_L produces an output signal the electric valve V_R closes AND simultaneously the motor M_0 starts to operate AND IF, due to a fault, the motor M_1 stops, THEN the motor M_2 also stops AND …*

All the previous demands, which are being produced from the type of the industrial process, constitute a set of dependencies that, without their formal satisfaction, the production of the final product is not feasible. The devices from these five categories, which are indicated in Figure 1.3, with their dependencies and the rest of the purely mechanical parts of the integrated machines or processes constitute the overall industrial system. This industrial system needs proper automation in order for the whole operation to be executed with a minimization of human intervention. The control of the devices that are characterized as ON and OFF states is denoted as automation, and is being carried out by the utilization of automation circuits; while the control of the devices that are analog is denoted as process control, and is being carried out by the utilization of automatic control systems.

1.2 Automation and Process Control

Following this introduction to the field of industrial control systems and automation, special attention should be focused on the differences in the fundamental meanings among the concepts of automatic control and industrial automation. Automatic control can be defined as the continuous control of a physical analog variable through the utilization of any kind of actuators, while industrial automation refers to the sequential or digital ON-OFF control of the two-state devices. As has been presented in Figure 1.3, among the discrete devices, a continuous time control device has also been included in the industrial system, in order to present the overall concept that in an industrial control system, multiple continuous time control units can be integrated and act in a cooperative way with the rest of the automation control units.

In the case of industrial control processes (batch processes), there are multiple process variables that, although we would like to have them set at constant values, show random variations, mainly due to multiple external disturbances during the production phases. The reduction and elimination of these variations can be achieved through the proper application of automatic control principles. In many cases, it is also desirable for a process variable to alter the set value from an existing converged one into another operating point, while certain specifications usually are amended to achieve this transition, e.g., a fast or slow transition time, a minimum control effort change, a low overshoot during the alteration of the set point, a fast convergence, etc. This problem can also be addressed by the theory of automatic control and by applying existing theoretical and applied approaches e.g., the theory of Proportional-Integral-Differential (PID) control, which is presented in Chapter 9.

In contrast to the automatic control principles, the theory of industrial automation focuses on physical variables and machines that are in one of two states, e.g., "a liquid flow exists or not", "the pressure has reached the desired value or not", or "the compressed air piston has been extended or not". Moreover, industrial automation refers to devices, machines, and circuits; and, in general, electronic, electromechanical, and electro-pneumatic integrated machines, where their operational principle is described from the Boolean logic and the corresponding sequential interconnections among the production stages. In the automation field, the action of control is restricted by being applied by two state actuators, and therefore the applied control action can only have the specific values of either ON or OFF.

In Figure 1.7, a simple process of controlling the level of a liquid in a tank is presented. In this process, it is assumed that the supply of the liquid in the tank is provided by an uncontrollable variable, while a valve is controlling the liquid's output flow from the tank. In the described setup, it is desired that the level of the tank be kept at a specific height h_0, independent of the liquid supply. To solve this problem, after the initial achievement of the specified height h_0, the output flow should be equal to the input flow. To implement this control law, due to the fact that in this example the input flow is not directly measured, the control scheme should be able to measure the

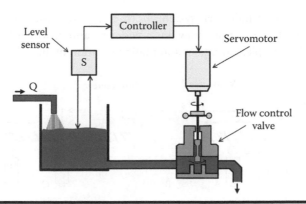

Figure 1.7   Example of an automatic control system.

level of the liquid in the tank by a level sensor and, subsequently, appropriately tune the outflow valve. In this case, the outflow valve has not only two states "Fully Open—Q_{max} flow" and "Fully Closed—0 flow", but it can take any kind of desired state value, thus allowing for a flow within the $(0, Q_{max})$ continuous space. In the era of classical industrial automation, this control scheme would have been implemented by analog circuits, whereas now it is commonly implemented by the utilization of computers and, more specifically, programmable logic controllers (PLCs), which are computational devices designed and configured for operating in industrial environments.

In Figure 1.8, a lead screw setup is presented, where the worktable can be translated by the proper connection and rotation of the lead screw into two directions (left and right). For this reason, the motor generating the rotation of the lead screw has two directions of rotation. Moreover, the motor has two rotation speeds, which means that the worktable can be translated in two speed profiles. With the help of the indicated position sensors (limit switches) and the provided motor, we can design an industrial automation with the following desired operation. As shown in Figure 1.8, the worktable should be continuously moving between the final positions A and D, while in the translation space from B to C the table should move at the fast speed and at the low speed for the remaining ones. In this setup, there is no continuous variable that needs to be controlled, while all the devices have two states of operation, ON and OFF. In this case, the controller is an automation circuit, which is responsible for implementing the described sequential (or Boolean) control logic, e.g., partially described in the following form:

> ....*When sensor B is energized, the motor should be set (start operating) and remain in the fast speed. When sensor C is energized, the motor should be set and remain in the slow speed. When the sensor D is energized the motor's direction of rotation should be inversed, without changing the speed of translation....*

At this point, it should be mentioned that the aforementioned translational system is not described by a specific transfer function, as in the case of automatic control systems, but from a Boolean function that expresses the desired operational logic.

Many books in the field of automatic control refer to the sequential control of two states as the fundamental form of industrial control, while providing minimum reference to this topic and concentrating on the analog and continuous time closed-loop control (feedback control). On the contrary, this book will focus on the methods needed for designing and implementing industrial automation systems, which cover a significantly larger set of the current trends in the area of industrial control systems.

After defining that, in an industrial system, both continuous and sequential control setups exist, the term "industrial automation" now has a wider meaning, which includes every kind of system being designed for implementing an automatic operation of an industrial process.

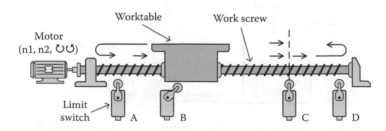

Figure 1.8   Example of an automation system.

1.3 Purpose of Industrial Automation

The industrial era was initiated by the efforts to automate existing industrial setups as a way to improve the quality of the produced products and the overall production volumes. Contrary to what is generally understood, industrial automation is not a discovery of the recent past, but it is rather as old as industry itself. From the beginning, the designer of an industrial production system has attempted to achieve an operation as autonomous as possible, always based on the available instrumental tools. The initial industrial production processes have based their operation on the workers' eyes, hands, and brain, as alternatives to contemporary sensors, actuators, and computational units. All of the current automated operations of industrial processes are based on these three factors. Through the sensors, the necessary signals and measurements are being gathered from the controlled process, as presented in Figure 1.9. Subsequently, this information is being analyzed by the control logic, running in a computational unit and, in the final step, the control actions are interacting with the controlled process, through proper control of the provided actuators.

In the beginning, efforts to automate industrial processes were focused on the replacement of human labor by independent machines, each one being able to accomplish a specific task and in a limited surrounding space, under the premise that you feed the machine with the raw material, and as an output you are receiving the complete product. The automatic operation of those machines were initially independent of each other, and thus there was a constant need for a human-centric coordination of these machines. Subsequently, through the evolution of multiple, related technologies and through the developments in the field of analog and discrete control in the era of microprocessors and PLCs, a transition took place from having small-scale, centralized, industrial automation, to a decentralized and large-scale one, fully controlled by numerous distributed PLCs, able to synchronize multiple industrial processes. The decentralization of industrial automation took place through the introduction of industrial networks, which are discussed in Chapter 8.

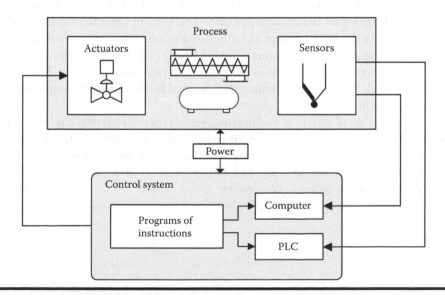

Figure 1.9   Basic elements of an automated system.

A fundamental motivation for the automation of an industrial process is that the viability of the enterprise, and particularly business profit, can be achieved through the following objectives:

- Production increase
- Cost reduction, mainly due to the reduction of human-related cost
- Improvement of the product's quality
- Improvement of the raw material utilization (reduced loss in materials)
- Reduction of energy consumption

Many other secondary benefits may be derived from the automation of an industrial process, i.e., plant safety, environmental pollution reduction, etc. The aims of industrial automation are frequently difficult to achieve for several reasons, such as inherent limitations of the plant, implementation costs and general situations in the marketplace. Regardless of these difficulties, there has been continual development of industrial automation from a control tools and methods point of view. Consequently, advancements in automation and control made possible the development of larger and more complex processes of various kinds (e.g., oil and gas refining, chemical, pharmaceutical, food and beverage, water and wastewater, pulp and paper, mining, iron and steel, cement, etc.), thus bringing numerous new technological and economic benefits.

1.4 Industrial Automation Circuits

In Section 1.1 it has been mentioned that the automation of an industrial system is being achieved through the utilization of automation circuits. These automation circuits will be analyzed in detail in Chapters 3 and 4, where their design principles and operations will be presented and through the utilization of numerous realistic examples. In Chapter 7, which focuses on PLCs, the main effort will be in transforming these automation circuits into proper software programs for PLCs. The design of the automation circuits, from the perspective of a functional design, consists of fundamental knowledge for all industrial automation engineers in order to understand the industrial operation; identify the needs; design and simulate the solution to the industrial automation problem; and, of course, produce the optimal solution from a cost-related perspective, either in the form of classical industrial automation or in the form of software for PLCs. In Chapter 1, we will refer only to the scope of automation circuits, their general form, and their alteration from other types of industrial circuits.

In the broad discipline of electrical engineering, many types of electrical circuits are involved, such as, for example, the electrical circuits of basic electrical components (R, L, and C), electronic circuits, power circuits, telephone circuits, integrated circuits, and many more. However, the electrical circuits being utilized for the study and implementation of an industrial manufacturing plant can be divided into the following categories:

- Power circuits
- Automation circuits
- Wiring diagrams

Power circuits (also called "main circuits") indicate the type of power supply for the utilized motors and all other related power devices. As an example, Figure 1.10 depicts: (a) the single-line,

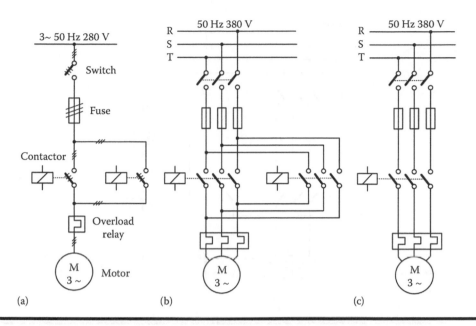

Figure 1.10   Examples of power circuits: (a) Single-line, three-phase circuit design, (b) complete multi-line design of circuit, and (c) circuit design of a direct starting motor.

three-phase circuit of a motor with two directions of rotation; (b) the complete multi-line circuit of the same motor; and (c) the power circuit of a direct starting motor. More details on the different types of motor starting, rotation inversion, etc., will be presented in Chapter 3.

Automation circuits (which can also be referred to as control circuits, auxiliary circuits, secondary circuits, or schematic circuits) represent the operational logic and control of the power devices, as indicated in Figure 1.11, for a start/stop operation of the previously depicted motor in Figure 1.10c.

Wiring diagrams are circuits representing both the power circuit and the automation circuit while, at the same time, representing the actual positioning of all the devices and components in the industrial installation, which is ideal information for the technician executing the wiring and overall installation. In Figure 1.12, the wiring diagram which is produced from the synthesis of the power circuit shown in Figure 1.10c, and the automation circuit shown in Figure 1.11, is displayed.

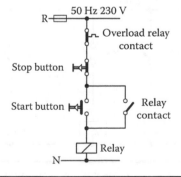

Figure 1.11   Example of a simple automation circuit.

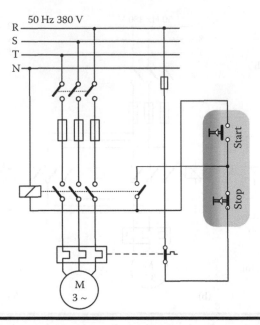

Figure 1.12 The actual wiring diagram of a direct start-up motor, including both power and automation circuits.

It should be noted that in the wiring diagram of Figure 1.12, the wires can intersect among each other and thus it is very difficult to follow the route of each wire and the overall functionality of the design, even in this very simple case, where we have only four devices (one relay, two buttons, and one overload protection). Having this in mind, it can be easily generalized how this complexity will grow in cases with more intersections, e.g., in wirings with 50 devices. In contrast to this, in the automation circuit of Figure 1.11, there are no wiring intersections and thus it is very easy to follow and understand the overall operation logic. For this reason, automation circuits provide an overview of the automation functionalities are most commonly utilized during the development, installation, and operation of an automation system.

Automation circuits are being developed in branches (sectors), which are presented in Figure 1.13. Each branch denotes the operational function of a corresponding relay, solenoid,

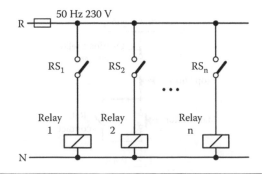

Figure 1.13 The simplest form of an automation circuit which is equivalent to a full manual operation.

or actuator, while the whole automation circuit denotes the operational logic of the overall industrial process automation. Each branch in the automation circuit can have multiple parallel sub-branches, depending on the complexity of the logical function being implemented. In Figure 1.13, the indicating branches have the simplest form. Each one implements the logic "If the rotary switch RS_i is closed, then the corresponding i motor will be in operation". An industrial system with an automation circuit of the form presented in Figure 1.13 has a full manual operation. In reality, this is not common, since the start and stop operations of an industrial process with similar machines are executed in an automated manner, based on the sequence of the sensing signals and the corresponding status of the machines. In these cases, which are dominant in industrial automation, automation circuits are becoming more complex and thus a proper methodology for designing such automation circuits is needed. Nowadays, industrial automation circuits have been transformed, as has been mentioned before, into a set of software programs for PLCs. However, it is of paramount importance to note that although the final implementation of the industrial circuits has been changed from a hardwire approach to a soft approach, the need for understanding and designing the electrical drawings for solving an automation problem cannot be replaced, except for cases where the focus is on very simple and small automation problems. Chapter 7 will present that the first step in writing the PLC program is to solve the problem, based on the methodologies that will be discussed subsequently, independent of the selected PLC or software language for the program implementation.

At this point, it should be mentioned that it was the authors' aim, when writing this book, to present all the necessary steps to the interested engineers or automation students for understanding the concepts of an industrial automation, mastering the procedures and the methodologies for developing the automation circuits, mastering the design methodologies for more complicated automations based on state machines, and understanding and mastering the principles of PLC programming and PLC networking. However, it should be highlighted at this stage that an industrial automation engineer is not a software programmer of a PLC, which is a mistake usually reproduced by lots of books in the field. The implementation of a fully functional and optimal industrial automation, as will be described subsequently, involves the understanding of fundamental principles in the area of sensors, actuators, electrical wiring, electrical machines, electrical circuits, process control, programming, and networking. However, conversely, a good programmer is not an industrial automation engineer, and thus programming of PLCs is just a small subset of the capabilities found in an experienced and professional automation engineer.

1.5 Computer-Based Industrial Control and Automation

The task of controlling an industrial process has evolved a lot over recent years, starting from a complete manual operation, continuing in the analog control and low-level automation era, and recently reaching a totally computer-based control and automation approach. Prior to the introduction of solid-state electronics, the designer of an industrial production process was attempting to make the automation operate as automatically as possible, based on the various instrumental tools. To enable the vision of a full automation technology and after the appearance of various digital processors, a rapid increase in process control computers and minicomputers took place, especially in small plants, which changed radically the situation in the field of industrial process control and automation.

Nowadays, an industrial control and automation system, from a hardware point of view, is a general term that encompasses several types of digital devices, such as industrial personal computers (I-PCs), programmable logic controllers (PLCs), programmable automation controllers (PACs), embedded PLCs, and other specific digital controllers. Furthermore, the larger control and automation system configurations include software and hardware platforms, such as supervisory control and data acquisition (SCADA) subsystems, distributed control subsystems (DCS), and industrial communication subsystems. At this point, it should be highlighted that the utilization of all of the aforementioned technologies in the industrial sector is of critical importance in order to achieve the desired performance and quality, while a proper mixture of all these computer-based solutions should always be considered.

After the introduction of the first powerful personal computers and PLCs, automation engineers have been divided into two groups. The first group was in favor of utilizing the PCs, equipped with proper input/output (I/O) hardware in order to accomplish a proper industrial automation functionality; while the second group rejected the PCs as inappropriate computational devices in an industrial environment, while promoting PLCs for the same purpose. However, these two categories have specific characteristics with certain advantages and disadvantages. PCs provide the user with the ability to utilize various software sets, spanning from simplified to extremely complicated software applications for implementing advanced control laws and industrial automations, providing extended graphical user interfaces and advanced interaction capabilities, increased computational power, and, in general, a simpler and more flexible programming environment for the user. From the other side, PCs are generally not suitable for a pure industrial environment. Even if the PCs can be equipped with the proper I/O hardware, they have the general disadvantage of not having been designed for installation in rugged industrial environments, and thus are characterized with generally reduced operational stability and durability. In contrast, PLCs have been designed specifically for industrial control and automation applications, are characterized by a high operational durability, and are equipped with a reconfigurable digital and analog I/O hardware that could be specifically tuned to the needs of the current application. Finally, PLCs provide fully optimized software for the exact needs of the industrial automation and process control, and nowadays this technology is considered as a standard solution in the industry. It could be left unattended and in continuous full operation for decades without operational errors or faults. For these reasons, PLCs are considered as the first choice of automation engineers, especially when compared with the classical PCs targeting a more home-based operation. From another point of view, PLCs are unable to support advanced control algorithms, are dedicated platforms for developing automation algorithms and have no support for other types of software. As a disadvantage, PLCs don't have a universal, standardized, and widely accepted way of communication with other types of devices from other vendors, thus restricting automation engineers in integrating products from specific vendors. Furthermore, after the finalization of the automation programming (the hardware connection to the I/O field devices and the initialization of the run state), a PLC operates as a "black box", without the ability to provide to the user any kind of online information, except for elementary information via optical light-emitting diodes, which indicate only the states of the digital I/Os. Regardless of these disadvantages, the PLC is still a very effective solution for general-purpose industrial control and digital I/O automation, mainly because of its reliability and transparent scope for which it has been developed.

The natural and acceptable competition among PLC vendors and the aforementioned industrial engineering groups, and the prevailing analogous situation in the marketplace of industrial controllers, were the reasons for various vendors to develop ways to remove boundaries between these two hardware technologies and add advanced functionalities, one of which has been the "industrial PCs". During the last few years, industrial PCs have been significantly expanded and improved in order to cover the existing gap between PCs and PLCs, but this category still has not replaced PLCs, nor has it been widely accepted and installed to a large extent. Additionally, industrial PCs have introduced multiple integration issues to engineers, due to the included multi-vendor hardware and software and the missing compatibility across different platforms.

The vendors of industrial automation systems for supporting the increased demands of the current industrial applications have developed industrial automation devices that could combine the advantages of PLCs for classical control and automation of a complex machine or of a process, with the advantages of the PC-based systems that provide the user with significantly high flexibility in configuring and integrating them into the industrial enterprise. Such a digital device has been established in the industrial world with the term programmable automation controller (PAC). A PAC is generally a multifunctional industrial controller, which can simultaneously monitor and control digital, analog, and serial I/O signals from multiple sources based on a single platform, while supporting multiple, built-in communication protocols and data acquisition capabilities. Although PACs represent the latest proposal in the programmable controllers' world at this time, the authors are not able to predict the future and the overall applicability of this technology. However, it is commonly agreed that PACs are an efficient and promising solution for complex industrial control and automation applications. In Figure 1.14, an overview of the available fundamental computational components for the implementation of industrial automation and control systems is presented.

In parallel with the developments in computational power in the control and automation devices, their ability to communicate, interact, and exchange information has also been developed in recent decades, thus leading to the introduction of industrial networks. Starting from a small number of industrial networks, and being introduced by three or four large vendors of industrial

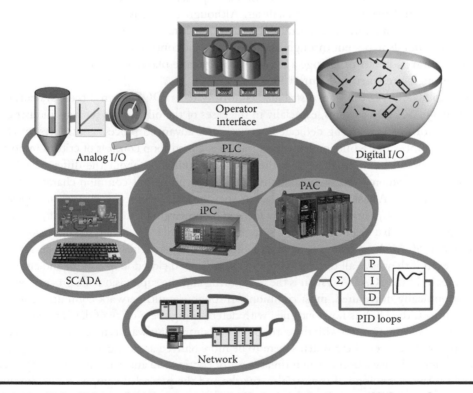

Figure 1.14 PLCs, iPCs, and PACs support control, communication, and other tasks.

automation equipment, there exist more than 20 industrial networks today, addressing all levels of industrial production, which are examined in Chapter 8. Industrial networks differ quite significantly from traditional enterprise networks due to their specific operational requirements. More specifically, industrial networking concerns the implementation of communication protocols between field equipment, digital controllers, various software suites, external systems, and graphical user interfaces. In general, by allowing the connection of digital industrial controllers, the industrial network offers mainly the possibility of sensing messages and control commands through a decentralized approach, which can be geographically spanned. Thus, today the controller of a specific production process could sample the information from another part of the factory automation or control the operation of a machine in a remote part of the industrial field. Since this concept can be fully generalized on the full automation floor, the ability to control the whole industrial process and to have a complete overview of the ongoing sub-processes has been made more achievable than ever before, and thus the concept of supervisory control and data acquisition (SCADA) has been introduced. Furthermore, today's industrial networks can interconnect industrial controllers from different producers, converging in a similar way as the well-known "open communications" demand.

The SCADA concept has been introduced from the real need to gather data and supervision-like control subsystems on a large industrial process plant in real time. Regardless of its initial definition, the term SCADA today represents a combined hardware and software system, including the remote field devices, the network, the central station equipment and the software platform. This software platform, in the case of SCADA, offers the user all the functionality required to receive or send data, represent data graphically, manage alarm signals, perform statistic calculations, communicate with other databases or software applications, schedule control actions, print various reports, and many other user facilities. Although the focus of SCADA systems is data acquisition and presentation on a centralized human machine interface (HMI), it also allows for high-level commands to be sent through the network to the control hardware, for example, for the command to start a motor or change a set point in a remote place. A characteristic example of a SCADA system is presented in Figure 1.15.

Similar to SCADA systems are distributed control systems (DCS), even if these systems existed before the era of SCADA systems, especially in the cases of the oil and gas refiners' industries. The DCS system consists of a strong dedicated network and advanced process controllers, often with very powerful processors, while implementing multiple, closed-loop controls of critical importance. In general, it should be highlighted that there is some confusion about the differences between these two types of automation systems, mainly due to the numerous common characteristics that these systems possess. A basic difference is the fact that the DCS is process-oriented, as opposed to a general-purpose software suite, and generally focused on presenting a steady stream of process information. This means that although the two systems appear similar, their internal operations may be quite different. SCADA systems, on the other hand, does not have the control of processes as a primary role, even if they have all the capabilities to apply limited closed-loop control and automation. The main focus of the SCADA system is the monitoring and the supervision of a process, which has been geographically distributed, most commonly through a multi-network communication structure. In contrast, the DCS is not concerned with determining the quality of data and visualization approaches, as communication with the corresponding control hardware is much more reliable. Even if the boundaries between these systems seem to be more blurred as time goes by, the computer and network technologies have become an intimate part of control and automation system engineering.

Based on the technology of industrial networks and the powerful computational automation units, the optimal implementation of the concept of computer integrated manufacturing (CIM)

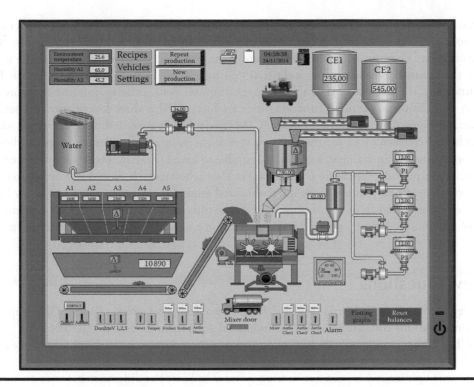

Figure 1.15 Typical example of a SCADA screen for process visualization.

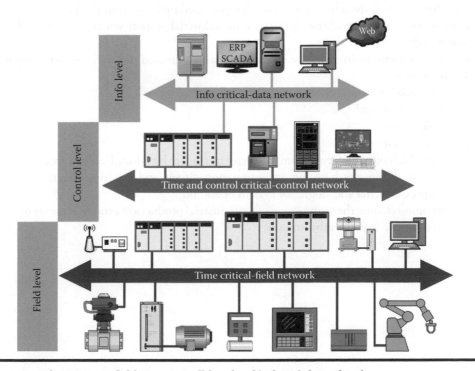

Figure 1.16 The CIM model integrates all levels of industrial production.

has been achieved, a concept that was initially introduced in the early 1970s. With respect to the CIM model, an industrial process can be organized in a three-layer hierarchical structure, where the lowest layer is comprised of sensors, actuators, and embedded micro-controllers. The middle layer is the control layer where the industrial controllers, industrial PCs, and industrial PACs are connected. The highest layer is the management level, where the mainframe computers for SCADA and resources planning functionalities are located. This three-layer generic structure is presented in Figure 1.16. Every level of the CIM model has its own dedicated network, with its own technical characteristics for the network speed, the number of nodes, etc., due to different operational goals. As an example, it could be desirable to have a real-time control loop in the lowest field level and a periodic supervisor control loop in the highest level. It should also be apparent that the networks in the various levels are interconnected among themselves in order to allow for the transfer of information from the bottom layer to the top ones, and vice versa. Consequently, the target of CIM in the industry process is the integration and utilization of the overall information.

Review Questions

1.1. Define the term "industrial system" from the automation point of view. Give important considerations concerning either the industrial equipment or its characteristics.

1.2. What is the "dominant variable" of an industrial production procedure? Give some examples of the dominant variable in concrete industrial applications.

1.3. In your opinion, what is an automation system? Indicate the difference between an automation system and an automatic control system.

1.4. What is the purpose of using sensors in an industrial system? Explain the difference between the information derived by sensors and that provided by indication devices.

1.5. What is the role of handling devices in an industrial system when this system has been automated?

1.6. The control terms utilized in the operation of an automated industrial system may include the following:
    a. Sequential control
    b. ON-OFF control
    c. Logic control
    d. Digital control
    Indicate which of these terms primarily refer to control based on Boolean theory.

1.7. Explain the differences between an automation circuit and a wiring diagram. Which of them expresses better the "logic" of system operation?

1.8. In your opinion, does the automation of an industrial production process increase or decrease unemployment? Explain substantially.

Chapter 2

Hardware Components for Automation and Process Control

2.1 Actuators

An actuator is a device that uses some type of energy and produces the required force, either providing motion to an object or actuating something. Actuators (independently of their shape, form, and size) produce the mechanical movements required in any physical process in a factory. It should be highlighted that in any industrial production line, if the actuators are removed, what will remain are only the "passive" metallic and plastic components, while the whole automation will lose its ability to alter or produce something meaningful. All actuators are controllable devices for performing the desired manufacturing operations, in order to have a well-controlled and automated production process. In general, there are various kinds of actuators that can be categorized based on the operation principle, such as thermal, electric, hydraulic, pneumatic, and micro-electro-mechanical (MEMS) ones. Figure 2.1 illustrates a number of different types of actuators.

Thermal actuators convert the thermal energy into movement based on various physical principles. As an example, a bimetallic strip is one type of thermal actuator made from two different metals, such as steel and copper. In this specific case, it is known that the two metals have different temperature coefficients, and when they are heated, their expansion occurs at different rates. Therefore, two similar strips from different metals, jointed together along their entire length, may generate motion at their free end. When heat is applied, the bimetallic strip bends in the direction of the metal with the smallest thermal coefficient, and deflects enough to energize an electric contact, for example. Such a bimetallic strip actuator is used in thermal overload relays, which are described in detail in Section 2.2.1.

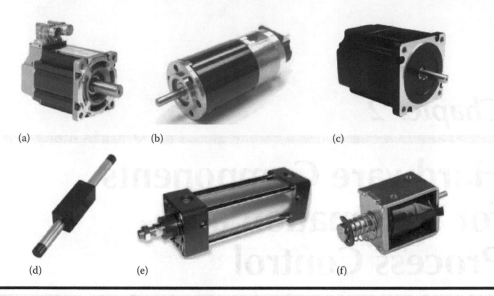

Figure 2.1 ✓ **Various types of actuators in industrial automation, (a) servo motor, (b) DC motor, (c) stepper motor, (d) linear motor, (e) pneumatic cylinder, (f) solenoid actuator.**

2.1.1 Electric Motors

Furthermore, electric actuators include the category of electric motors of all kinds, such as the stepper motors, servomotors, linear motors, and solenoids. However, in the industrial world, the term "actuator" is usually connected to low-power actuating devices and not to high-power electric motors. For an industrial engineer, the motor is a separate category itself and is not always straightforwardly connected to the actuators, although its definition includes them. Thus, the rest of this book will adopt the following approach: separate the electric motors for high mechanical power production from other types of actuators. In general, electric motors have the capability to convert the electrical energy into mechanical or kinetic energy. All electric motors*, AC (alternate current) or DC (direct current), use the principle of electromagnetic induction and the subsequent interaction of two magnetic fields to generate torque on a rotational element called a "rotor" inside a stationary housing called a "stator". In Figure 2.2, an indicative internal construction for the case of AC and step motors is provided, where the existence of the coils gives a first impression of the generated magnetic forces that are responsible for achieving motor rotation.

More analytically, most DC motors operate by electric current flowing through a number of coils at the rotor (depending on the number of poles in the motor), which are positioned between the poles of a permanent magnet or electromagnet of the stator. The interaction of the two magnetic fields, one created by the rotor and the second one due to the stator, causes the rotor-shaft to rotate. To reverse the motor, it is needed to change the polarity of the supply voltage to either field winding or armature winding, but not both, since this will cause no change in the direction of rotation.

In the case of AC induction motors, only the stator has coils by means of a three-phase winding circuit, which produce a rotating magnetic field. This field induces an alternating current in the rotor, which consists of a cylindrical laminated core with slots that can carry conductors from

---

* P. Vas, *Electrical Machines and Drives: Space Vector Theory Approach*, Oxford Science Publications.

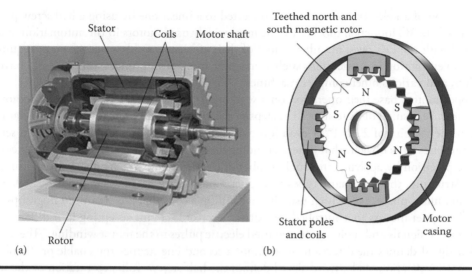

Figure 2.2   Internal view of an AC motor (a) and step motor (b).

copper or aluminum bars. Since these conductors are directly shorted by an end ring, they form the rotor winding, which cuts the stator rotating magnetic field, causing the flow of electric current through them. The attraction and repulsion between these two magnetic fields, according to Lenz's law, causes the rotation of the rotor. For the case of a three-phase induction motor, switching two out of the three input voltage lines causes motor rotation in the opposite direction, as indicated in Figure 2.3. At this point, it should be highlighted that the alteration of three phases at the same time will result in no alteration of the magnetic flux, and thus the rotation of the motor will not change.

In both types of motors, the rotational movement of the rotor is transferred to a shaft and, subsequently, to a series of spindles, gears, pulleys, and smaller shafts in order to increase the output torque or to transform the rotary motion to a linear or reciprocating one. For example, the

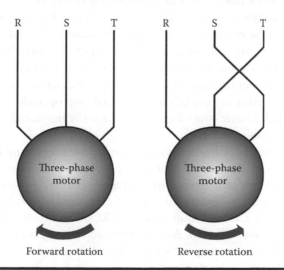

Figure 2.3   Forward-reverse rotation of a three-phase motor.

rotary motion of an electric motor can be converted to a linear one by using a ball screw pair and guide rails. The AC induction motors are the most widely used motors in the automation industry compared with the DC ones, mainly because of their efficiency and less maintenance required. It is the simpler solution in applications such as machine tools, fans, pumps, compressors, conveyors, extruders, and various other complex machines.

Stepper motors base their operation on a working principle similar to that of DC motors and can rotate in very small discrete steps. The steps of a stepper motor represent discrete angular movements in the vicinity of 2° or 1° or even less, which are performed successively due to a series of digital impulses. It is obvious that a stepper motor can perform any number of rotation steps with the same precision by applying an equal number of electrical pulses to its phases. Regarding their internal structure, there are many types of stepper motors (such as unipolar, bipolar, single-phase, two-phase, multi-phase, etc.) which usually have multiple coils that are organized in groups called "phases". Stepper motors are controlled by a driver electronic circuit accepting four different pulse digital control signals and applying the required electric pulses to the motor windings. The one-step function signal defines the direction of rotation, a second one defines the enable or disable state of the motor operation, and a third signal defines the half-step or full-step rotation of the motor. Finally, a pulse train signal causes the rotation of the motor. Each control pulse causes the motor to rotate by one step, while the speed of the rotation is determined by the frequency of the pulse train. These control signals may be produced by a programmable logic controller that will be described in Chapter 10, where step motor applications will be examined. In general, stepper motors provide precise speed, position, and direction control in an open-loop fashion, without requiring encoders or other types of sensors which conventional electric motors require. A stepper motor does not lose steps under normal conditions of mechanical load, while the final position of the stepper motor's rotor is determined by the number of performed steps and expresses the total angular displacement. This position is kept until a new pulse train is applied. These properties make the stepper motor an excellent actuator for open-loop control applications, for low to medium power requirements.

When higher torque demands precise control, servomotors are then the best solution to be used. Servomotors are not a specific class of motors and the term "servomotor" is often used to refer to a motor suitable for use in a closed-loop control system. A servomotor consists of an AC or DC electric motor, a feedback device, and an electronic controller. In the case of a DC motor, this can be either a brushed or brushless type. Typically, the feedback device of a servomotor is some type of encoder built into the motor frame to provide position and speed feedback of the angular or linear motion. The electronic controller is a driver, supplying only the required power to the motor, in the simplest case. A more sophisticated controller generates motion profiles and uses the feedback signal to precisely control the rotary position of the motor and generally to control its motion and final position, thus accomplishing the closed-loop operation. Since the servo motors are driven through their electronic controllers, it is quite easily interfaced with microprocessors or other high level programmable controllers.

Figure 2.4 provides some fundamental torque-speed curves regarding the selection of the AC (a) and stepper (b) motors, characteristics that can be found in each motor's manual and contribute also in the comprehension of its respective operation. For AC motors, at rest the motor can appear just like a short-circuited transformer and, if connected to a full supply voltage, draw a very high current known as a locked rotor current (LRC). The motors also produce torque that is known as locked rotor torque (LRT). As the motor accelerates, both the torque and the current will tend to alter with the rotor speed if the voltage is kept constant. The starting current of a motor with a fixed voltage will drop very slowly as the motor accelerates, and will only begin to fall significantly when the motor has reached at least 80% of the full speed. The actual curves for

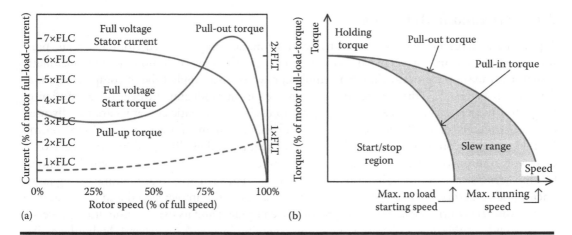

Figure 2.4 **Torque-speed curves facilitate the selection of (a) an AC or (b) stepper motor.**

induction motors can vary considerably between different types of motors, but the general trend is for a high current, until the motor has almost reached full speed. The LRC of a motor can range from 500% of full-load current (FLC) to as high as 1400% of FLC. Typically, good motors fall in the range of 550% to 750% of FLC. The starting torque of an induction motor with a fixed voltage will drop a little to the minimum torque, known as the pull-up torque; when the motor accelerates it will then rise to a maximum torque, known as the breakdown or pull-out torque, at almost full speed; and then it will drop to zero at the synchronous speed. The curve of the start torque against the rotor speed is dependent on the terminal voltage and the rotor design. In the case that the load curve is added, the intersection of the load curve with the torque and voltage curves will define the operational point of the motor.

For the case of a stepper motor, the characteristic torque speed curves are the following ones. The pull-out torque curve is the curve that represents the maximum torque that the stepper motor can supply to a load at any given speed. Any torque or speed required that exceeds this curve will cause the motor to lose synchronism. Holding torque is the torque that the motor will produce when the motor is at rest and rated current is applied to the windings. Slew range is the region where the stepper motors are usually operated. A stepper motor cannot be started directly in the slew range. After starting the motor somewhere in the self-start range, the motor can be accelerated or loaded while remaining within the slew range. The motor should then be decelerated or the load should be reduced back into the self-start range before the motor can be stopped. As in the previous case, the intersection of the motor's characteristic curves with the load curve will indicate if the size of the selected motor is sufficient for the envisioned application.

At this point, it should be highlighted that it is beyond the scope of this textbook to present the details of any type of electric motors, since the objective is to get an understanding of the basic principles of operation of the various actuators and to study their use in automation systems. Therefore, the objective of this chapter is to provide an overview of the basics of other common actuators (particularly the non-electrical ones) such as pneumatic actuators. Finally, regarding electric motors, we will need to distinguish their control task from their automation task. The control task refers to a closed-loop control scheme for the regulation of their speed, angular position, and torque output. The automation task refers to the sequential steps of power relays (energizing or de-energizing) in order for an electric motor to change the direction of rotation or to startup according to a star-delta configuration.

2.1.2 Pneumatic Actuators

A pneumatic actuator converts and transmits pneumatic energy derived from a central, pressurized air source into mechanical movement energy. The most common type of pneumatic actuator is the simple piston-cylinder assembly connected to a supply tube of compressed air, as depicted in Figure 2.5. The air pressure acts on the piston producing a direct linear motion of the piston rod's free end. Since air is highly compressible, pneumatic actuators are usually used for the movement of small and lights objects and are not suitable for accurate position control. Many operations on automated production lines require actions like pushing an object on a conveyor, transporting a machine component between several closely placed assembly stations, feeding adhesive across a straight path on a surface, etc. For such or similar manufacturing tasks, pneumatic actuators are ideally suited, since they can be easily implemented and also require a simple, discrete control logic. Since the pneumatic equipment and its use in automation processes is an extended subject, it will be presented separately in Chapter 5. In general, hydraulic actuators are operated by a pressurized fluid that usually is oil. Their operation logic is similar to that of the pneumatic ones, but different in terms of construction. Since fluids are non-compressible, hydraulic systems are generally used when high forces and accurate control are required. Finally, it should also be highlighted that an engineer, who possesses the technology of the pneumatic systems, will assuredly be able to face up to any issue in automation applications using hydraulic actuators.

2.1.3 Micro-Electro-Mechanical Systems

Micro-electro-mechanical systems (MEMS) are a process technology based on techniques of microfabrication, and are used to create devices that have the ability to sense, control, and actuate on the micro scale, while generating effects on the macro scale. Over the past few decades various micromechanical actuators (or microactuators) began appearing in numerous industrial products and applications. These miniaturized actuators are MEMS devices, which range in size from a few micrometers to millimeters, that convert one form of signal or energy into another form. MEMS microactuators have a wide variety of actuation mechanisms producing very small forces or displacements, and can be categorized in four basic groups: electrostatic, piezoelectric, magnetic, and thermal. MEMS microactuators are used in various industries, such as the automotive, electronics, medical, and communications. Some typical examples of microactuators are micro-valves for control of gas and liquid flows, micro-grippers for robotic surgery, and focusing micro-mechanisms for cameras in mobile devices.

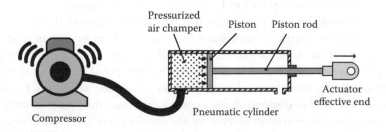

Figure 2.5 The pneumatic cylinder as linear motion actuator.

2.1.4 Relays

Since electric motors are supplied with electric power through the utilization of relays or contactors, their operational principles and characteristics are presented in this chapter as a specific kind of actuator. Another reason for describing relays in this chapter is the need to know their structure and operation in order to utilize them in designing and synthesizing automation circuits, which will be presented in Chapter 3.

In general, a relay is a binary actuator as it has two stable states, either energized and latched or de-energized and unlatched, while a servomotor, for example, is a continuous actuator because it can rotate through a full 360° of motion.

Some synonymous terms, such as "magnetic contactor", "relay", "remote operated switch", and "tele-operated switch" are used for an electromechanical device that opens and closes electrical contacts, and should not create confusion to the reader, since all of these terms refer to an electrically operated switch that has the same principle of operation, and these terms are often used interchangeably. A soft distinguishing between the terms "relay" and "contactor" is that the second one is used for powering large electric motors. In the rest of the book, the term "relay" will be adopted as it is the most widely used one. There are many types and categories of relays, each one developed to satisfy the various specific needs of several applications. From a construction point of view there are two types of relays, the power relays and the general purpose or control relays, both having the same principle of operation described below.

2.1.4.1 Relays' Operation Principle

Relays consist of two parts; a first one containing the various main or auxiliary electric contacts, and a second one containing the electromagnet-based mechanism that creates the motion required for the operation of the electric contacts. Figure 2.6 shows a simplified form of a relay, including the coil-core electromagnet, the movable arm, the electric contact consisting of two parts (the fixed and the movable one), and the return spring. The relay's contacts are electrically conductive

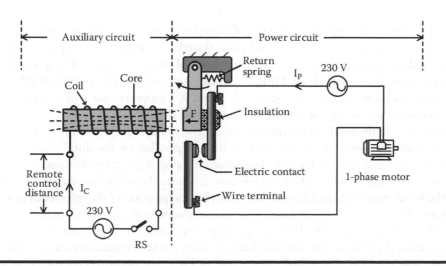

Figure 2.6 A simplified structure of a relay.

pieces of metal that, when contacted together, complete a circuit and allow the circuit's current to flow, just like a simple switch. The operation of the relay is achieved by forming two independent and isolated circuits, the power circuit and the auxiliary or the control one. The power circuit refers always to the electrical device usually called "load", which is powered through the relay contact, while the auxiliary circuit refers always to the coil of the relay. Since the auxiliary circuit, which is explained in Chapters 3 and 4, performs the automated operation of a machine, it is also called the "automation circuit". The very simple form of the auxiliary circuit shown in Figure 2.6 should not lead the reader to false conclusions. As is explained in Chapter 3, the complicated and interdependent operation of many relays in a complex machine requires usually very complicated automation circuits. As a kind of definition, the auxiliary circuits of a large number of relays in a complex industrial machine or production line, embedded all in one common circuit, constitute the automation circuit.

The power circuit consists of the relay's electric contact, the electric source that provides the electric power and the load. Generally, in the considered industrial applications, the electric source will be always the public electric power network and the powered load a motor that, in this case, is a single-phase motor. The auxiliary circuit consists of the coil, the electric power source (usually different and independent from the corresponding source of the power circuit), and the hand switch RS. When the switch RS closes, 230 V is applied on the coil and the generated magnetic field attracts the movable arm due to the generated force F to the direction shown in Figure 2.6. The movable arm carrying on the insulation material and the movable contact part, causes the closing of the electric contact, thus permitting the electric current to flow in the power circuit and hence the motor to operate. If the hand switch RS opens, the magnetic field is nulled and the attraction of the movable arm stops. The return spring then brings back the movable arm, the relay's electric contact opens and the motor stops. The main result is briefly summarized as follows:

RS is closed → Relay is energized → Motor operates

RS is open → Relay is de-energized → Motor does not operate

Therefore, the possible operational states of the relay are two: the first one corresponds to the voltage applied on the relay coil; then it is said that the relay is energized or simply the relay is ON; and the second one corresponds to the voltage not applied on the relay coil, then it is said the relay is de-energized or simply the relay is OFF. These two states of the relay operation ON and OFF can be corresponded to the digital logic signals 1 and 0, a property that will be invoked later in Chapter 4 concerning the logic design of the automation circuits.

The current I_C in the auxiliary circuit is generally low, of the order of a few tens of mA. Instead, the current I_p of the power circuit may be very high, depending on the durability of the electric contact and is the basic parameter that determines the size of the relay. Various types of power relays exist in the market where their electric contacts have the ability to rate up to 2000 A. In general, the hand switch RS could be far from the physical position of the relay, even at a distance of a few tens of meters, while the distance limit is introduced by the permitted voltage drop across the lines of the auxiliary circuit. This voltage drop, and hence the maximum permitted remote control distance, depends on the characteristics of the wires of the auxiliary circuit, the nominal coil voltage, and the kind of supplied voltage (AC or DC). The nominal coil voltage of a relay is selected from a set of standardized voltages that are the most usual in the market, as shown in Table 2.1. The power absorbed by a coil (self-consumed power of a relay) depends on the relay size

Table 2.1 Standardized Values for Nominal Voltage Selection of a Relay Coil

Values of Relay Coil Nominal Voltage	
AC 50 Hz	24, 42, 48, 60, 110, 230, 380 V
DC	12, 24, 36, 48, 110, 230 V

and is usually of the order of a few watts. This absorbed power is greater during the activation of the relay, when the inertia torque of the movable arm should be overcome and is reduced when the activated state of the relay has been achieved. The time required for the activation and deactivation phases of the relay depends on the type of relay and is usually rated up to 10–50 msec. Since the described relays include mechanical and electrical components, they are also called "electromechanical relays" in order to be distinguished from the "solid state relays" consisting only of electronic and semiconductor components (transistors, thyristors, and triacs) and circuits.

2.1.5 Power Relays

Power relays are made in order to feed the various kinds of electric motors with the required electric power. In proportion to the typical motors' powers, these power relays are made in several sizes, from the smallest of nominal power 5.5 KW, to the largest of 500 KW under 660 V, while it is obvious that the power relays and particularly their electric contacts must withstand a "switching under load". This means simply that an electric contact of a power relay should have the mechanical strength required in order to open while the nominal current is passing through it. The reason that causes damage to the contact material is that an electric arc is created during the opening or closing of the electric contact. Thus, it is obvious that the electric contact consisting of two thin metallic plates, as shown in Figure 2.6, does not have the strength to break a high electric current. Therefore, the construction of a power relay leads to the increase of the contact surface and subsequently to larger dimensions of the metallic plates. This, in turn, leads to the need for stronger attraction of the movable arm and hence to larger dimensions in the coil and the corresponding coil's core. In conclusion, a power relay has the same principle of operation as the one described in Section 2.1.1, but the higher its nominal power, the larger its size is.

Power relays are characterized from both the nominal current of its electric contacts and the nominal power of the electric motor that supplies it. These magnitudes are different for the various categories of power relay use, such as for example AC1, AC2, etc. (IEC 158-1, BS 5424, and VDE 0660). Figure 2.7 shows the side cross section of a large size power relay, while the internal structure of the power relay includes all the components described in the "principle of operation" (Figure 2.6), which are the coil, the fixed iron core, the movable iron core or arm, the electric contacts, and the arc chamber for protection. It should be noted that the rotary motion of the movable contact found on small relays, shown in Figure 2.6, is replaced in the case of power relays with parallel motion, resulting in the realization of the electric contact at two different points and the so-called double-break contact. To reduce the effects of contact arcing, modern contact tips are made of, or coated with, a variety of silver-based alloys (silver-copper, silver-cadmium-oxide, silver-nickel, etc.) to extend their life span. Since power relays switch their rated loads, their electric contacts are characterized by an electrical and mechanical life, expressed in thousands or millions of operations, while the usual values of electrical or mechanical life expectancy may be one million, 10 million, or even 100 million operations. The electrical life is usually lower in comparison

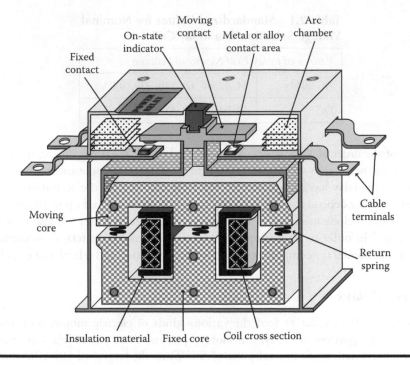

Figure 2.7   Typical internal structure of power relay.

to the mechanical one because the contact life is application-dependent, such as when a set of contacts switches a load of less than rated value.

The power relays, except for their three main power-contacts (one for each phase of the three-phase network), may be equipped with one or more pairs of auxiliary contacts, which are used in the implementation of the automation circuit. Figure 2.8 shows a typical form of a power relay

Figure 2.8   Typical external view of a power relay by Siemens.

with power contacts and a block of auxiliary contacts on its right side. Usually, two to four maximum blocks of auxiliary contacts, whose supply is optional, can be adapted to the body of a relay, in order for the required mechanical coupling to be achieved. The auxiliary contacts of power relays are rated up to 15 A, because they are used only in automation circuits and don't supply any kind of load.

The possibility of selecting the coil's nominal voltage is also valid for the case of the power relays. The coil nominal voltage should be the same as the operation voltage of the whole automation circuit. It should be also noted again that the coil nominal voltage is quite independent of the power circuit voltage. Therefore, it is possible with a coil of a nominal voltage (e.g., 24 V DC) to control the operation of a power circuit (electric motor of a machine) of a nominal voltage (e.g., 660 V AC).

2.1.6 General Purpose Relays

General purpose relays are usually miniature relays, used either as auxiliary components for the implementation of automation circuits, or as switches for supplying very small electric loads, such as electric valves, micromotors, small fans, alarm sirens, etc. Their use as auxiliary relays or as auxiliary contacts of a power relay means they are necessary components for the implementation of the Boolean logic functions described in the automation circuit. This issue will be further analyzed in Chapter 4, where the logical design of the automation circuits is presented.

In general, there is a huge variety of general purpose relays regarding their size, shape, number of contacts, coil voltage, and mounting methods. All of the relays of this type have the same operation principle as that of power relays and an example of their typical form is shown in Figure 2.9. These relays are mounted either on a specific base for accepting electric wires (easy replacement of a relay), or directly on a PCB board as a component of an electronic circuit. Figure 2.10 shows the two side views of a general purpose relay in its physical dimensions, where the basic parts of this relay are evident as are the coil, the movable armature and contacts, the double fixed contacts, and the return spring. General purpose relays are protected by a plastic case to protect the mechanisms from dusty and corrosive environments. The movable contact arm, called common terminal (C), has two contact tips and it is located between two fixed contacts, which forms a normally

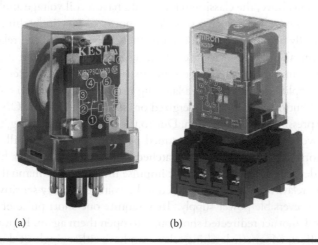

(a)   (b)

Figure 2.9   Typical form of a general purpose relay without (a) and with (b) a mounting base.

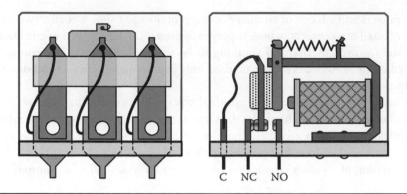

Figure 2.10   Two internal side views of a general purpose relay.

closed contact (NC) and a normally open one (NO). This type of electric contact is usually called "changeover two-way contact", and all these types of electric contacts and their nomenclature are presented in Section 2.1.10.

The number of two-way contacts varies and depends on the type of the general purpose relay. For example, the relay in Figure 2.10 has three sets of changeover two-way contacts. The electric contacts of a general purpose relay are passing a current rated up to 15 A approximately, except from the very small ones (with dimensions of $1 \times 1 \times 2$ cm for PCB boards) that have a nominal current of 1 A. The operation of a relay is normal and stable when the coil voltage is exactly the nominal one. Therefore, care is required to apply the correct coil voltage because if it is larger or smaller than the nominal one, the temperature rise of the coil or the insufficient holding force of the movable contact or other electromechanical malfunctions may be caused. Such situations reduce the relay contact life expectancy and tend to cause fusing of the contacts.

2.1.7 Latching Relays

A latching relay, after its activation, maintains its contact position although the coil power supply has been removed, and therefore has two relaxed states as an electronic, bistable flip-flop. As we have seen from the description above, the classical relay should have a coil voltage applied to it at all times that it is required to stay energized. Such a situation is not necessary in latching relays, where their contacts are mechanically or magnetically locked in the ON state until the relay is reset manually or electrically. Mechanical latching relays use a locking mechanism to hold their contacts in their last set position until commanded to change state, usually by means of energizing a second coil. Figure 2.11 shows a simplified schematic of a latching relay operated by two coils, each one with a corresponding control button. The relay is energized or "set" by pressing the ON button and deactivated or "tripped" by pressing the OFF button. Due to the mechanical latching, the locking strength will not be reduced over time and the self-consumed power of the relay is null. Some conventional power relays (contactors) can be converted into latched contactors by adding a block containing the mechanical latching device with electromagnetic impulse unlatching or manual unlatching.

Magnetic latching relays are typically designed to be voltage polarity sensitive and hence can be driven directly from a reversible power supply. They require one short pulse of coil power to close their NO contacts, and another redirected short pulse to open them again. Repeated pulses from the same input have no effect. Magnetic latching relays can have either single or dual coils. On a single coil device, a permanent magnet is designed to hold the contacts in the energized position, while the

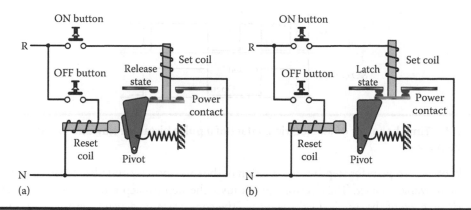

Figure 2.11 **Schematic diagram of latching relay: (a) relay tripped and power circuit open, and (b) relay set-latched and power circuit closed.**

relay will operate in one direction when power is applied with one polarity, and will reset when the polarity is reversed, thus overcoming the holding effect of the permanent magnet. On a dual coil device, when polarized voltage is applied to the reset coil, the contacts will return to their rest state. Latching relays are used in situations where the energized status is held for a long period or when the relay condition should be held invariant during a breakdown or power supply interruption.

2.1.8 Pulse Bistable Relays

Pulse bistable relays are small electromechanical devices, which have the ability to open and close their contacts in a preset sequence. Although they are called "relays", their operation principle, construction, external form, and dimensions are quite different from that of the standard relays described in Section 2.1.1. Their use in automation systems is limited, but in the cases where this is required, it facilitates immensely the implementation of an application. A pulse bistable relay consists of a cam mechanism, an electromagnet, and the electric contacts as shown in Figure 2.12. The cam is a double

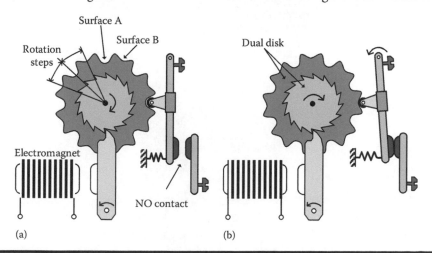

Figure 2.12 **Typical internal structure of a pulse bistable relay in its two possible states: (a) a movable contact in front of type-A disk surface, and (b) after one pulse, the movable contact comes in front of the type-B disk surface and the electric contact closes.**

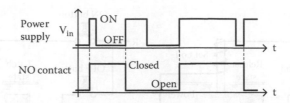

Figure 2.13   Time behavior of an electric contact of a pulse bistable relay due to a series of four successive pulses.

plate disk, one tooth plate for step-by-step rotation of the disk and a second plate for setting the position of the movable contact. The electromagnet causes the step-by-step rotation of the disk at each positive rising edge of the supplied voltage pulse. The contact status depends on the position of the second plate, particularly if the surface A or surface B is in front of the movable contact. When the coil is pulsed, the relay armature moves a lever that, in turn, rotates the disk to position B stepwise in the sequence, closing the NO contact. This position will remain independent of the pulse duration and as long as another pulse is not supplied to the coil. Figure 2.13 shows the behavior of the electric contact due to a series of successive pulses. The first pulse closes the contact, the next opens it, and so on, back and forth. It is obvious that this relay has two discrete states, which can be retained without the existence of a coil voltage, justifying the term "bistable". A typical use of the pulse bistable relay is remotely starting and stopping a machine from a single momentary push button.

2.1.9 Solid State Relays

Solid state relays (SSRs), like the one presented in Figure 2.14, are electronic devices with no mechanical contacts capable of switching various AC or DC loads, such as heating elements,

Figure 2.14   Typical external view of a solid-state relay by Siemens.

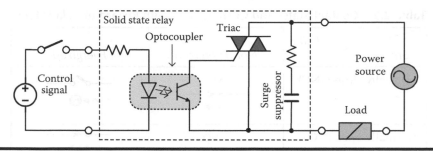

Figure 2.15   Simplified circuit of a photo-coupled solid-state relay.

motors, and transformers. SSRs perform the same switching function as the electromechanical relays, however their structure is quite different, consisting of semiconductor switching elements, such as thyristors, triacs, transistors, and diodes. Also, SSRs include a semiconductor-type opto-coupler to separate the input circuit from the output circuit, offering complete isolation of the input and output signals. As shown in Figure 2.15, the control signal (which corresponds to the coil voltage of the electromechanical relays) is applied to a light-emitting diode (LED). The light, or infrared radiation, is detected from a phototransistor that triggers the triac, which switches on the load current supplied from an external source. In order to avoid temperature rise in the triac junction, SSRs are usually equipped with a heat sink made by a heat-conductive metal and integrated in their body in order to dissipate the heat to the surrounding air. Because of their electronic structure, SSRs provide high-speed or high-frequency ON-OFF switching operations and generate low noise. Furthermore, due to the absence of movable contacts, they do not produce electric arcs and hence are suitable for use in hazardous areas. In chemical, petrochemical, mining, and many other industries where combustible materials are transported, stored, or processed, potentially explosive atmospheres are inevitable. In such cases, the use of SSRs as load switches is mandatory, according to national laws and regulations, since they ensure reliable operation and safety for personnel and machinery, even in extremely explosive environments.

2.1.10 Electric Contact Classification

An electric contact of a relay may be open or closed when the relay is de-energized, i.e., no supply voltage connected to the relay coil. Such a status of an electric contact, when the relay is not energized, is called "normal" and hence may be "normally open" or "normally closed", abbreviated as NO or NC correspondingly. The same relay of any kind may contain NO and NC contacts simultaneously, independent of one another. When the relay is energized, all the NC contacts open and all the NO contacts close. The automation circuits are always designed for the normal status of the relays and other devices that participate in the automation system. Table 2.2 shows the possible states of an electric contact in relation to the energized or NO state of the relay and the corresponding possible current flow.

The electric contacts shown in Table 2.2 (and Figures 2.6 and 2.7) permit the current to flow only one way or "throw", as this term is widely accepted and utilized in the field of industrial automation. An electric contact of a relay or switch may have one or more throws. In other words, the number of throws indicates how many different output connections each electric contact can connect its input to, which is called a "common terminal". For example, if an electric contact has two throws, the common terminal can be connected to one of two possible terminals. The two most common types are the single-throw and the double-throw contacts. A double-throw contact

Table 2.2   Contact States and Current Flow with Regard to Relay State

Contact / Relay	Non-Energized	Energized
Normally closed (NC)	NC —o—o— I ⇝	⟋o— I ⇝●
Normally open (NO)	NO —⟋o— I ⇝●	—o—o— I ⇝

is also called a "changeover contact". On the other hand, relays or hand switches can be made up of one or more individual electric contacts with each "contact" being referred to as a "pole". The number of poles refers actually to the number of separate circuits that the relay contact or the switch may control, while a single-pole relay or switch controls just one circuit. A double-pole relay or switch controls two separate circuits. Relays, hand switches or even sensors may have electric contacts with various combinations of poles and throws defined as "single-pole, single-throw", "single-pole, double-throw", "three-pole, double-throw", "four-pole, double-throw", and so on, which are abbreviated as SPST, SPDT, 3PDT, and 4PDT, respectively. Table 2.3 shows the possible states of the simplest changeover contacts (SPDT and DPDT) in relation to the energized or non-energize state of the relay, with Figure 2.16 presenting the most basic electric contacts and their abbreviations. The electric contacts of the relays are numbered on convenient locations of the relay body, according to a standard numbering system, the basic rules of which are:

1. The power contacts are numbered with one-digit numbers (1, 2), (3, 4), and (5, 6) for the three phases R, S, and T, respectively.
2. The auxiliary NC contacts are numbered with two-digit numbers (X1, X2), where X=0–9.
3. The auxiliary NO contacts are numbered with two-digit numbers (X3, X4) where X=0–9.

Table 2.3   Possible States of Two Simple Changeover Two-Way Contacts

Contact / Relay	Non-Energized	Energized
SPDT	I → NC I → I → ⟋o NO	I → ⟋o → I
DPDT	I_1 → NC I_1 → NO I_2 → NC I_2 → NO	I_1 → ⟋o → I_1 I_2 → ⟋o → I_2

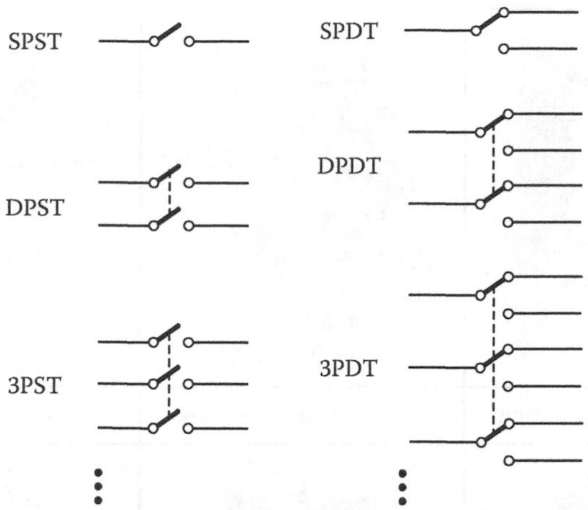

Figure 2.16   Single-throw and double-throw electric contacts for several numbers of poles.

The issue of selecting the suitable relay for an application depends on many parameters that concern the kind of construction; operation principle; size; and electrical characteristics, such as coil nominal voltage, the contact's rated current, the number of contacts, and many others. The manufacturers of relays offer detailed catalogs containing tables with all mechanical and electrical characteristics of the relays to customers, as shown in Figure 2.17, which summarizes the required information for general purpose relays of plug-in socket, or PCB mounting type. For power relays, the same technical characteristics are available, but their selection is mainly based on the rated current of the main contacts. Furthermore, the power relays are also characterized by the power being able to feed to the electric motors.

We will close the section on relays with some general conclusions:

■ Relays are used either as power switches or as auxiliary logic components.
■ Relays permit a low-voltage circuit to control another high one or in general different voltage one.
■ Relays can be used as current "amplifiers".
■ Relays provide complete electric isolation between the control signal and the power signal, that is between the auxiliary control circuit and the power circuit.
■ By using relays with multiple poles of contacts, it is possible with one low-voltage signal to control the operation of many loads, each one with a different voltage.

2.1.11 Solenoid Linear Actuators

A general purpose solenoid actuator is an electromagnetic device that converts electric energy into a mechanical pushing or pulling force or motion. Most solenoid actuators produce a linear motion called therefore "linear solenoid actuators"; however, rotational solenoids are also available. The linear solenoid actuator works on the same basic principle as the electromagnet of the electromechanical relays that causes the required movement of contacts. Usually, the general purpose linear solenoid actuators produce small movements and apply low-range forces, capable though of opening or closing valves, activating latches or similar mechanisms, and generally moving light mechanical elements. The linear solenoid actuators, as shown in Figure 2.18, consist of an electric

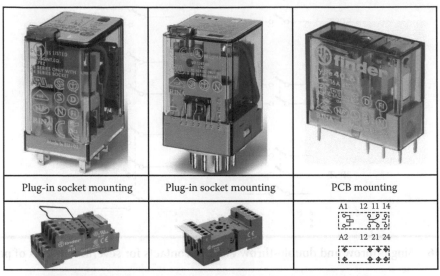

Technical information for general purpose relay type 55.32 by finder, (above left)	
Contact specification	
Contact configuration	2 CO (DPDT)
Rated current/maximum peak current A	10/20
Rated voltage/maximum switching voltage V AC	250/400
Rated load AC1 VA	2500
Rated load AC15 (230 V AC) VA	500
Single phase motor rating (230 V AC) kW	0.37
Breaking capacity DC1: 30/110/220 V A	10/0.25/0.12
Minimum switching load W (V/mA)	300 (5/5)
Standard contact material	AgNi
Coil specification	
Nominal voltage (U_N) V AC (50/60 Hz) V DC	6-12-24-48-60-110-120-230-240 6-12-24-48-60-110-125-220
Rated power AC/DC VA (50 Hz)/W	1.5/1
Operating range AC/DC	$(0.8 … 1.1)U_N/(0.8 … 1.1)U_N$
Holding voltage AC/DC	0.8 U_N/0.5 U_N
Must drop-out voltage AC/DC	0.2 U_N/0.1 U_N
Technical data	
Several other technical data, such as mechanical life, electrical life, operate/release time, dielectric strength, ambient temperature range, etc., are also given by manufacturers of general purpose relays.	

Figure 2.17   An indicative table with technical characteristics of general purpose relays.

coil winding around a cylindrical tube, a ferromagnetic "plunger" that is free to move or slide IN and OUT of the coil's body and optionally a return spring. In the case of the absence of a return spring, an external return action is necessary. In general, solenoid actuators may be used either as a holding mechanism under continuously supplied voltage or as a latching mechanism under pulse-type (ON-OFF) supplied voltage. Linear solenoid actuators are available in a variety of types, forms, applied forces, voltages of operation, and other attributes, while the selection of the most suitable and efficient types is dependent on the kind and the specific characteristics of the application. Some linear solenoid actuators, designed to perform a specific task in circuit breakers, will be presented in the section on specific industrial equipment.

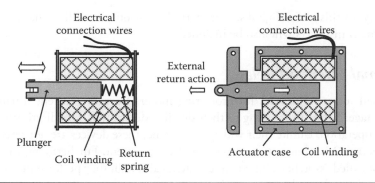

Figure 2.18   General purpose solenoid actuators with/without a return spring.

2.2 Sensors

Almost any industrial automation system includes sensors for the detection of the various "states" of the controlled manufacturing process and actuators as outputs for real-time acting and achieving the desired behavior of the production procedure. Following the presentation of actuators in the previous section, sensors (by means of their kinds, properties, implementation and the basic theory behind them) are described in this subsection. Sensors are devices that, when exposed to a physical phenomenon (temperature, pressure, displacement, force, etc.), produce an output signal capable of being processed by the automation system. The terms "transducer" and "meter" are often used synonymously with sensors, while simultaneously some sensors are combined with the term "switch", causing confusion about the correct terminology. Furthermore, some writers consider that "sensor" is only the sensing element that detects the physical magnitude and not the whole device that, together with the sensing element, transforms the physical variable into a form of electrical signal. Let's define the meaning of these terms as they will be used in this textbook.

In general, sensors transform the variation of a physical quantity into an electrical output signal, which may be an analog or digital one. In the first case, the sensor produces a continuous output signal that is proportionally varied to the sensed parameter. For example, a pressure sensor may produce a 4–20 mA DC, or 0–10 V DC output signal for a 0–725 psi pressure variation. In the second case, the sensor produces a discrete output signal in the form of an ON or OFF, usually causing a SPDT contact to change state when the physical quantity gets over a predefined value. Sensors with analog output may also be called transducers, while sensors with discrete (or binary or digital) output are called switches, e.g., "proximity sensor or switch". When transducers include an analog-to-digital converter (ADC), then they are called digital transducers, since their output can be directly fitted to a digital controller, and should not be confused with the binary or digital sensors of a switch operation type.

A conventional automation circuit cannot manipulate the output of transducers because of its analog form, and thus it may be combined only with sensors of a switch-type output. On the other hand, the modern automation systems are realized by programmable logic controllers, which can accept both digital type signals (ON-OFF) and analog type current or voltage signals, varied into some standard ranges. It is obvious that for the same physical variable (e.g., temperature), there are in the marketplace both sensors with switch-type output and transducers with analog-type output, while the selection between them depends on the kind of application. It should also be noted that some transducers, including a signal-conditioning unit, might offer a second switch-type output, except those from an analog-type output, responding to a manually defined value of the detected physical quantity. In this sense, digital sensors with discrete outputs will be mainly covered in this section, but some transducers will also be included.

2.2.1 Thermal Overload Relay

Thermal overload relays are devices for protecting motors from overcurrent situations that may cause them damage. Before proceeding to their detailed description, let's discriminate some issues regarding their operation. Unlike their widely used name, these devices are not relays like the ones described in Section 2.1.1, but sensors detecting electric current. Furthermore, these relays could be considered or called "switches", but are not switches although they perform indirectly the action of switching. In fact, thermal overload relays can interrupt the auxiliary or control circuit of a power relay—that is, the real switch—causing its deactivation. From this point of view, a thermal overload relay is an integral part of any power relay supplying an electric motor. It is mechanically and electrically coupled with the power relay, and both of them constitute together a unified power device that is inserted in the power circuit of an electric motor. The relay acts then as a switching mechanism and the overload relay as a sensor detecting the motor current. For this reason, all the power relays manufacturers produce also the corresponding thermal overload relays.

Every motor is characterized by the nominal current $I_{nominal}$ absorbed during the motor's operation and under normal conditions. The thermal overload relay detects the possible overloading situation of the motor, expressed by the condition $I_{real} > I_{nominal}$, where I_{real} is the actual current of the motor. If the motor draws more current than $I_{nominal}$ for an extended period of time, then it will be damaged. In order to avoid such a fault, the thermal overload relay protects the operation of the motor indirectly by deactivating the power relay or the contactor, as is described subsequently. The output of the thermal overload relay is, in most cases, a DPST contact, as shown in Figure 2.19, which helps stop the motor operation.

The principle of the thermal overload relay operation is based on the known behavior of the bimetallic strip, which consists of two dissimilar metals by means of two oblong metallic pieces bonded together. The two metals have different thermal expansion characteristics, as for example the brass and the nickel-iron alloy, and therefore the bimetallic strip bends at a given rate when heated. In Figure 2.20a the simplified form of a bimetallic strip is shown, where the motor current flows through it and hence it is directly heated, while in Figure 2.20b the strip is indirectly heated through insulation winding around the strip. When the motor current has its nominal value or lower, the bimetallic strip doesn't bend, and the contacts have their normal status. As motor current rises for any reason, the bimetallic strip bends, pushes the trip lever, and mechanically changes the state of the two contacts, as presented in Figure 2.20c. The NC contact is inserted in the control circuit of the power relay supplying the motor and, in such an overload condition,

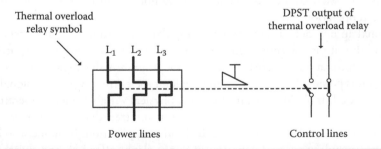

Figure 2.19   Schematic symbol of a thermal overload relay.

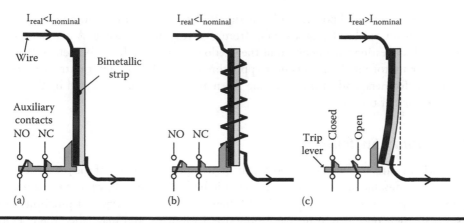

Figure 2.20 **Operation principle of the thermal overload relay with bimetallic strip, (a) during normal operation (direct heating), (b) during normal operation (indirect heating), and (c) during fault case operation.**

causes the deactivation of the power relay and hence the overall stopping of the motor. The corresponding automation circuit, with a thermal overload protection of a motor, was given in Figure 1.11 as an example of control circuits. Its operation explanation and other examples are described in more detail in Chapter 3. The NO contact of the thermal overload relay is used to activate the fault indication of the condition "overload relay tripped".

Furthermore, Figure 2.20 shows a single bimetallic strip, while the thermal overload protection relays are of a three-pole type, and hence it contains three bimetal strips mounted together with a uniform tripping mechanism in a housing made of insulating material, as shown in Figure 2.21. Each bimetallic strip detects the motor current of the corresponding phase, making the overload relay also sensitive to the phase imbalance or losses. Each thermal overload relay covers a standard range of the motor current, e.g., 4–6 A or 6–8 A (there are about 20 such ranges), and is equipped with a small dial for the ampere constraint setting into the standard range by the displacement of the trip mechanism relative to the bimetal strips. After an overload situation, the bimetallic strip

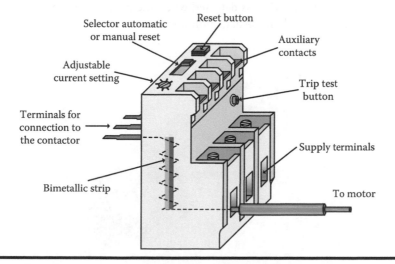

Figure 2.21 **Typical external view of a thermal overload protection relay.**

cools and returns to its initial position either manually by a reset button or automatically by a self-resetting mechanism, the last being selectable from a corresponding switch. A manual test button is also provided for testing the operation of the overload relay's auxiliary contacts in the automation circuit. For motor rated currents over approximately 80 A, the motor current is conducted via current transformers, and in this case, the thermal overload relay is heated by the secondary current of the current transformer.

2.2.2 Proximity Switches

Proximity switches are, in general, sensors for the non-contact detection of the presence of various objects, whether metallic or not, in front of the switches' effective area, while the form of a typical proximity switch is shown in Figure 2.22. There are three basic types of proximity switches with respect to their operating principles: inductive, capacitive, and magnetic. The operation of inductive proximity switches is based on the variation of the magnetic inductance. The capacitive proximity switches base their operation on the variation of capacitance and the magnetic proximity switches on the variation of the magnetic flow. The most widely-used proximity switches in industrial applications are the inductive and the capacitive ones. The substantial advantages of using proximity switches over other detectors (e.g., limit switches) are as follows:

- Long electrical life
- Durability to vibrations, accelerated motion, and toxic environments
- Operation under DC or AC voltage
- Effective response in objects' high speed approaching and high frequency excitation
- Absence of any kind of movable elements

The basic parameters of the proximity switches on the basis of which their selection is performed are the following:

Detection or sensing distance is the maximum detection distance between the target object and the effective area of the switch, as shown in Figure 2.23. The presence of an object in front of the effective area of the switch is detected when the distance between the object and the effective area is not larger than the detection distance.

Figure 2.22 Typical external view of a proximity sensor.

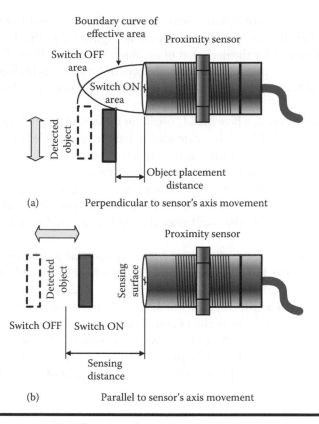

Figure 2.23   Object movement in front of the sensing area of a proximity sensor perpendicular to the sensor's longitudinal axis (a), and parallel to the sensor's longitudinal axis (b).

Object placement distance: The motion of an object in front of the effective area of the proximity switch may be perpendicular to the switch axis, as shown in Figure 2.23a, or parallel, as shown in Figure 2.23b. The object placement distance has meaning only in the first case and is defined as the constant distance between the passing object and the effective area of the proximity switch. The object placement distance should always be less than the detection distance. Usually, the object placement distance must be precisely adjusted during mounting of the switch. For this reason, the cylindrical type proximity switches have a threaded cylindrical surface for micro-adjusting their location.

Power supply is the supplied voltage range that the proximity switch will operate at. This voltage should not be confused with the operating voltage of the automation circuit where the output of the proximity switch, usually a SPDT contact, is inserted.

Response time is the time elapsed between the object detection instant and the switch output activation.

Operating frequency is the maximum number of ON/OFF changes per second that the proximity switch, and particularly its SPDT output, is able to perform.

Residual current is the current that flows through the proximity switch when it is not activated by an object. The residual current has a particular importance when the proximity switch is connected in series with the load and, furthermore, when it is connected as input in a programmable logic controller, as is explained in Chapter 6.

The main types of proximity switches, with respect to their external form and shape, are the cylindrical, rectangular, and the slot ones, as shown in Figure 2.24. The double cycle area in all of them is the effective area for the detection of an object's presence. From the internal structure point of view and the used operating principle, the proximity switches include three basic stages, the LC oscillator, the Schmitt trigger unit, and the amplifier-output switching circuit, as shown in Figure 2.25.

Inductive proximity switch is where the LC oscillator creates, through the projected coil, a high frequency electromagnetic field which is extended in front of the whole effective area. If a metallic object enters inside the zone of the electromagnetic field, then eddy currents are generated in the object's body. These eddy currents generate their own electromagnetic field that opposes the field of the proximity switch and draws energy from the oscillating circuit, reducing thus the oscillation amplitude. The rise or fall of the circuit oscillation triggers the amplifier and the output switching circuit, where the oscillation change is converted to a SPDT contact-output change.

Capacitive proximity switches are similar to inductive proximity switches, since they have the same external form and size but a different operating principle. In general, the capacitive switch produces an electrostatic field instead of an electromagnetic field, and can detect metal as well as nonmetallic materials such as paper, glass, wood, liquids, and granular materials. In the capacitive proximity switch, the capacitor of the LC oscillator, shown in Figure 2.25, takes the place of the coil, just behind the effective surface of the cylindrical switch. The capacitor is formed by two concentrically shaped metal electrodes like two circular sectors (plates) as shown in Figure 2.26. It is known that the capacitance is proportional to the surface area of the electrode plates and the

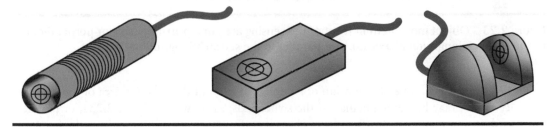

Figure 2.24 Basic exterior forms of proximity sensors for various applications.

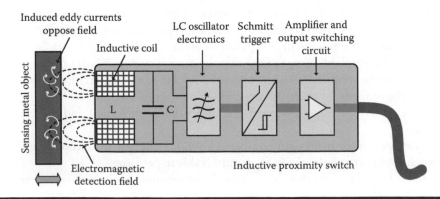

Figure 2.25 Internal elements of an inductive proximity sensor and object-electromagnetic field interaction.

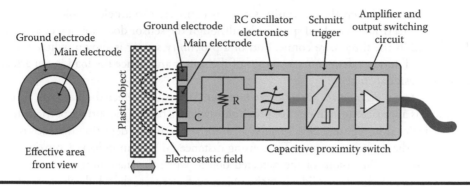

Figure 2.26   Internal elements of a capacitive proximity sensor and object-affecting dielectric constant.

dielectric constant of the material between them, while it is inversely proportional to the distance between electrodes. When an object approaches the sensing surface, it enters the electrostatic field of the electrodes and changes the capacitance of the LC oscillator circuit, since the object plays the role of a new dielectric material. As a result, the oscillator begins oscillating and continues until the object is removed. The existence or absence of the oscillation is subsequently converted to the SPDT contact-output change, in a similar way to the case of the inductive proximity switch.

Electric connection of proximity sensors: Inductive and capacitive proximity switches can be connected to an AC or DC power supply source in a series or separately with a load, the so-called two-wire or three-wire connections. The connected load may be the coil of a power relay, the electronic circuit of a control unit, or the digital input of a programmable logic controller. The two-wire connection of a proximity switch is shown in Figure 2.27a, where the load is connected in series under a 24 V DC power supply. In the two-wire connection of an electronic-type proximity sensor, a residual current flows through the load, even when the sensor is in the non-activated state. Due to the residual current, a rotary switch (RS) is necessary for the load isolation during the rest period. When the sensor is in the activated state, a voltage drop occurs among it and the flowing current increases to a level that is being determined by the load. Attention should be paid

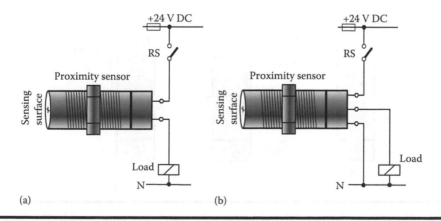

Figure 2.27   Electric circuit of a proximity switch with (a) a two-wire connection, and (b) a three-wire connection.

to the residual current, when the proximity sensor is connected to an electronic circuit, since it may activate the electronic-type load spuriously, although the sensor does not detect any object.

In Figure 2.27b the three-wire connection of a proximity sensor is shown, where the load is driven by an SPST independent contact output of the sensor and hence the load is not affected in any way by the residual current.

Effects of the target size and the metallic environment: When the detected object has an uneven-surface, while it is approaching an inductive proximity switch, a part of the surface participates in the activation, as depicted in Figure 2.28a. In such a case, the maximum detection distance is reduced proportionally to the surface curvature. The sensing distance (SD) of an inductive proximity switch is also affected by the dimensions of the detected object, while the maximum detection distance, provided by the manufacturer, is valid when the object surface covers the whole effective area of the proximity switch. In general, smaller dimensions of the detecting object lead to the reduction of the corresponding sensing distance, as shown in Figures 2.28b and 2.28c correspondingly. The material of the object also affects the maximum detection distance of a proximity switch. Particularly, for the case of inductive proximity switches, the sensing distance is usually given for carbon steel St37 material.

In general, the inductive proximity switch may be shielded or not. The non-shielded inductive switches are affected by the presence of a metallic environment, due to the electromagnetic field scattering as shown in Figure 2.29a. This means that the switch may be activated by the metal-lic environment and not by the target object. In such a case, the mounting of the non-shielded inductive switch should satisfy some safety distances, provided by the manufacturer from the adja-cent surfaces, as presented in Figure 2.29b. The shielded inductive proximity switches create an electromagnetic field that is restricted just in front of the sensing area, as shown in Figure 2.29c. Therefore, these types of proximity switches can be mounted in a metallic environment without risk of pseudo activation.

Magnetic proximity switches have a different and simpler operating principle from that of the inductive and capacitive ones. They consist of a permanent magnet with a projected pole piece, a pick-up coil around a pole-piece, and its required housing. When the magnetic flow through the coil is varied, then a current by induction is produced, and thus a voltage output appears at the coil terminals, as shown in Figure 2.30. The variation of the magnetic flow may be caused by any ferrous material moved through the magnetic field. The magnetic proximity switch usually works with a ferrous gear, while its rotation produces a series of pulse-type voltages. If the gear is

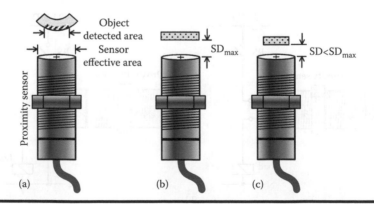

Figure 2.28 The sensing distance (SD) of a proximity switch that is affected by the form and the dimensions of the detected object: (a) spherical object in front of the sensing surface, (b) the object covers the whole of the effective area, and (c) the object has smaller dimensions than needed.

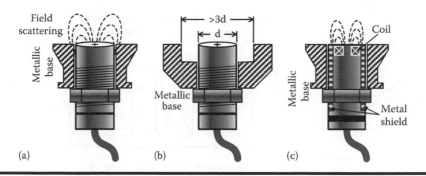

Figure 2.29 **The metallic environment affects the operation of a non-shielded inductive proximity switch: (a) non-shielded switch presenting field scattering, (b) safety distance for field scattering avoidance, (c) shielded switch in metallic environment without scattering.**

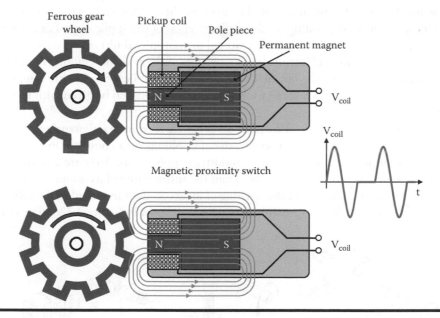

Figure 2.30 **Typical variable reluctance magnetic proximity switch.**

mounted on the shaft of a rotated machine, then the magnetic proximity switch can act as pulse generator, exactly synchronized with the rotation of the machine.

Magnetoresistive proximity switch: Another kind of proximity switch that bases its operation on magnetic field sensing technology is called magnetoresistive, because of the low or high resistance in a specialized chip element, depending on the existence or not of a magnetic field correspondingly. The magnetoresistive element (MR) is just behind the effective surface of the proximity switch, as shown in Figure 2.31, and is combined with a trigger circuit and an amplifier for outputting a digital signal. Magnetoresistive proximity switches usually detect the magnetic field of a permanent magnet and are able to detect very weak magnetic fields. Since the magnetic fields are able to pass through many non-ferromagnetic materials, the detection process can be triggered without the need for direct exposure to the target object. Magnetoresistive sensors react to both axially and radially magnetized magnets, have significantly smaller physical size, are vibration resistant, and have superior noise immunity.

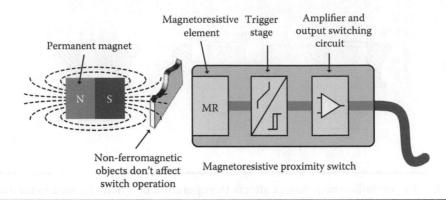

Figure 2.31 Magnetic proximity switch based on magnetoresistive detection.

Various other types of proximity switches exist in the market and may give the industrial engineer a better solution, depending on the distinctiveness of the application. For example, there are ultrasonic proximity sensors which use reflected or transmitted ultrasonic waves to detect the presence or absence of a target object. A deep knowledge of the available equipment in the market is a fundamental path toward the optimal solution to each application. In this case, optimality should be considered not only as a correct solution, but it should also be given with respect to the corresponding cost.

Applications of proximity switches: Proximity switches are used in manufacturing processes, e.g., to detect the position of machine components, to count objects on a conveyor, to monitor normal operation of a machine, and numerous other industrial applications. They are also used in robotics in order to monitor the distance of a robot from the surrounding objects and in safety systems, where a simple application can detect the opening of an access door in an industrial cell. The most common industrial applications of proximity switches are shown in Figure 2.32. Some of the most

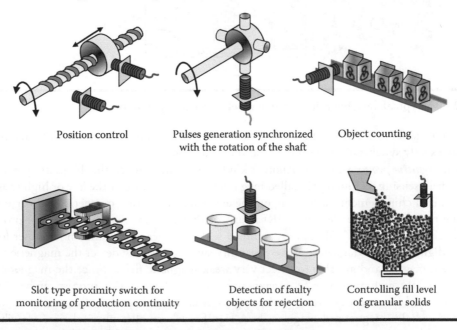

Figure 2.32 Typical applications of proximity switch usage.

typical applications of proximity switches are the precise position detection of a moving object in a machine tool; the number of produced objects as they are traveling on a conveyor belt; and the pulse generation for synchronizing purposes, which are also performed by suitable proximity switches. A slot type proximity switch can monitor reliably the continuity of production of lightly connected objects. As an example, the utilization of a proximity switch could also check the presence of the closures on boxes, in order to reject the faulty boxes from the production line. Furthermore, capacitive proximity switches can be used in specific applications, as in the case of level detection of a granular solid in a tank. The market offers a wide range of proximity sensors to meet different types of applications and many others. Finally, it should be noted that the lack of physical contact with the target object and the absence of mechanical parts increase the life of a proximity sensor and make it more reliable in comparison to other mechanical proximity sensors, such as "limit switches".

2.2.3 Photoelectric Switches

Photoelectric switches, called also photocells, are solid-state sensors that are able to detect the presence of an object at long ranges and use an output transistor to change the state of their SPDT-type digital output. The non-contact detection is performed by a beam of light, visible or infrared, which can be interrupted or ignored by the detected object. In general, the interruption of the light beam emitted by the emitter component, due to the presence of an object, is detected by the corresponding receiver and causes the activation of the sensor output. A typical exterior form of a photoelectric switch is presented in Figure 2.33. In general, there are three basic types of photoelectric switches, regarding the utilized reflection type of the beam, which can be categorized as the "through-beam", "retro-reflective", and "diffuse".

Through-beam photoelectric switches: Any photoelectric switch consists of an emitter and a receiver of the light beam. In a through-beam type of photoelectric switch, the emitter and the receiver are separate units, contained in different housings and positioned opposite each other, as

Figure 2.33   Typical exterior view of a photoelectric sensor.

shown in Figure 2.34. When an object breaks, the light beam causes the receiver to change the state of the output contacts, which have been inserted in the automation circuit. The sensing distance is the maximum operating distance between the emitter and receiver, and is achieved when the two units have perfect alignment and hence the maximum amount of the emitted energy reaches the receiver. This means that the placement angle of the emitter that is relative to the receiver must not exceed a predefined value given by the constructor. This type of photoelectric switch has the advantage of a longer sensing distance, in comparison to other types of photoelectric switches.

Retro-reflective photoelectric switches: In this type of photoelectric switch, the emitter and receiver are embedded in the same housing unit, and a reflector is used for the reflection of the light beam, as shown in Figure 2.35. Any object situated between the sensor and the reflector, interrupting the light beam from reaching the receiver, activates the output of the photoelectric switch, such as in the case of the through-beam mode of sensing. Reflectors are usually made up of many small, corner-cube prisms that reflect the beam back to the receiver, almost in parallel to the entering beam. A simple mirror can also be used as reflector, but precise positioning of it is needed in order for the emitted beam to strike almost perpendicular to the mirror surface, while reflectors may have a small deviation angle. When the detected object is opaque with a highly reflective

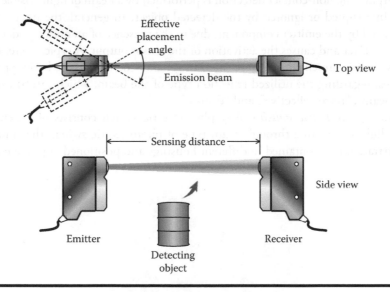

Figure 2.34 Through-beam photoelectric switch.

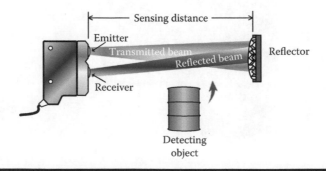

Figure 2.35 Retro-reflective photoelectric switch.

surface, it is possible to play the role of an artificial false reflector which is normally an unwanted faulty situation. In such a case, polarizing filters can be placed in the light beam in order to prevent the sensor from false triggering due to non-polarized light signals.

Diffuse-reflective photoelectric switches, which base their operation on the reflection of the light beam directly on the surface of the detected object, are also called "direct reflection" photoelectric switches for this reason. Obviously, in this case the emitter and the receiver are embedded together in one housing unit, as in the retro-reflective type. The emitter emits a beam of light that is not returned by reflection to the receiver. When the target object is inserted in the light beam trajectory, the beam is diffused in many directions, one of which is reflected back to the receiver, as shown in Figure 2.36. Since the reflection of the beam is performed on the detected object, the color and the type of its surface affects the operation of the sensor. Light colors usually have a better behavior, offering the maximum of sensing distance, while shiny opaque objects affect the reflection of the beam by type and quality of the surface rather than by color.

Photoelectric switches come in a variety of designs, sizes, and technical characteristics, each type being suitable for a specific application. The terminology that has been presented for proximity switches is also used in the case of photoelectric switches in a similar way. The detection distance for the through-beam and retro-reflective types of photoelectric switches is defined as the maximum distance between the emitter and receiver, or between the emitter and reflector correspondingly. In the case of diffuse reflection types, it is the maximum distance between the photoelectric switch and the detected object. The detection distance varies with the type, size, and model of the photoelectric switch and ranges usually from less than 10 cm up to 1500 cm. Photoelectric switches have a residual current, which is necessary to power the sensor, while their frequency or ON-OFF output cycles per second depend on the AC or DC voltage of operation and may range from a few Hz to 700 Hz or more. A new term, used only in photoelectric switches, refers to their "Dark On" or "Light On" operations. The "Dark On" operation means that the photoelectric switch is energized when the beam of light is interrupted, or simply when the switch is "in the dark". Instead, the "Light On" operation means that the switch is energized when the beam of light reaches the receiver or simply when the switch is "in the light". All cases for the three basic types of photoelectric switches are presented in Table 2.4.

The photoelectric switches can be connected to an AC or DC power supply source through a two-wire or three-wire connection, similarly to the proximity ones. The output of photoelectric switches is usually an SPDT contact that changes state when the switch is energized. The NO or NC contact change is transformed into a digital signal that can be sampled by an external control unit (e.g., a motion controller, PLC or an automation circuit) in order to trigger a variation in the operation of the overall controlled system. In addition to the above-described photoelectric switches, various other types, such as fiber optics types or laser types, may be used for specific applications.

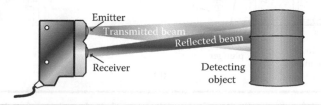

Figure 2.36   Diffuse-reflective photoelectric switch.

Table 2.4   The Dark-on and Light-on Operation of a Photoelectric Switch

Photoelectric Switch Type	Light ON	Dark ON
Through beam		Target
Retro-reflective	Reflector	
Diffuse reflective	Target	

2.2.4 Limit Switches

Limit switches are purely mechanical constructions, where the only electrical parts they have are their electric contacts. These types of switches have the ability to detect the "end or limit of a motion" through the rectilinear movement and force of a plunger that is transferred to the internal mechanism and converted to open or closed electric contacts, as shown in Figure 2.37a. The return spring resets the switch to its initial normal position, when the cause of the plunger movement is eliminated. The switch housing may be made from aluminum or plastic material, but in all cases, it protects the internal mechanism with superior electrical insulation and mechanical strength. Figure 2.37a shows the manual actuation of the limit switch by hand in a similar way as a push button switch, but in reality, limit switches are mechanically actuated by a movable object, e.g., a piston rod or a movable shaft. Whereas the movement of the plunger during actuation is possible in only one direction, the motion of the object actuating the switch may have any direction with respect to the plunger. From a practical point of view, it is very difficult for the two movements, of the plunger and the actuating object, to be identified. For this reason, an actuator head is adapted to the body of the limit switch, as shown in Figure 2.37b, in order for a

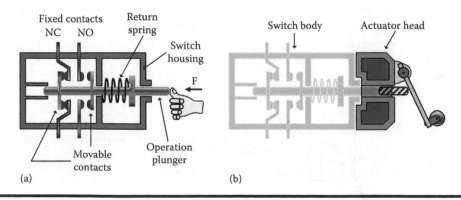

Figure 2.37   Typical form of a limit switch with (a) an interior view of switch body, and (b) a switch body equipped with an actuator head.

specific motion of the actuating object to be acceptable with safety from the limit switch. A wide range of limit switches, with various actuator heads, exist in the market, thus being able to detect the motion of any kind of moving objects. In all these types, the switch body is the same, with the only difference being the type of sensor head. Some basic types of limit switches are shown indicatively in Figure 2.38.

Limit switches are cheaper in comparison to proximity or photoelectric switches, and thus may be preferred in any application where noncontact actuation is not required. There is also a wide range of limit switches regarding their size, from subminiature size for use inside small machines to heavy-duty ones for large mechanical systems. Furthermore, limit switches are available in a variety of rotary arm styles, operating forces, reset forces, over-travel distances, operation requirements, and environmental factors (including moisture, contamination, temperature, shock, and vibration). Limit switches have contacts with electrical ratings that are usually around 250 V AC and between 10 and 15 A. Some manufacturers of plug-in style limit switches offer switch bodies that have a LED or neon lamp status indicator. As with all other sensors, the actuation of a limit switch

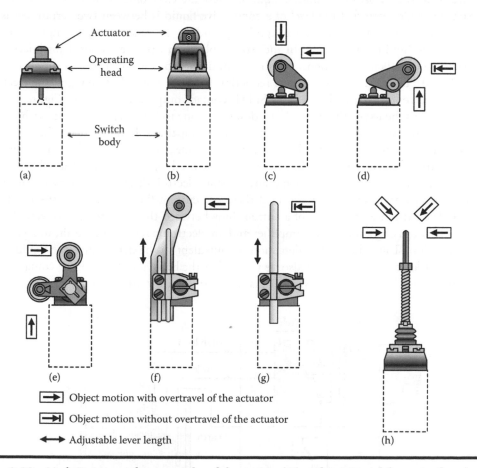

Figure 2.38 **Various types of actuator head for transmitting the external force to the change-over mechanism suitably and thereby engaging the movable contact's (a) push plunger, (b) roller plunger, (c) roller lever plunger (motion from the front or side), (d) roller lever plunger (motion from the back or side), (e) fork type, (f) adjustable roller arm, (g) adjustable rod, and (h) spring flexible rod.**

provides an electrical digital signal (change of contact state) that causes an appropriate control system response.

2.2.5 Level Switches

Level switches or level sensors are devices of various kinds, from fully electronic to purely mechanical types, with a general usage in detecting if the level of a liquid has a definite height. The detection of the liquid level inside a tank is a common need in any industrial process, large or small. For example, from large tanks in a petrochemicals refinery to small tanks in a brewery or water treatment plant, it is quite necessary to know if the liquid level has reached a high or low limit (level switches) or to measure continuously the value of the liquid level (level transducers). The same need exists for various granular materials of any kind inside a silo in many industrial applications, while in the following, the most representative basic types of level switches or sensors are described.

Electronic level switches for conductive liquids: These are electronic devices that can, by using immersed electrodes, detect if the level of a conductive liquid is between two certain levels that the engineer can define according to the application needs. There are many examples of electrically conductive liquids encountered in industry applications, such as water, fruit juices, milk, beer, sewage, acids, alkaline solutions, etc. The main device consists of the power supply circuitry (230 V AC, 50 Hz) including the voltage reduction to the required low voltages, the electronic circuit, the electrodes' terminals, and the SPDT contact output, as shown in Figure 2.39. Its operation requires the use of three rod electrodes, where one rod (E_3) represents the earth connection and the other two electrodes (E_1, E_2) represent the upper and lower limits of the liquid level respectively. If the liquid tank is metallic, the earth electrode can be connected to any point of the metallic construction, and therefore does not necessarily need to be immersed. The voltage applied to electrodes is an alternating one to avoid any possible electrolytic effect and very low for safety purposes. The electronic circuit can detect if there is a current flow between electrodes through the liquid. The existence or absence of a current flow between the electrodes is converted by the electronic circuit, using a measuring amplifier and an electronic self-holding unit, to a switching contact alteration. Thus, the current detection is equivalent to the detection if the liquid level is at the upper or lower level. It should also be noted that the behavior of the SPDT contact output is different during the level's rise versus fall. As shown in Figure 2.40, when the level is rising and

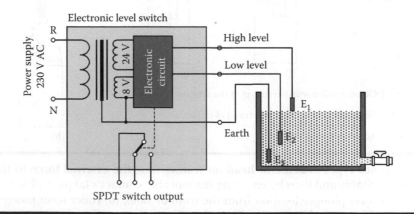

Figure 2.39 Electronic level switch with immersed electrodes for conductive liquids.

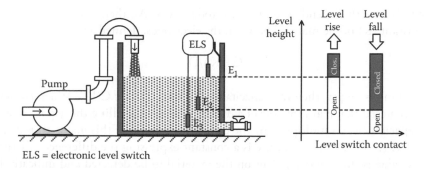

ELS = electronic level switch

Figure 2.40 Differential behavior of an electronic level switch output during level rising and falling.

reaches the upper electrode (E_1), then the contact output is activated. If the liquid level starts to fall, the output remains activated and will be deactivated only when the liquid level falls just below the lower electrode (E_2). When the electronic level switch is combined with a pump, either filling or emptying a tank automatically with its so-called "differential" behavior, the frequent and rapid start-stop repeated operation due to ripples on the surface of the liquid is avoided. Because of this differential behavior between the two-level limits and their electronic structure, the electronic level switches are sometimes called "level controllers". Since the conductivity of the liquids can differ considerably, the response sensitivity of the electronic circuit detecting the current through the liquid is usually adjustable.

Capacitive level switches: It is obvious that the electronic level switches described in the previous section cannot be used for the case of non-conductive liquids or solid materials. The measuring system of level switches is capable of detecting liquids, as well as bulk solid materials, and is based on the capacitance measuring method. The capacitive level switch is a unified device that is side-mounted entirely on the tank walls, exactly at the ideal point the level should reach. As shown in Figure 2.41, the device body is on the outside of the tank, while the projected set of the two capacitive electrodes is on the interior of the tank. The first electrode is the sensitive one, meaning that it senses the presence of material, while the second electrode is called insensitive and usually is permanently connected to the earth (tank wall). The two electrodes are electrically isolated from one another by a suitable polyacetal material and form a capacitor with dielectric material, the air. The operation principle of the capacitive level switch is based on the variation of the capacitance due to the presence of liquid or bulk solid material around the electrodes. In general, the

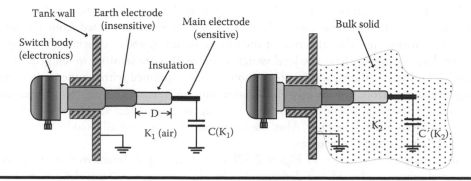

Figure 2.41 Capacitance type level switch for direct contact with bulk solids and liquids.

capacitance C is directly dependent on the electrodes areas (A), their distance apart (D) and the dielectric constant (K) of the material between the electrodes, defined as:

$$C = KA/D$$

When the tank is empty, then the capacitance (air dielectric) is $C(K_1)$. When the level switch is covered by material that plays the role of a dielectric substance filling the gap between the electrodes, the capacitance gets multiplied by the dielectric constant of this material (K_2) and specifically varies to $C'(K_2) = e\ C(K_1)$, where e is a constant expressing the difference of two dielectric constants. Its value is directly dependent on the material to be detected and indicatively has the value 2–3 approximately for corn and 70–80 for water.

The variation in capacitance is subsequently translated into a switching output by an oscillating circuit, the frequency of which is dependent on the value of capacitance. The oscillation or stabilization of the circuit corresponds to the two states of the SPDT contact output and hence to the existence or absence of a material in the tank. The rod-type electrodes of a capacitive level switch are strong against the buried situation, and therefore are suitable not only for fine powder but also for bulk solid materials. Since the temperature, moisture content, humidity, and density of the process material can change its dielectric constant, the capacitive level switches are equipped with a sensitivity regulating mechanism for calibration. If more than one capacitive level sensor needs to be mounted in the same tank, a minimum distance between them should be provided by the manufacturer in order to avoid interference of their electromagnetic fields. Capacitive level switches can be used for level detection in silos, tanks, and bunkers in all areas of industry for conductive or non-conductive liquids, as well as for bulk solid material with a dielectric constant greater than that of air.

Ultrasonic level sensors: Ultrasonic level switches and transducers work both on the same basic principle of generating and receiving after a target reflection of ultrasonic waves, but are different in their output type. Ultrasonic level switches have a digital output signal of a SPDT contact type, while ultrasonic level transducers are capable of non-contact measuring of the level through a microprocessor-based circuit and producing an analog output signal, usually 4–20 mA. Some ultrasonic level transducers offer both switch and current outputs, which means they can be used either as a level switch or as a level meter. Ultrasonic level sensors can detect liquids, sludge, and solid materials.

An ultrasonic wave is a high-frequency acoustic wave that cannot be heard by someone. In general, people can hear an acoustic wave or sound if it is within the range 20 Hz to 20 KHz. The transmitter of an ultrasonic level sensor emits acoustic waves usually within the range 30 KHz to 200 KHz. The emitted ultrasonic waves hit the liquid, sludge, or solid surface and are reflected back to the sensor, as shown in Figure 2.42a. The level is then calculated from the time lag between the emission and the reflection of the ultrasonic wave, and is converted to a digital or analog signal accordingly. Ultrasonic level switches and sensors are sensitive to temperature, pressure, and humidity conditions, and for this reason they are equipped with compensation units for reducing measuring errors. On the other hand, ultrasonic level sensors have some basic advantages over other technologies of level detection. For example, they can detect various materials that are quite far away, even more than 15 m. Also, the ultrasonic waves are not affected by the color of the target surface and its possible changes.

The ultrasonic switch shown in Figure 2.42b uses a slightly different method to detect the presence or absence of a liquid at a designated point. It contains two piezoelectric crystals, one transmitting a high-frequency (about 2 MHz) sound and one receiving the previous sound, which

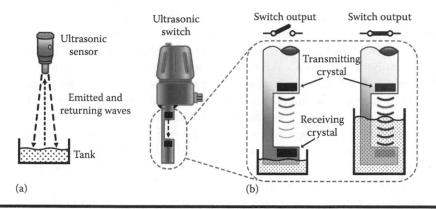

Figure 2.42 **Ultrasonic sensor for non-contact level measurement (a), and ultrasonic switch for contact level detection (b).**

are mounted opposite each other at a small distance of a few millimeters. The ultrasonic switch uses the different behavior of sound transmission in air and liquid to detect the liquid presence. When there is no liquid in the gap between the two crystals, the receiver accepts a weak signal, due to the sound transmission in air which presents attenuation. When liquid is present, the sound retains almost all of its signal strength and the receiver accepts a strong signal. Subsequently, the electronics detect this difference and switch an SPDT contact output accordingly.

Radar type level sensor. Since it is difficult for the lag time-based method described in the previous section to give very accurate measurements for such small time intervals, the frequency modulated continuous wave method is used in radar-type level sensors. A radar signal that is emitted via an antenna toward the liquid surface is a microwave signal with a continuously varying frequency. When the reflected signal returns to the receiver, it is compared with the outgoing signal. Since the transmitter continuously changes the frequency of the emitted signal, there will be a difference in the frequency between the transmitted and the reflected signals. The distance of the level from the sensor location (taking into account the dead zone) is then calculated by measuring the proportional frequency difference, as illustrated in Figure 2.43.

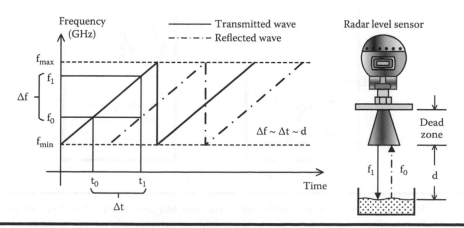

Figure 2.43 **Radar level sensor operation principle.**

Float level switches: The simplest level switches from a construction point of view are the float-based level switches. In Figure 2.44 some basic types of float switches are shown. The float switch shown in Figure 2.44a requires sidewall mounting, while a similar one shown in Figure 2.44b is suitable for a vertical placement. Both types use a sealed-in-glass magnetic switch (called "reed relay", which is presented in Section 2.3.4) and a floating part that contains a permanent magnet. When the liquid level is low, the permanent magnet is far from the magnetic switch, and thus its SPST contact is open. When the liquid level increases, the floating magnet is moved toward the magnetic switch. Once the magnetic switch is reached, the floating magnet activates the SPST contact, which subsequently closes. Therefore, the floating magnet follows the changes of the liquid level, and at a designated height, the contact output changes status and remains there as long as the level height does not change.

Another type of a float level switch is shown in Figure 2.44c where the two metallic plates play the role of an electric contact, and the metallic ball consists of the medium closing the contact. All the mentioned components are inside a plastic floating box, the orientation of which, up or down, depends on the liquid level height and determines the status, closed or open, of the contact.

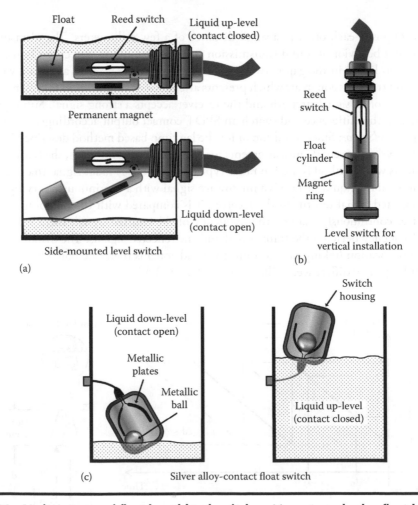

Figure 2.44 Various types of float-based level switches: Magnet–reed relay float level switch for (a) sidewall mounting, (b) vertical placement, and (c) metallic plate–ball float level switch in two possible states.

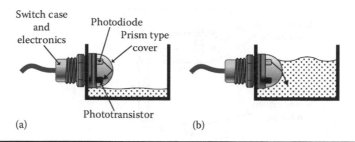

Switch case
and
electronics

Photodiode

Prism type
cover

Phototransistor

(a)   (b)

Figure 2.45   Optical level switch is non-refraction state (a), and refraction through the liquid state (b).

Optical level switch: The optical level switch can detect the level of transparent liquids, while its operation principle is based on the refraction of infrared light through the liquid. Due to its very small dimensions, it is suitable for mounting in small devices for general use or containing small vessels of water or any other transparent liquid. Typical examples include switches used in drink vending machines, medical machines and devices, in motor vehicle technology, etc. As shown in Figure 2.45a, the optical level switch consists of plastic housing, a transparent hollow hemisphere, an infrared diode, and a light-sensitive semiconductor. The infrared diode acts as a transmitter of a light beam and the semiconductor as a receiver of the light beam. When the switch is not covered by liquid, the infrared light beam is fully reflected on the surface of the hemisphere and is guided to the receiver. When the switch is covered with a transparent liquid, the infrared light is mostly refracted into the liquid and less light reaches the receiver, as shown in Figure 2.45b. The incidence or irregularity of the infrared beam on the receiver is then converted to a switching output that can be used suitably in an automation system.

2.2.6 Flow Switches

In many industrial applications, it is necessary to detect if there is liquid flow inside a pipe. For example, the cooling system of the bearings of a heavy-duty machine requires flow monitoring for ensuring water circulation. When the cooling fails due to malfunction of the water pumps, the switching signal of the water flow detector causes the machine to be switched off, first protecting the machine, and second, saving the water pumps from possible catastrophic dry running. Furthermore, it enables a simple light or visualized indication of the fault location. The flow switches can detect a minimum of flow rate, while remaining unresponsive below their cutoff value. Although the following description of two flow switch types will refer to liquid flow detection, in general, flow monitors can be used in both liquids and gases.

Paddle-type liquid flow switches are purely a mechanical construction, while the only electrical part relies on its electric contacts. As shown in Figure 2.46, the switch operates with the use of a paddle made from stainless steel or brass that is inserted in a pipe. When the liquid flow being detected pushes against the paddle, the paddle-arm system swings away. This movement changes the position of a stick tangent to the paddle cam, and thus activates the SPDT contact output. When the flow is interrupted, the paddle moves back to its initial rest position. Some types of flow switches use a permanent magnet for actuating and a magnetic switch as contact output. Paddle-type flow switches can be mounted on both horizontal and vertical pipe lines. The same flow switch of some manufacturers can accept paddles of various sizes according to the pipe size.

The *thermal dispersion flow switch* operates based on the principle of thermal conductivity. The term "calorimetric" flow switch is also used as a detector of the thermal energy that is removed

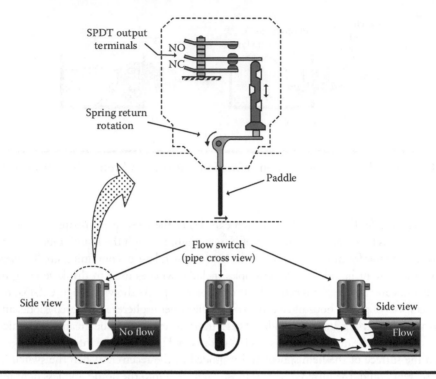

Figure 2.46 Paddle-type liquid flow switch.

from a heating source due to a medium flow. This sensing technology can be reliably applied almost in any liquid or gas. As shown in Figure 2.47a, the flow switch consists of two resistor temperature detectors (RTD) acting as temperature sensors and the required housing with electronics. The unheated RTD 1 sensor measures the temperature of the fluid where the probe is immersed. The RTD 2 sensor is heated from an embedded, constant, low-power heater and hence measures the temperature of the heater. This creates a temperature difference between the two sensors, which varies according to the existence or absence of flow, since the flowing medium cools the heated sensor. Particularly at high flow rates, the temperature difference is lower as flow removes more heat; at low flow rates, the temperature difference is higher, since flow removes less heat. The

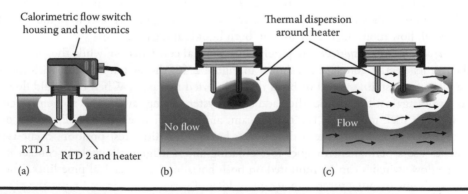

Figure 2.47 Calorimetric flow switch mounted on a pipe (a) and thermal dispersion around its heater probe for fluid at rest (b) and for existence of flow (c).

thermography around the sensor RTD 2 when the flowing medium is at rest or the medium is flowing is shown in Figures 2.47b and 2.47c correspondingly, and illustrates the heat dispersion principle. Therefore, the temperature difference is inversely proportional to the flow rate and provides a primary signal that is converted to an SPDT relay switching output.

A basic advantage of the thermal dispersion flow switch is that it has no mechanical moving parts providing precise and fast detection of flow and no flow. Compared to the traditional paddle type flow switch, thermal dispersion flow switches offer higher sensitivity and no limitation of installing locations, while are capable of responding under low flow rates.

2.2.7 Temperature and Pressure Switches

Temperature detection in an industrial process (and by extension, its monitoring and control) is a very common problem for an engineer of automatic control systems and automation. The production of heat is described in the industry either as a positive (desirable) phenomenon or as a negative (undesirable) phenomenon. In the first case, electric energy is converted into heat through resistors; for example, in an industry producing plastic objects, where the raw material is heated in order to be fluidized, hence it is suitable for processing. In the second case, mechanical friction (or the Joule phenomenon) results in rising temperatures; for example, in the case of a heavy-duty gearing mechanism or in the interior of a power transformer. In both cases, independently of the causation, it is necessary to check the temperature continuously; when it rises over a predefined limit, a proper control action needs to be triggered. The temperature detection may be very simple, in the form of an ON-OFF thermostatic signal (temperature switches), or a precise measurement in the form of an analog variable (temperature sensors).

There are many different types of temperature sensors, all with different characteristics, depending upon their actual application and principle of operation. The methods used for temperature measuring can be subdivided into contact and non-contact temperature detection methods. In contact temperature detection techniques, it is required that the sensor and the part containing the temperature-sensitive element are in physical contact with the object being sensed, while the thermal conduction property is used to detect changes in the temperature. In non-contact detection techniques, heat radiation and convection properties are used to detect changes in temperature via the energy exchanged between the detected object and the sensor. These two basic types of temperature sensors can be further subdivided into many other categories of sensors, according to the applied principle of physics. Due to many principles of physics that have been used for temperature detection, there is a wide variety of temperature sensors and switches on the market today, including thermocouples, resistance temperature detectors, thermistors, infrared, and semiconductor sensors or switches. Due to the large number of these sensors, it is not possible to give a detailed overview of all of them and thus, subsequently, the most widely used types of temperature switches only will be presented, keeping in mind that the use of switching contact outputs in an automated system has a similar principal of operation across the board.

Bimetal temperature switch (thermostat): The bimetal temperature switch is a contact type electromechanical temperature detector whose operation is based on a bimetallic strip described in Section 2.2.1. The bimetallic strip of the temperature switch, as shown in Figure 2.48a, consists of two different and similar bonded metals, usually made from nickel, copper aluminum, or tungsten. Simultaneously, the bimetallic strip constitutes an electric contact in a closed status under a normal temperature condition. When the strip is subjected to heat, as shown in Figure 2.48b, a mechanical bending movement is produced, opening the electric contact. Therefore, the bimetallic electric contact inserted in an automation circuit can interrupt the operation of a machine when the temperature rises.

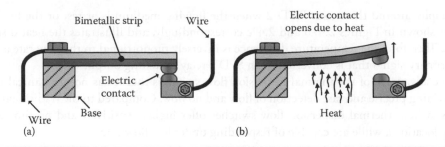

Figure 2.48   Bimetal temperature switch: (a) electric contact is closed at normal temperature, and (b) electric contact opens at high temperature.

With additional mechanical or electrical equipment, such as a regulator of the temperature interruption set-point or a pre-interruption switching signal, more complex control actions can be facilitated. Due to their simple and inexpensive detection mechanism, bimetal temperature switches are suitable and preferable for many industrial applications.

Thermocouple: The thermocouple is a type of temperature sensor or switch that is made by joining two dissimilar metals or alloys at their one end, as shown in Figure 2.49. The two disjointed ends of these dissimilar metals represent a junction that is kept at a constant temperature called a "cold junction", while the other junction, called a "hot junction", constitutes the sensing or measuring element that is in contact with the object being detected. The operating principle of the thermocouple is based on the well-known "Seebeck phenomenon" according to which a small voltage (few millivolts) is created between the two terminals of the cold junction when there is a temperature difference between the two junctions (cold and hot). If both junctions are at the same temperature, the potential voltage across them is zero. Since the resulting voltage is very small, an amplification is created by an operational amplifier (A) in order for the thermocouple output to produce anything more, such as an On-Off switching circuit.

Various combinations of metals are used in thermocouples according to the desired, detected temperature range. For example, the nickel chromium–nickel aluminum combination (Type K) is suitable for a –200 °C to 1250 °C temperature range to be detected. Two other types of widely used thermocouple materials are the iron-constantan (Type J) and copper-constantan (Type T) that have been recognized internationally as standards. One noticeable advantage of thermocouples used in temperature switches as sensing elements is their very small size, allowing thermocouples to be inserted into very narrow spaces.

Resistance temperature detectors (RTD): A well-known physical law of electricity is the resistance variation (to the flow of an electric current) with temperature in metallic materials. Its operation is based on this effect, and the resistance temperature detector is constructed from high-purity

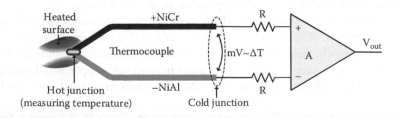

Figure 2.49   Thermocouple as a sensing element in a thermal sensor or switch.

conducting metals, such as platinum, copper, or nickel. The corresponding temperature switch uses an RTD as a sensing element, characterized by a linear positive change in resistance with respect to temperature. There are two common forms of construction for RTD sensing elements: Wire-wound RTD elements are manufactured by winding a small diameter of wire on a ceramic bobbin, as shown in Figure 2.50a. Another form is the thin-film RTD sensing element, which consists of a very thin layer of metal deposited onto a flat ceramic substrate, as shown in Figure 2.50b.

From the above-mentioned conducting metals, platinum is the most commonly used in RTD manufacturing, and the PT100 temperature detector is widely available in the market. It has a standard resistance value of 100 Ω at 0 °C and about 140 Ω at 100 °C, and it can cover a temperature range from –200 °C to 850 °C. There are also PT500 and PT1000 RTDs for higher measuring resistances.

Since the RTD elements are constrainedly connected to lead wires for extension purposes, their line resistances add errors, due to the increased total resistance value. On the other hand, since RTD is a resistive element, an electric current is needed to pass through it in order to detect the resulting voltage and to have the corresponding temperature. This creates self-heating of the resistive wires, causing additional errors. To avoid these errors, the RTD element is connected to a Wheatstone bridge, the other branches of which have a compensation action, as shown in Figure 2.50c. A voltage source excites the bridge, and the indicative voltage across the bridge output is proportional to the resistance of the RTD. This circuit is called a "two-wire connection" of the RTD sensing element. Better results from an accuracy point of view are given by the "three-wire" and "four-wire" connections in combination with a constant current source, connections that are not going to be considered further in this chapter. Figure 2.51 shows an external typical form of an RTD temperature sensor or switch, whose probe is inserted in the space where the temperature is being detected.

There are many other types of temperature sensors and switches, some of which will be mentioned briefly here.

The *thermistor* is similar to the RTD temperature sensor, whose resistance changes with temperature, having either a negative or positive temperature coefficient. The thermistor is made from ceramic type semiconductor materials and presents a large resistance variation for a small temperature change.

The *infrared temperature sensor* is a non-contact electronic detector of thermal radiation. It measures the infrared energy emitted by a stationary or moving object as a result of its thermal state.

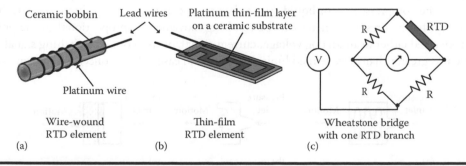

Figure 2.50 **Resistive temperature detectors and their incorporation in a Wheatstone bridge circuit: (a) wire-wound RTD element, (b) thin-film RTD element, and (c) RTD element connected to a Wheatstone bridge.**

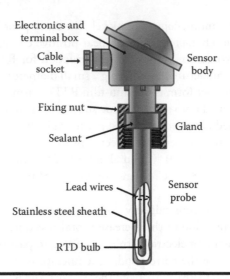

Electronics and terminal box
Cable socket
Sensor body
Fixing nut
Gland
Sealant
Lead wires
Sensor probe
Stainless steel sheath
RTD bulb

Figure 2.51 Typical form of an RTD temperature sensor or switch.

The *semiconductor sensor* of temperature is an integrated circuit fabricated in a similar way as all others. These sensors are classified into different types like voltage, current, digital, diode, and resistance silicon. The most common semiconductor temperature sensor is based on the fundamental temperature and current characteristics of the transistor. If two identical transistors are operated at different, but constant collector current densities, then the difference in their base-emitter voltages is proportional to the absolute temperature of the transistors.

Pressure and temperature are the two most commonly measured or simply detected quantities in industrial processes. As happens for temperature, there are many ways of sensing fluid pressure. Pressure sensors and switches consist of a mechanical part sensitive to pressure and an electrical part producing the output signal, analog (measurement) sensors, or digital (switching contact) switches. In general, there are different sensing elements, each of which prescribes the design of the mechanical part. The goal of all the pressure sensing elements is to produce a movement as a result of the presence of fluid pressure, while the most common of them base their operation on a diaphragm and piston configuration, as presented in Figure 2.52. The air or liquid pressure at the inlet port acts on a movable surface (e.g., a flexible membrane or piston base surface). The force applied to the movable surface depends on the area of the membrane or piston base and the pressure of the compressed air or liquid. Since the area is constant, the produced force is directly proportional to the pressure. Therefore, the resulting movement is proportional to the pressure, and is converted either to an analog voltage, current signal, or to a contact switching signal, when movement is greater than a predefined limit. In the second case, which concerns a pressure switch,

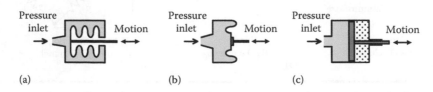

Pressure inlet — Motion   (a)
Pressure inlet — Motion   (b)
Pressure inlet — Motion   (c)

Figure 2.52 Pressure sensing elements for motion production: (a) Bellow, (b) Diaphragm, and (c) Piston.

the pressure set-point at which the switch is activated may be constant or adjustable. The above-described relation between the movement and the pressure is valid when the pressure on the other side of the membrane is the atmospheric pressure. In the case that the pressure switch accepts two different pressures, it is called a "differential-pressure switch" and the movement is proportional to the difference between the two pressures. The term "differential" should not be confused with the differential behavior of the pressure switch output during rising and falling of the pressure, as shown in Figure 2.53a. The difference between the switch operating point on rising pressure (P_{max}) and the switch operating point on falling pressure (P_{min}) is called the "dead band", and may be adjustable either independently in the setting points or as a fixed range.

A pressure switch with a piston as a sensing element is shown in Figure 2.53b. This is the simplest form of a pressure switch that is able to activate the SPDT contact, when a hydraulic or pneumatic system has reached a defined pressure value. In this case, the pressure enters through the connection inlet port and acts on the piston base. If the force resulting from the inlet pressure is greater than the downward force of the pre-loaded spring, then the piston is moving upward. The piston rod is also moving upward and the lever changes the state of the electric contacts. The trip-setting regulator is a nut that compresses the spring in order to increase the pressure set-point value where the switch is activated.

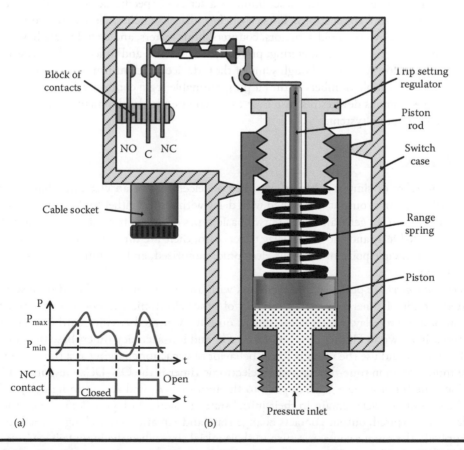

Figure 2.53 **Random pressure variation and NC switch-contact differential behavior of a pressure switch (a) and a non-differential piston type pressure switch (b).**

It should be clear to the reader that, concerning the characteristics of various industrial sensors, a minimum representative set of sensor types and detecting methods have been covered in Chapter 2, since there are hundreds of other kinds and types of sensors for all physical quantities and it is beyond the scope of this book to provide a complete list of these sensors (e.g., optical laser sensors for the detection of very small objects, fiber optic sensors using optical fiber as sensing element, linear position sensors based on the magnetostrictive technology, piezoelectric accelerometers for sensing vibration, inclination sensors, and vibrating level switches). Even with this huge variety in industrial sensors, all of them participate in an automation system in exactly the same way, which is the insertion of their digital or analog output into the process controller, which is described in Chapters 3 and 4. The industrial engineer should be in a position to have a comprehensive overview of the sensing elements, as well as be aware of the current trends and new, upcoming sensing devices, in order to select the best sensing technologies for a specific application; a selection that should satisfy both the performance requirements, as well as the cost constraints.

2.3 Timers, Drum Switches, and Special Components

This chapter is completed with the description of a series of specific devices used in industrial automation and control applications. Timer switches, drum switches, electric-network monitoring switches, reed relays, solenoid actuators, counters, hour meters, and encoders are basic devices, usually of small size, performing metering, protection, actuation, and time-based processing in an industrial automation system. Although some of these devices (such as timers and drum switches) are contained in multiple numbers within a programmable logic controller (PLC), they are still available as autonomous devices because they can facilitate the development of either a small or economical automation system.

2.3.1 Timers

A time delay relay or simply a "timer" is called an electronic or an electromechanical switch, where its SPDT contact output is activated with delay with respect either to the start or stop of the switching operation. In general, there are two basic types of timers from the time-function point of view: the ON-Delay and the OFF-Delay timer. Also, there are three basic kinds of timers from the operation principle point of view, the electronic, motorized, and pneumatic timers, which are presented next.

Electronic timers are pure electronic circuits with an operation that is based on a solid state, integrated circuit of timer type and a number of other electronic components that control the operation of a micro-relay, which is the timer's output. A printed circuit board of such an electronic timer is shown in Figure 2.54. Some low-cost and low-accuracy timers base their operation on an RC circuit and use the initially linear behavior of capacitor charging as a time meter.

The more common time-function of an electronic timer is the ON-Delay response of its output. When the input voltage V_{in} is applied to the terminals A_1 and A_2, a timing delay T begins while the output contacts remain in their initial state, as shown in Figure 2.55. When the preset time delay has expired, output contacts change state and remain there as long as the input voltage is applied. The output contacts return to their initial state when the input voltage is no longer applied. Of course, this output response or timer behavior is repeatable as many times as desired.

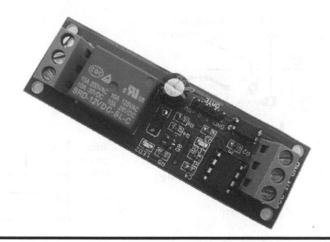

Figure 2.54   Printed circuit board of an electronic switch timer.

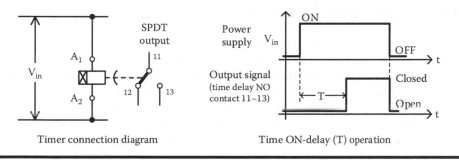

Figure 2.55   Connection diagram and ON-delay response of an electronic timer.

An electronic timer is able to cover a large time range, and the time delay is adjustable within this range. There are various types of electronic timers with several time ranges (for example, 1–10 s, 1–30 s, 1–6 min, etc.), offering a broad choice of timing ranges from a second or even less to several minutes, hours, or days. Also, there are several ways of timing adjustment from external knobs, dual in-line (DIP) switches, and thumbwheel switches to recessed potentiometers. Electronic timers can accept a variety of input voltages from 12 to 230 V AC or DC, while the switching capability of the output contacts is usually up to 12 A.

Pneumatic timers: The most common pneumatic timers are designed to operate in conjunction with a power relay or a contactor. They are mounted on top of a relay of the same manufacturer and base their operation on restricting the flow of air through an orifice to a rubber bellow, diaphragm, or small air cylinder. Figure 2.56a illustrates the principle of operation of a simple air cylinder timer versus the construction design. When the relay is activated, the piston is moved downwards, overcoming the spring force; the air enters into the cylinder, while the check valve and the output contacts remain in their initial state. At the end of this movement, the piston rod is released from the relay and the piston is allowed to return back, due to the spring force. The check valve is then closed, forcing the air to escape through the orifice, the size of which is adjustable via a suitably mounted knob. The rate at which the air is permitted to be exhausted (adjusted by knob) is proportional to the time interval (time delay) needed for the piston to return to its initial

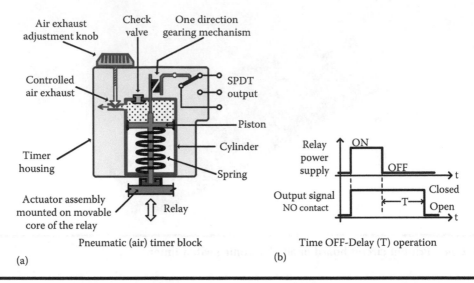

Figure 2.56 timer simplified internal view (a), and an OFF-Delay timer response (b).

Figure 2.56 Pneumatic timer simplified internal view (a), and an OFF-Delay timer response (b).

position. At the end of the piston's back movement, the projected part of the rod piston causes the output contacts to change state. The output contacts remain in this state until the relay is de-energized, when another mechanism (not shown in the figure) returns the contacts to their normal state. The described behavior of the pneumatic timer corresponds to the ON-Delay time function of the electronic timers. In a similar mode of operation, pneumatic timers can offer an OFF-Delay time function of output contacts. With an OFF-Delay pneumatic timer, timing begins when the relay is de-energized, as shown in Figure 2.56b. The output contacts change state immediately after the relay activation and return to their normal state after the time delay T has expired. Some pneumatic timers are designed to offer both time functions (ON-Delay and OFF-Delay) and permit the timer to be changed from one to the other by a simple mechanical micro-switch. Electronic timers with an OFF-Delay time function are also available, but they need one more input signal for triggering the off-event time instant and an uninterrupted power supply.

Pneumatic timers are popular throughout the industry because they have some basic advantages. For example, they are unaffected by variations in ambient temperature or atmospheric pressure, they are adjustable over multiple time ranges, they have a single turn time-adjustment knob, and they are simply convertible from ON- to OFF-Delay and vice versa. But the most significant advantage, in the authors' opinion, is that the pneumatic timer is the only device that can produce a kind of electric signal (electric contact change) with a time delay after the general failure of the central power supply system. In such case, the electric contact can be combined with a battery bank to activate with time delay, or a siren or a flashing light for indicating the existence of a fault, a functionality that cannot be achieved with other types of timers.

Motorized or electromechanical timers are motor driven timers with single or multiple time ranges, accurate and quick recurring operations, and instantaneous and/or delayed contacts. They use either a clock quartz motor or a synchronous motor to rotate a set of gears through which a mechanical activation of the electric contacts is achieved. If an electromechanical timer has multiple time ranges, the selection is performed by changing the gear ratio through a small recessed switch and a friction-clutch mechanism. Figure 2.57 shows two electromechanical timers from different manufacturers, on the front of which the time adjusting knob is visible. In timers with motors, the adjusting knob is back-rotated during timing, thus offering a moving pointer function.

Figure 2.57 Motorized electromechanical timers from Panasonic Corporation and A.G. Engineering Enterprise.

Several additional types of timers are available from the time-function point of view. For example, the delay-on with pulse activation timer, the flasher-type with an ON-OFF alternate change timer and the two-time-constants timer (T_1 for ON and T_2 for OFF) are some of them. More types of timers are described in Chapter 7, where PLCs contain multiple sets of timers each with multiple time functions. The use of timers in an automation system will be clear in Chapters 3 and 4, where the synthesis of automation circuits including timers is presented.

2.3.2 Cam Timers

Cam timers, or motorized drum switches, are electromechanical sequencers used in industrial applications with repetitive processes of a finite number of steps. They are considered an old-fashioned type of timer, due to the current PLCs where such timers can be easily programmed, but for small automation applications they are an appropriate solution, even today. Furthermore, their operation knowledge helps students to better understand the industrial processes, including a sequence of successive and interrelated ON-OFF operations of different machines. Figure 2.58 shows a general view of a multiple cam timer consisting of a motor, a gearbox, a setting disk, dual cams, and limit micro-switches. Details of its construction are shown in Figure 2.59 for a five-cam timer. The synchronous motor rotates the camshaft continuously through a gearbox with a constant cycle time period, usually of 1 min. This means that a full rotation (360°) of the camshaft takes 1 min. Several other cycle times, from 12 s to 24 hrs, are available from manufacturers.

The camshaft carries a number of fix-mounted dual cams with the ability to adjust their initial position. The number of dual cams may vary from 1 to 12 or even more, according to the application needs. Each dual cam has two separated sides, each of which can slide concentrically with respect to the other. The periphery region of each cam-side is properly formatted in order to form a groove angle in each cam by sliding, as shown in Figure 2.59. Obviously, the value of the groove angle corresponds to a defined constant time interval per cycle of operation. In front of dual cams there are an equal number of leaf-spring-mounted limit micro-switches, each with DPST electric contacts that are wired to devices on a machine for ON-OFF control. Therefore, adjustable hand cams can turn the micro-switches on and off, and the switches can be adjusted at specific times within an operation cycle, hence they can "program" any sequence of operations that will be continuously repeated. Multiple cam timers are often referred to as time-step programmers,

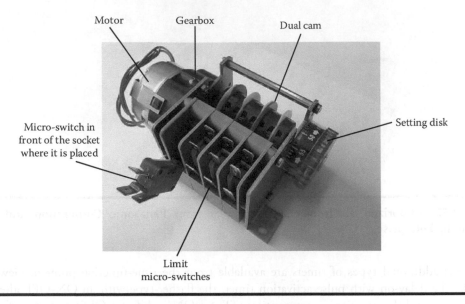

Figure 2.58   Electromechanical switch cam timer-programmer.

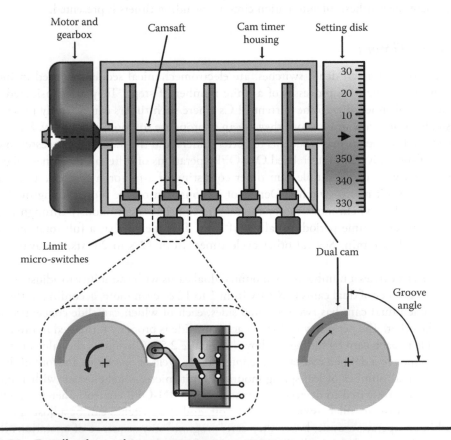

Figure 2.59   Details of cam timer programmer components.

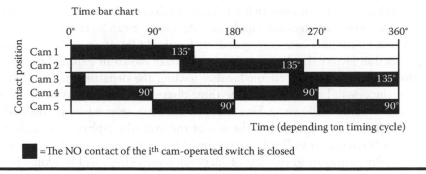

Figure 2.60   Timing bar chart of a five-cam timer.

sequentially energizing and de-energizing various electrical devices. A setting disk, calibrated in degrees, facilitates the user to adjust the groove angles and the relative position between them. In order to denote the groove angle for each cam with clarity and present the contact status with respect to the cycle time, a time bar chart should be designed, such as in the example of a five-cam timer shown in Figure 2.60. As mentioned above, the various cam angles correspond to equivalent time durations depending on the cycle time. For example, in the case of a 1 min cycle time, the groove angles of the first three cams correspond to a time duration of 22.5 s, while the closed status of the NO contacts creates an overlapping of 2.5 s with each other.

2.3.3 Three-Phase Monitoring Relays

Three-phase monitoring relays are electronic devices that monitor various basic electrical quantities of the three-phase power network. A typical utilization is for the measurement of the current values of the electrical quantities and their deviations from the nominal values, where the monitoring relay is able to activate their output(s) when these deviations exceed a predefined limit. Their output is usually an SPDT contact, suitably inserted in the automation circuit in order to cause the interruption of the load operation, as shown in Figure 2.61a. The interruption of the operation

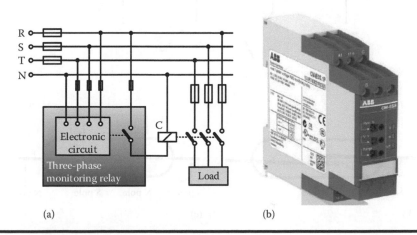

Figure 2.61   Connection diagram of a three-phase monitoring relay (a), and a single phase monitoring relay from ABB Ltd. (b).

is a protective action, since it is known that all electric devices can be damaged when they operate continuously under abnormal electrical conditions. As a load, it can be considered either a simple machine with one motor (this case corresponds to Figure 2.61a) or an entire department of an industrial production factory. In the second case, the relay shown in Figure 2.61a is substituted by the electromagnetic three-pole circuit breaker feeding the department, and particularly by the trip coil of the circuit breaker. When the three-phase monitoring relay is connected to the side of a power network, as shown in Figure 2.61a, then it is referred to as "network monitoring". Otherwise, it may be connected to the side of the load (three-phase line feeding the load) and in this case it is referred to as "load monitoring". Obviously, an eventual interruption, due to the monitoring-relay activation in the case of "network monitoring", will be global for the entire department, while in case of "load monitoring", it will concern only the monitored load.

Several important parameters of a three-phase power network can be monitored reliably and continuously by a three-phase monitoring relay. For example, it can monitor a phase failure, a phase sequence, a phase imbalance, over- and under-voltage conditions, over- and under-current conditions, a power factor, a frequency deviation, a broken neutral wire, and combinations of them. Therefore, there is a wide variety of monitoring-relay types in the market, mainly depending on the number and kind of monitored electric parameters. In general, the utilization of such a multifunctional three-phase monitoring relay ensures a trouble-free production procedure, while a single-phase voltage monitoring relay is shown in Figure 2.61b used for the protection of sensitive equipment and critical control systems.

2.3.4 Reed Relays

Despite their widely used name "relay", reed relays are switches that are simple in structure and magnetically actuated. As shown in Figure 2.62a, where a reed relay is depicted, it has a quite different form from the relay. Reed relays or switches are manufactured with two ferromagnetic reeds that actually are electric contact blades, partially enclosed in a hermetically sealed glass capsule.

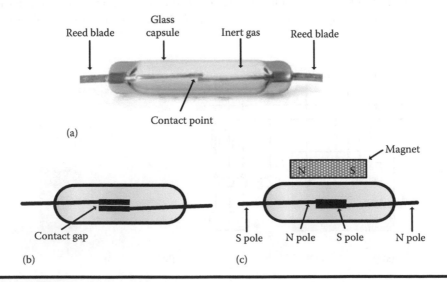

Figure 2.62 A simple no-cased form of a reed switch (a), NO status of the reed switch contact (b), and the permanent magnet that closes the contact (c).

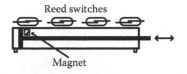

Figure 2.63   Example of reed switches used as position sensors.

The two opposing blades are overlapped at their free ends inside a capsule with a gap between them, as shown in Figure 2.62b, while the other two ends are the terminals of the reed switch that are able to be soldered. The contact area of the two nickel-iron blades is coated with special metals, such as ruthenium for protection from arcs. The glass capsule either has an internal vacuum (for high voltage switching) or is filled with an inert gas to exclude contaminants and to prevent oxidation of the contacts.

When the magnetic field from a permanent magnet or a wire coil is close to the reed relay, then poles of opposite polarity are created among the blades and the contact closes, as shown in Figure 2.62c. Obviously, the field strength must be suitable so that the magnetic force exceeds the self-spring force of the reed blades. When the permanent magnet is moved away and the force between the blades is less than the restoring force, the contact opens. An electromagnetic coil for the creation of the required magnetic field (instead of a permanent magnet) may be part of a unified construction together with the reed switch. From the above description, it is advisable to distinguish the reed switch from the reed relay in order to remove any confusion. A reed switch is only the glass encapsulated two-blade contact, while a reed relay is a reed switch with an embedded electromagnetic coil.

Reed relays and switches have several advantages, for example their low cost, long life (billions of switching operations), simple construction, fast switching time, absence of moving parts except for the elementary motion of blades, etc. But they also have some disadvantages; for example, the low current rate, sensitivity to other magnetic fields, sensitivity to vibrations, etc. However, both devices are ideally used for switching and sensing applications in industry, instrumentation, security installations, vehicle manufacturing, and many others. In fact, a reed switch is the simplest sensor of a magnetic field and, in combination with a permanent magnet, it can be used for liquid level detection, position detection, and counting. For example, a double-action pneumatic cylinder usually has two piston position states (in or out) that may be equipped with an internal permanent magnet for reed switch activation, as shown in Figure 2.63. Then, by installing a number of reed switches outside the cylinder, it is possible to stop the piston in intermediate positions corresponding with the reed switches.

2.3.5 Specific Solenoid Actuators

General purpose solenoid actuators were described in Section 2.1.11, where their operation principle was explained. These actuators are able to convert electric energy into a mechanical pulling force able to open or close valves or similar devices. In this section, solenoid actuators of the same operation principle and produced force are presented, but for specific use in power circuit breakers. Circuit breakers for industrial use are automatic, three-phase switches that can interrupt load currents under fault conditions. They are equipped with a complex over-toggle mechanism, operating either manually (handle) or remotely (motor and gear train) for closing or opening their contacts. In both cases, a strong spring is compressed and stores energy that can be released during

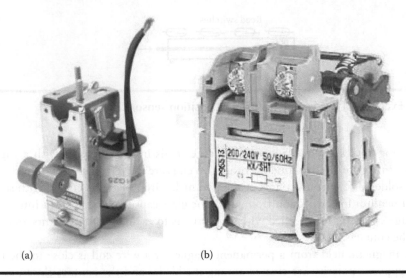

(a)   (b)

Figure 2.64 Solenoid actuators for current breaker operation, shunt trip coil (a) and under-voltage release coil (b).

the opening of the circuit breaker. In order to enable this action, a solenoid-produced force with the capacity to open a circuit breaker is needed. Such a solenoid is called a "shunt trip coil" and is shown in Figure 2.64a. It is mounted inside a circuit breaker to perform the trip operation. There are many reasons why a trip operation or disconnection is desirable. Any abnormal occurrence or critical fault may cause more destruction if the circuit breaker remains closed. For example, in the case of a factory fire, the automation circuit detecting fire via a corresponding sensor must activate the trip coil in order for a general power interruption to take place. In such a case, the automation circuit should be powered by a remote separate source, such as a battery.

The under-voltage release coil, shown in Figure 2.64b, is also a solenoid actuator that performs a similar trip to the circuit breaker, due to the low voltage of the supply network. It is known that a low voltage may destroy all electrical devices (mainly electric motors) if operated for a long period of time. The under-voltage release coil monitors the voltage of the supplying power line and automatically trips the circuit breaker when the voltage falls below 70% of its nominal range. If the under-voltage coil mechanism is not energized, it is impossible to close the circuit breaker, either manually or electrically. Both trip coils may have an additional break-make contact for signaling purposes in the automation circuit. In Chapters 3 and 4 dealing with the automation circuits' design, how the automation circuit of a power transformer activates the trip coil of a circuit breaker is demonstrated.

2.3.6 Counters and Hour Meters

In many industrial applications, it is necessary to count pulse-type events such as strokes, rotations, "product passing", ON-OFF changes of devices, etc. The simplest way to achieve this is to use an electromechanical counter, like the one shown in Figure 2.65a. Counting only has an indication purpose, unlike counting-value processing. Conventional automation systems are unable to process counting values. However, in the case of PLCs that contain many digital counters, it is possible to both indicate and process counting values. Electromechanical counters are often referred as "totalizing" counters, since they display a total quantity. For example, in counting bottles, the

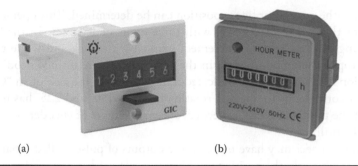

(a)   (b)

Figure 2.65   Typical view of an electromechanical counter from General Industrial Controls Private Limited (GIC) (a), and hour meter from Camsco Electric Co. Ltd. (b).

counter value will express the total number of passed bottles in front of a photocell application. Electromechanical counters accept pulse input signals from sensors, switches, and relay contacts. When the output contact of a sensor closes, the automation circuit supplies the nominal voltage to the counter input, causing one increment per voltage pulse. The counters usually have a 3–6 digit display, which may have presetting capabilities and accept manual or electric reset.

Hour meters are used in applications where time duration is critical. To make a process more effective and safe, hour meters are used to determine how long a machine has been running in order to schedule machine maintenance or to start another process at a predetermined point in time. Electromechanical hour meters are designed for the determination and monitoring of the operating time of electrically driven machines, and are therefore connected in parallel to corresponding relay coils. Obviously, an hour meter will only work during the operation of a machine or device. In Figure 2.65b, a panel-mount hour meter for industrial use is shown. In general, there are a wide range of electromechanical hour meters in all common voltages (AC or DC) and in various mounting styles.

2.3.7 Encoders

Encoders are specific sensors that generate square pulses in response to a wide range of motion tasks. In many manufacturing and production processes, there is a need for precise shaft rotation positioning or linear motion position measuring. For this purpose, there are shaft or hollow shaft encoders that respond to rotation (rotary encoders) and linear encoders that respond to a linear movement along a straight-line motion. As electromechanical devices, encoders are able to follow the detected motion through mechanical coupling and provide a motion control system with feedback information, concerning position or distance, rotation speed or velocity, and direction. Linear and rotary encoders are both available with two main types of output interpretation: the absolute encoder and the incremental encoder.

The output of absolute encoders is a multi-bit digital code that indicates the actual position of the moving mechanical object directly. This means that an absolute encoder has a reference position. Even in the event of a power outage, it has the ability to maintain a record of its absolute position. After restarting, the motion system can resume motion immediately without the need for rehoming. On the other hand, the output of incremental encoders generates a series of pulses proportional to the rotation of the shaft (rotary encoder) or distance traveled (linear encoder). By counting the number of pulses per a desired period of motion and based on the resolution of the encoder—i.e., the number of pulses per revolution or millimeters per unit travel given by the manufacturer—the angular or linear position can be determined. This operation is presented in Figure 2.66 for the case of an encoder with three digits of accuracy. In the left part of this figure the encoding of the encoder is presented, marked with black and white regions that can be detected by proper photodiodes, while in the right part, the resulting encoding and the corresponding rotation measurements are indicated. For example, in the case of an "ON-OFF-OFF" measurement that corresponds to line 4, a rotation between 180° and 225° has been achieved. If more accuracy in the positioning task is needed, alternative types of encoders with more digits of accuracy should be utilized.

An incremental encoder may have one or more outputs of pulses called "channels". A single-channel output can give only the position, and hence is not used in applications where direction of movement is required. A two-channel encoder produces two pulse trains that are 90° out of phase with each other. Then, the direction of movement can be determined through the phase relationship between them. Depending on the number of the rising and falling points of the two pulse trains that are counted, the counting resolution may be single, double, or quadruple. Single resolution corresponds to rising edges of one channel only. Double resolution is obtained when both rising and falling edges of one channel only are counted. Quadruple resolution, the most precise motion detection, is achieved when both rising and falling edges are counted from both channels. Some incremental encoders can also include an additional channel known as the "0-index pulse" or "Z channel", producing one pulse per rotation for a rotary encoder or at specific, precisely-known positions for a linear encoder. Since, the incremental encoders don't have a constant reference point (as absolute encoders do), the 0-index signal can be used for the homing procedure at startup; for example, the operation of an incremental encoder, shown in Figure 2.67, for the case of an encoder with dual channel and zero index output. From this figure, the resulting pulse trains for the channels A and B are also depicted, that are also different for different rotation directions and the resulting coding for the clockwise and the counter-clockwise operation. Based on the accuracy of the encoder and the number of district pulses per rotation, the rotation achieved can be measured. The bottom part of this figure also indicates a 2D representation of the pulse trains, and a zero index in the case of a quadrature encoder.

Absolute and incremental encoders can base their operation on optical, magnetic, or laser technology. Independently of their operation principle and kind (which is beyond the scope of this book to analyze), encoders are used in industrial applications in exactly the same way. It is

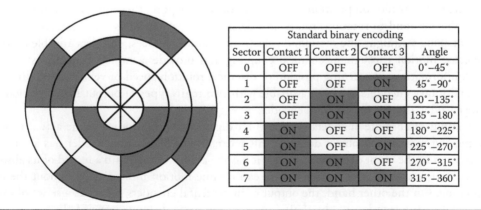

Standard binary encoding				
Sector	Contact 1	Contact 2	Contact 3	Angle
0	OFF	OFF	OFF	0°–45°
1	OFF	OFF	ON	45°–90°
2	OFF	ON	OFF	90°–135°
3	OFF	ON	ON	135°–180°
4	ON	OFF	OFF	180°–225°
5	ON	OFF	ON	225°–270°
6	ON	ON	OFF	270°–315°
7	ON	ON	ON	315°–360°

Figure 2.66 Binary encoded output for a full rotation of an encoder.

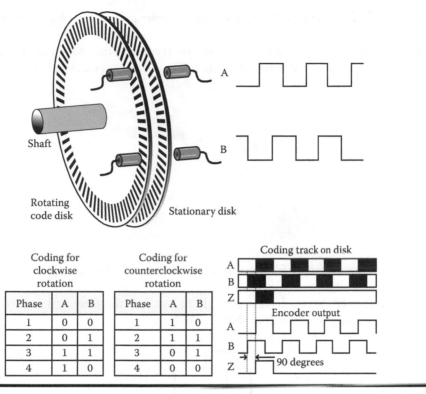

Figure 2.67 Pulse trains of a two-channel encoder and zero index for resulting coding data.

noticeable that encoders cannot be used with a conventional automation circuit, except if a specific electronic unit accepts the pulses and produces some digital contact-type outputs inserted in the automation circuit. On the contrary, they can be easily used with PLCs equipped with suitable counting modules for positioning tasks. Such applications are met in robotics for controlling and positioning joints, in computer numerical control (CNC) machines, drilling machines, assembly machines, cutting machines with a measuring wheel-encoder, and many others.

Review Questions

2.1. What kind of materials can detect an inductive proximity switch and why?

2.2. How many categories of relays exist regarding to their use in industry? Explain each one of them.

2.3. Is a thermal overload relay really a switch breaking a load circuit?

2.4. Can a latching relay be considered as an electromechanical type of memory? If yes, how many bits of data can be stored in it?

2.5. In your opinion, can the electronic level switch with immersed electrodes be used for detecting the level of a tank containing kerosene?

2.6. Explain if capacitive type proximity switches can detect metallic objects or not.

2.7. Describe two different ways to detect the end position of a pneumatic cylinder.

2.8. What is the role of auxiliary relays in an automation system and what are the corresponding power relays?

2.9. What kind of level switch should you select in order to detect the level of barleycorn inside a silo?

2.10. Explain the operation principle of a thermal overload relay and the role of its electric contacts.

2.11. What kind of sensor is shown in the figure? What does it detect and how does it operate?

2.12. A pump is automatically filling a tank via an electronic level switch with immersed electrodes. The tank is emptying from an unpredictable consumer. At a random time interval, you are informed that the level is at height L. Based on this information only, can you conclude if the pump is operating at this time instance?

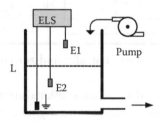

2.13. A production machine produces metallic objects slightly connected to each other in the form of a chain. What kind of sensor you would use to monitor the continuity of chain and where exactly you would position the sensor?

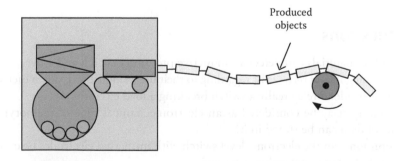

2.14. Give the time function response of the SPDT output of an On-Delay type and Off-Delay type of timer, both adjusted at 30 s in relation to the supply voltage V_{in} shown in the figure.

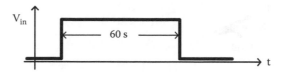

2.15. Give the time function of operation of relay C for the continuous operation of the cam switch the time constant of which is ½ rpm. The time diagram must contain two complete rotations of the cam.

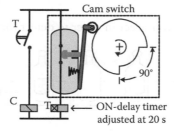

2.16. Suppose that you have a power relay with a coil of nominal voltage 24 V DC and a proximity switch of nominal operation voltage 230 V AC with an output of SPDT type. Explain if it is possible to combine the two devices, relay and sensor, in order to achieve an automatic operation of a machine which is fitted via the relay.

2.17. In the three applications in the figure, we want to detect if the corresponding objects A, B, and C have reached the imaginary line XY. What kind of sensors you would select in each case?

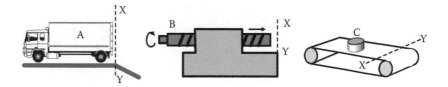

2.18. The same voltage V_{in} is applied to the coil of a common general purpose relay and a pulse bi-stable relay. Complete the state-behavior of their NO contacts in the diagram.

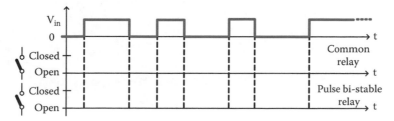

2.19. The figure shows a sensor or switch for industrial use. Identify the type of the sensor and its possible features, and give the connection circuit of a load.

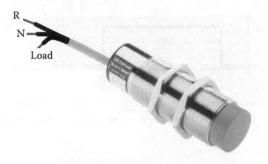

2.20. What kind of sensor would you select in order to detect the lack of label on the bottle?

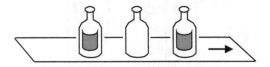

Chapter 3

Industrial Automation Synthesis

3.1 Introductory Principles in Designing Automation Circuits

3.1.1 The Latch Principle

Let's consider a machine operating with the help of an electric motor. The motor is a direct start type and is being controlled by the utilization of a relay C. In Section 1.1 it has been analyzed that the automation circuits are actuating on the power relays, supplying subsequently the motors or other industrial devices with the appropriate electrical power. In this case, the automation problem can be stated as follows: *What automation circuit can we select to energize or de-energize the relay C and thus directly control the operation or non-operation of the motor?*

A straightforward solution to this problem is the electrical circuit in Figure 3.1a. In this design, with the help of the RS switch having two switching positions, we can achieve the permanent operation of the motor in position 1 and the permanent non-operation in position 0. The term "permanent" means that if the operator leaves this control panel, the motor will continue to operate or shut down according to the selected state. This does not happen in the case of the automation circuits shown in Figure 3.1b and 3.1c, where the pressing of a button energizes the relay C. In the case that an NO button is utilized (Figure 3.1b), there is the option of a permanent stop but instant operation of the motor, as long as the button is pressed. In contrast, if an NC button is utilized (Figure 3.1c) there is the option of a permanent operation, but with an instant stop, as long as the button is pressed. The term "instant operation" does not only mean a moment in time but also contains the meaning of the time duration. This property comes from the construction of the button and the mechanism of the (spring based) automatic return in the relaxation position and depends on how long the operator keeps pressing the button. In the most common cases, the buttons are being pressed instantly and for a very short time interval. If for any reason the button (e.g., the NO button in Figure 3.1b) remains pressed for a long time interval, then the operation of the motor will last for a corresponding amount of time, but for simplicity this is characterized as "instant".

Therefore, the circuits in Figures 3.1b and 3.1c are not suitable for the usual and practical cases of machine operation in general. As a result, the following problem is formulated: *Is it possible to*

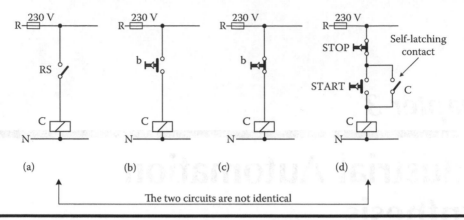

Figure 3.1 **Elementary manual ON-OFF control of a machine. (a) Permanent stop or operation of a machine (C), (b) Permanent stop, instantaneous operation of a machine (C), (c) Permanent operation, instantaneous stop of a machine (C), and (d) Permanent stop or operation of a machine (C).**

utilize only buttons to achieve the same operation depicted in Figure 3.1a, which means the potential of a permanent stop or permanent operation of a machine in general?

The solution to this problem is provided by the automation circuit in Figure 3.1d, which operates as follows. When the circuit is idle (there is no voltage among the power lines R and N), the relay C is de-energized. By pressing the START button, the voltage R-N is applied on the relay's coil, and thus the relay C is energized. After this, the NO contact of relay C (which is connected parallel to the START button) closes, and thus an alternative route to the current is provided. In this way, when the START button has been released and its contact is open again; the relay C still continues to be energized, based on the previously formulated alternative route of the current. Due to the fact that the relay C remains constantly energized from the flow of current through its own contact, this situation is called as "self-latch" for relay C, while the NO contact of C is called a "self-latching contact". In this energized state, the relay C remains as long as it is desired from the automation specifications. By pressing the button STOP, the application of voltage to the relay's coil is interrupted, and thus the relay is de-energized. After releasing the button STOP, the relay C is not energized again, since the contact of the button START and the self-latching contact remain open. Thus, the electric circuit shown in Figure 3.1d is able to provide a permanent STOP or START operation of the machine. However, it should be highlighted that the electric circuits in Figure 3.1a and 3.1d do not have the same operation, since they are characterized by a very important difference, which is going to be explained in the following.

Let's assume the switch RS in Figure 3.1a is in the 1 position, which means that the machine is operating. If a general shutdown of the power supply network is assumed, then the automation circuit (and the power circuit also) will be de-energized and the machine will stop. In the case that the power supply network is restored, the machine will operate by itself, without any intervention by a human operator, which is a case that could be totally catastrophic or dangerous for specific kinds of machines or industrial processes. For example, the smashing machines of various materials are not supposed to start when filled with the materials for smashing. In this case, these machines should initially start operating and subsequently the materials to be smashed should be supplied to the machine, otherwise the machine will be blocked and may be destroyed by the overload. In another case, it could be assumed that the multiple machines are operating sequentially due to corresponding commands from a human operator. In this case, after the event of a power loss, all the sequential operations of the machines will be terminated. After the restoration of the power loss, and for the case of the circuit shown in Figure 3.1a, all the machines will start operating exactly at the same time instant, an issue that could be totally catastrophic for the industrial process under consideration. In contrast to this undesirable operation, the automation circuit shown in Figure 3.1d does not allow the automatic re-operation of the machine after an impermanent power loss. In order for the machine to be able to operate again, the human operator should press the START button, which would provide full control of the machine's operations, while allowing the operator to consider specific safety actions in the case of a fault, e.g., the considered power loss.

Like in the circuits shown in Figure 3.1, two horizontal lines will represent the electrical supply of the automation circuit. In more detail, the power line marked by R will indicate the wire that is carrying the voltage (phase), while the wire marked by N will indicate the neutral phase. Between wires R and N we assume a voltage of 230 V, but this can be any kind of standardized AC or DC operation voltages for the relay's coils. For example, in the case that the coils of the relays have a nominal operation voltage of 24 V DC, then the upper horizontal power line will be marked as the positive pole (+24 V) and the lower horizontal power line as the ground pole.

For all cases of automation circuits, the following remarks should always be considered:

- Every automation circuit, simple or complex, is secured by a general safety fuse, especially in the cases where the automation includes control panels for human operators.
- In the case of control panels for human operators, all the automation circuits should be of low voltage in order to ensure the operator's safety in case of a short circuit, current leakage, faulty wires, etc.
- The automation circuits are being designed in parallel branches between the R and N lines, while there is no restriction in their "length", meaning the total number of branches.
- Every branch of an automation circuit can be simple or complex, which means that it can consist of a series, parallel, or a mixed type combination of switching contacts.
- Every branch of an automation circuit can contain an unlimited number of switching contacts, but always only one coil should be able to get energized.
- It is never permissible to have serial connections of relay coils in the same branch, while it is permissible to have a parallel connection of two or multiple coils in the same branch.

3.1.2 The Principle of "Command"

In the wiring diagram of Figure 3.1a, the operation of the machine "C" is controlled by the human-based actions on the switch RS, while this operation is presented more clearly in Figure 3.2a. However, it is possible that the operation of the machine is controlled in an autonomous manner, not from a human-based operation, but through the utilization of a switching contact, e.g., a contact of a light sensor, as presented in Figure 3.2b. In this example, when the photocell is not detecting an object, it has its internal contact open and thus the machine cannot operate. When the photocell detects an object, the previously open contact closes and thus the operation of the machine is allowed. In this case and in all other similar ones, the operation of the machine is based on the "command" from the utilized sensor. In both cases depicted in Figure 3.2, the machine is being operated through a contact, either by the switch RS or the sensor switch. In the second case, the definition of "command" is describing the absence of a human operator. An additional difference in the described cases is the meaning of the distance. In the first case, in Figure 3.2a, the operating switch is usually very close to the rest of the automation circuit, either

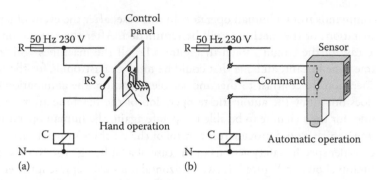

Figure 3.2 Machine operation (a) manually or (b) automatically via sensor command.

in the same electrical cabinet containing the automation circuit or in a nearby control panel. In the second case, in Figure 3.2b, the sensor cannot be placed inside or close to the cabinet with the automation circuit, since this device is placed in the environment needing the measurement of a specific quantity and thus proper wiring should be performed from the sensor to the automation cabinet. Hence, the definition of "command" also includes the meaning of the distance or the meaning of a control command from a distance.

From an automation circuit design perspective, it should be mentioned that the points in a circuit, where the "command" from a sensor is being utilized, are denoted by "Ø". In most cases, we are not designing the sensor device in the automation circuit, but we are simply noting the kind of the command and switching contact as it will be analyzed subsequently. Finally, the machine operation control through a sensor with an SPST output switching contact is commonly referred as a "two-wire command".

3.2 Step-by-Step Basic Automation Examples Synthesis

3.2.1 Motor Operation with Thermal Overload Protection

In Section 2.2.1, the thermal overload relay has been described, which has a specific utilization in protecting the motor form an overcurrent. As has been explained, the thermal overload relay is not directly breaking the power circuit, but it is happening indirectly and through the utilization of a supporting switching contact in the automation circuit. Today, all the motors are being supplied through an overcurrent thermal protection relay and thus the utilization of the corresponding thermal protection auxiliary contact in the automation circuit will be considered an obligatory component. The corresponding automation circuits shown in Figures 3.1a and 3.1d with thermal overload protection are depicted in Figure 3.3. The NC contact of the thermal relay is always included in the automation circuit. As long as the motor is consuming the nominal current (in which the thermal relay has been tuned), the automation circuit allows the operation of the motor. In the case that the consuming current becomes bigger than the nominal one, the thermal overload protection relay will react and will open the contact "e". Since contact "e" is connected in series to the coil of the power relay, its opening will cause the deactivation of the relay and thus it will lead to the overall stop of the motor.

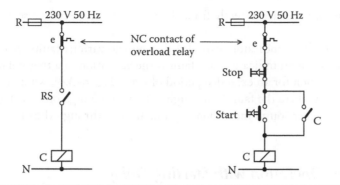

Figure 3.3 Motor operation with thermal overload protection.

3.2.2 Operation and Fault Indication

The indication that a machine is in operation can be achieved simply either by the sound produced from the operated machine or by a visual inspection of its moving parts. However, it is straightforward that both ways for indicating the operation of the machine are not accepted, not only because they are old-fashioned but also because these approaches are ineffective, e.g., in the case that the operator is away from the operating machine. As a design regulation, the operation of the machine is indicated through the utilization of indicative lamps that are placed on the control panel. This control panel can be located either on the machine or at a distance from it.

A lighted indicator is able to inform the operator that the corresponding machine is in operation. However, in reality, the exact information that the light indication is providing is that "the power relay in energized", as displayed in Figure 3.4. For example, in a case where the power cable from the power relay to the motor is cut, independently of the fact that the light indication will be on, the motor will not be in operation. In the majority of applications, the light indications of the machines' operation are implemented in the form of Figure 3.4. In special cases, where the

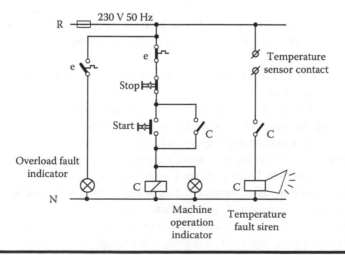

Figure 3.4 Signaling of machine operation or faults.

operation information is needed directly from the motor itself, special sensors can be utilized to provide this.

Different types of light indicators can also display the various faults in the operation of a machine. The most common indication of a fault is the activation of a thermal overload relay due to an overcurrent situation for an extended period of time (Figure 3.4), which is called a "thermal overload trip". In cases where the fault is of a high risk or very dangerous, the light indication can be accompanied by proper sound indications (sound alarms) through the utilization of a horn or a buzzer.

3.2.3 Machine Operation with Starting Delay

The most common cases of starting a machine take place directly after the application of the START signal in the automation circuit. This START signal can be provided in the automation circuit by a human, either through the pressing of a button or with the activation of a switch. Moreover, this command can be provided based on a sensing device, through the closing of an NO contact, when the sensing requirements are being met. However, there are specific cases where it is ideal that the machine starts after the elapse of a specific time delay T, from the time instant of applying the START command. At this point it should be mentioned that the logical question: "Instead of using the overall operation mentioned, the reason why someone cannot apply the command and signal START after the same time delay (T)" is justifiable but not correct. To clearly understand the importance of starting a machine with a time delay after the provided starting command, the following two examples will be presented.

Example 1: A specific category of pumps, called "hydro-lubricated", need to be filled with water through a supporting tank before starting. The water filling of the tank can be completed in a short time interval of 1–4 minutes. In the case that an automatic operation of the pump is needed, the following procedure should be followed:

1. In the beginning, the pump water filling mechanism should be set in operation, without the pump needing to be in operation. Let's define "T" as the required time interval for completing this process.
2. After the elapse of time T, the pump should be set in operation.
3. In parallel with (2), the operation of the water filling mechanism is terminated.

An alternative implementation of this operation can also be assumed, where in the beginning a command can be provided to start the water filling mechanism and afterwards, when we identify that the pump is full of water and ready to start, to give the second command for starting the pump and terminating the water filling process. In this implementation, it should be noted that there is no automatic operation of the pump but a fully manual one.

Example 2: Let's consider a parts processing station. The parts are being transported to the station through a conveyor belt. During the part processing, a proper initialization of the station (before) and the conveyor belt (after) can start. The initialization of the station could include the heating of specific area, the reception of complementary parts, and specific initialization tasks that are required to prepare the station for accepting the upcoming parts. This preliminary process of station preparation lasts a time interval T. It is more than straightforward that the start of the conveyor belt should be made automatic and, of course, after the elapse of the T time interval, from the moment that the station has been put in operation.

In both these examples, the problem of starting a machine with a time delay, after the start of the operation command, is evident. In these cases, where the time is included in the automation as a parameter of the physical process for implementing the automation circuit, the utilization of timers (time relays) is needed. A simple automation circuit that achieves the starting of a machine with a time delay is presented in Figure 3.5. In this case, the timer T is an electronic type time relay, which can be set at the desired delay time T, while the relay C corresponds to the machine that we would like to control. For a better understanding of the overall automation circuit and the utilization of the time relay, the time response of the automation circuit components (T, C) is also presented in Figure 3.5.

The characteristics of the presented automation circuit are the ability to offer a permanent operation of the machine with a time delay in starting; a permanent stop through the switch RS; the ability of repeating the operational sequence as many times as desired; the re-operation; and the permanent supply of voltage to the time relay during the whole operation. The last characteristic is of paramount importance, especially if it is considered that the time relay is an electronic device that should not be under voltage, in case that it is not utilized. Moreover, the time relay after the elapse of time T is no longer utilized in the automation circuit and thus it is desirable to have a redesigned automation circuit where, after the elapse of time T, the time relay will no longer be under voltage. Such a circuit is presented in Figure 3.6, where the time response of the voltage V_T at the terminals of the timer is also presented for a better understanding of the overall

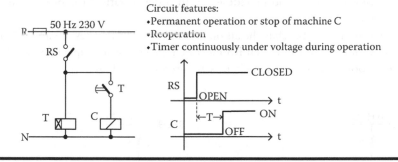

Figure 3.5   **Machine operation with a start-up delay (circuit No.1).**

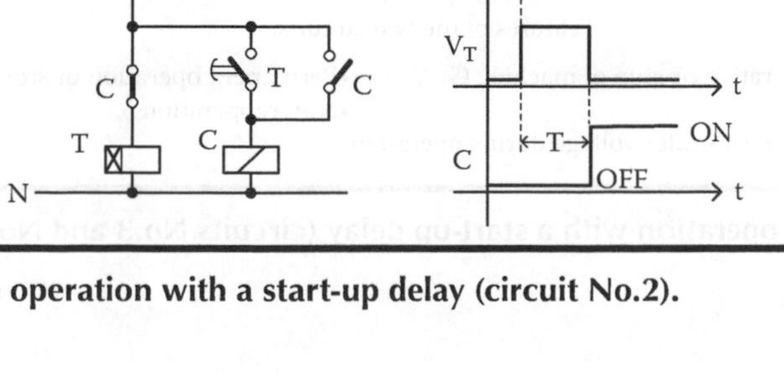

Circuit features:
•Permanent operation or stop of machine C
•Reoperation

Figure 3.6   **Machine operation with a start-up delay (circuit No.2).**

functionality. As can be observed, in this case, the voltage V_T is set to zero after the elapse of time T.

The automation circuit shown in Figure 3.6 contains a generic potential problematic situation, which can be found in many cases, which we will consider subsequently. This problem consists of the synchronization of two or more switching contacts of the same relay, during the change of their state. For further comprehension of this problem, let's consider that the NO contact of C delays to change its switching state, when compared with the NC contact of C. This means that the NC contact will open first and before the closing of the NO contact. In this case, the timer will be deactivated, the opening of the timer's contact T will happen before the creation of the alternative current route, through the NO contact of C, and finally this will result in the de-energizing of relay C. Such an operation of the automation circuit is wrong, and thus a proper redesign should be provided for dealing with this issue. In normal operations, such as a problematic delay of one contact with respect to the others does not exist. However, the problem still remains, even in the cases of simple automation circuits with usual relays and thus the question "will the NO contact of a relay close before the opening of the corresponding NC contact?" arises. In this question, a clear and generic answer cannot be provided. Due to the inertia of the relay's mechanical parts and their acceleration during the relay operation, usually the synchronization of the contacts' state change is satisfied. In the general purpose miniature relays, the printed circuit board (PCB) relays, or in the case of low-cost relays where the moving parts are of a low mass and with a short translation of the movable contacts, this problematic synchronization might be evident. In these cases, the automation circuit should be redesigned in order to satisfy the proper sequential logic of instances and states.

In the case that it is desirable that the automation circuits of Figures 3.5 and 3.6 do not present the feature of reoperation, the switch RS must be replaced by a couple of START and STOP buttons. The corresponding automation circuits are presented in Figure 3.7.

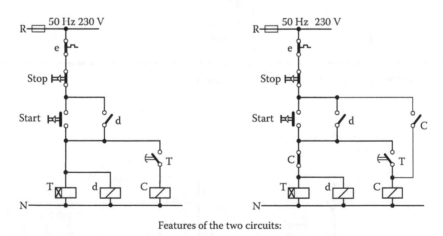

Features of the two circuits:

•Permanent operation or stop of machine C   •Permanent operation or stop of machine C
•Non reoperation   •Non reoperation
•Timer continuously under voltage during operation

Figure 3.7   Machine operation with a start-up delay (circuits No.3 and No.4).

3.2.4 Machine Operation with Stopping Delay

Let's assume a complex machine is receiving parts in a continuous manner, processing them for a certain time, and subsequently, placing them on a conveyor belt. If we would like to stop the machine, there are three possible ways to do so:

1. By pressing the STOP button, the machine will stop independently of the processing stage that it currently is in. In this case the most likely result would be to have an unfinished part inside the complex machine.
2. By monitoring the operation of the machine and detecting the termination of a processing cycle for a part, the STOP button can again be pressed quickly before the start of the next part.
3. The automation circuit can be designed in a way that the pressing of the STOP button does not stop the machine directly but will allow the operation to continue until the finalization of the processing of the current part.

The first two solutions describe unwanted automation operations, since in the first case problematic parts can be produced, while in the second case there is no automation. The third solution indicates the exact case of a machine operation with a stopping delay, or a delay in stopping after the application of the STOP command. For a constant delay time T, the proper automation circuit is presented in Figure 3.8. It should be highlighted that stopping the machine is performed by the NO button, while the termination until now had been achieved with the pressing of an NC button. Since the time relays are equipped only with "delayed" contacts, the auxiliary relay d has been utilized for providing the latch contact d. The NO contact C in the branch of the time relay T is being utilized in order to avoid the energizing of d and T when the machine is not in operation and someone is pressing the button STOP.

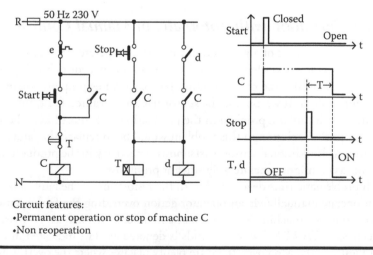

Circuit features:
•Permanent operation or stop of machine C
•Non reoperation

Figure 3.8   Machine operation with a stop delay.

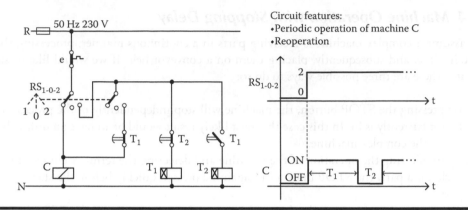

Figure 3.9 Periodic operation of a machine.

3.2.5 Periodic Operation of Machine with Two Time Constants

In Section 2.3, basic time relays (also called timers) were presented, while, apart from those, there is a big variation of more specific time relays. One of them is the time relay with two time constants T_1 and T_2, which can be utilized for achieving the periodic operation of a machine, where the machine will continuously operate for a T_1 time interval and will stop for a T_2 time interval. The same functionality can also be achieved through the utilization of two simple time relays of ON-Delay type. The corresponding automation circuit and the time response of the relay C, which controls the machine, are indicated in Figure 3.9. The selective switch RS_{1-0-2} allows the normal operation of the machine without pauses in position 1. In position 2, the machine is operated for a time interval T_1 and stops for T_2 repetitively, while in position 0 the machine stops. The second pole of switching contacts of the selective switch have been introduced in order to isolate properly the time relays and to avoid their unwanted operation, while the machine is in normal operation.

3.2.6 Machine Operation with Automatic or Manual Control

Let's consider a machine, as part of a production line, which operates through a command from a sensor (e.g., a pump being controlled by a level sensor, or a conveyor belt being controlled by a limit switch). For the time that the corresponding sensor operates properly, there is no problem with the existing automation. If we assume the sensor has an unwanted fault, then the machine is not able to operate, which causes a problem in the production line (most likely the whole production line will be stopped). A solution to this problem would be to replace the faulty sensor as soon as possible assuming such additional sensor exists; however, a delay in the production line may still occur, which on occasion might generate a significant profit loss.

To avoid such problematic time delays, and until the restoration of the fault, it is acceptable that the machine will operate manually via an operator action overriding the faulty sensor. The ability to control the operation of a machine in an automatic manner is denoted as "A" (based on a sensor) or through the classical START-STOP commands is denoted as "H", indicated in the automation circuit shown in Figure 3.10a. Moreover, there are plenty of cases where the overriding of the sensor command is desired, for example in Figure 3.10b, where in the corresponding circuit a time relay with a 24-hour period is being utilized instead of a sensor. With the switch RS_{1-0-2} in the A position, the machine will operate automatically in the time periods that we have initially programmed.

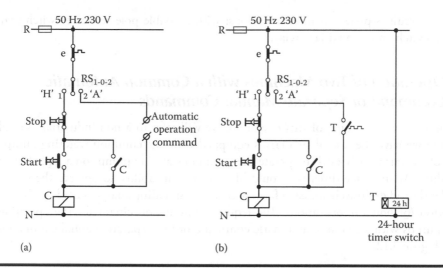

(a) (b)

Figure 3.10 Machine operation with: (a) a selectable auto-manual mode, and (b) with a time relay.

3.2.7 Operation of Two Machines with a Common Manual Command or Separate Automatic Commands

In an industrial environment, there are a lot of cases where the same process is supported by two machines. For example, consider a liquid extracted from a tank by the utilization of two pumps, parts that are being forwarded from one station to another through a double conveyor belt system and two machines that are packaging parts that are coming from a process station at the same time. In the case of an automatic operation of two such machines, it is straightforward that these machines will operate through commands provided by the utilization of sensors. Moreover, in case of faults, it is ideal to have the ability to operate these machines in a manual mode, overriding the sensory commands. In this particular case, we will also assume that there is a demand for having the two machines operate each other with a separate automatic command or both with the same manual command, after an initial selection of the desired state. In Figure 3.11, the corresponding

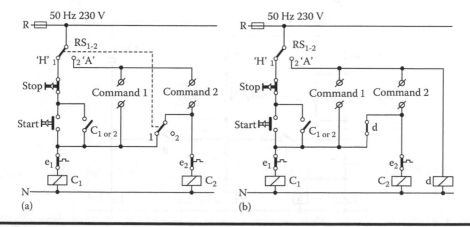

(a) (b)

Figure 3.11 Operation of two machines with selectable common manual commands or separate automatic commands. (a) Automation based on additional contacts of a switch, (b) automation based on an auxiliary relay.

automation circuit is presented in two versions: with a double pole selective switch and with a single pole switch and an auxiliary relay.

3.2.8 Operation of Two Machines with a Common Automatic Command or Separate Manual Commands

For the proper understanding of this operation, we will refer to a non-industrial example. Let's assume that we have the case of two farms equipped with one common watering pump station. For both of the farms, it is ideal to operate the common watering pumps on a daily (24 h) repetitive schedule. At the same time, we would like to have the ability to operate the water pumps independently and exclusively for each farm with a corresponding charge, based on the individual consumption (or hours of operation). The required automation circuit for the operation of the watering pumps with a common automatic command or two separate manual commands is presented in Figure 3.12.

3.2.9 Operation of a Machine from Two or More Points

The need for operating a machine from two or more control points is commonly found in industrial automation, especially in cases where the machine is being dimensionally extended or if there is a central control station. The automation circuit for the control (START and STOP) of one machine from N control points is presented in Figure 3.13. In the installation of such an automation circuit, the number of wires needed from one control position to another should be mentioned. In the control points of Figure 3.13, there is no operation indication, and thus three wires are needed for the electrical installation. In case that the operation indication is needed, then an extra wire is required.

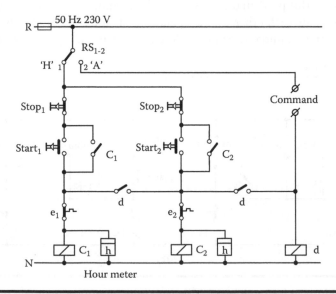

Figure 3.12 **Operation of two machines with selectable common automatic commands or separate manual commands.**

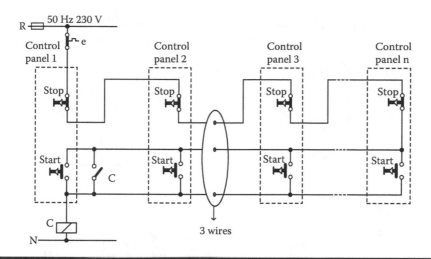

Figure 3.13   Machine manipulation from multiple control panels (without operation signaling).

3.2.10 *Control Panel for Operating n Machines*

It is generally acceptable in the most ideal cases to have multiple independent machines controlled from the same control station or, as it is commonly stated, to have a common control panel. This control panel can be physically located on the complex machine, placed somewhere in the area of the machines, or even placed remotely in a special area for control and supervision. Figures 3.14a and 3.14b indicate the concept of having a decentralized installation of the control panel (control buttons and switches) far from the switchgear with power relays, thermal overload switches, etc. The desired automation circuit for the n machines is presented in Figure 3.14c, while in this circuit it is worth studying the number of the desired wires for completing the automation. It is straightforward for this implementation that three n wires are needed, as shown in Figure 3.14d. This number can be reduced to 2n+1 if the contacts of the thermal overload relays are placed between the relays C_i and Neutral. In both cases, the number of wires is increased by 1 if a visual indication is desired in the control panel, which is the common practice in most automation installations. This additional wire is needed for connecting the neutral phase, while the terminals 3, 6, 9, ..., 3n will supply the corresponding light indications.

3.3 The Meaning of the Electrical and Mechanical Latch

Many times, the operation of a motor prevents the operation of another one, mainly due to the fact that the executed physical operations can be in contrast. For example, in the case that a motor creates the insertion of a robotic arm in the interior of a car during a car assembly production line, then the operation of the corresponding motor for the movement of the cars in the production line should be prevented. In these cases, the automation circuit should mutually exclude the operation of the second motor, even if such an operation is commanded by mistake. This functionality is achieved through the insertion of an NC contact of a power relay in the activation branch of the other power relay, in which the mutual exclusion should be achieved, as indicated in Figure 3.15. This action is called "electrical latch". Through the electrical latch, the activation of the relay is prevented based on an electrical approach, without determining that it is impossible to have a mechanical activation of both relays.

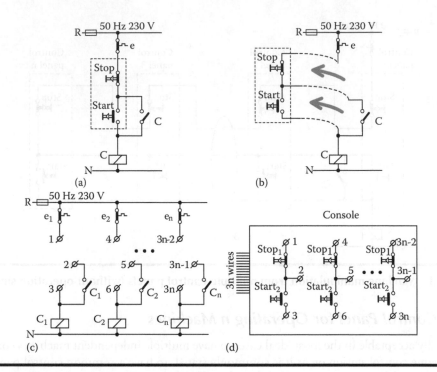

Figure 3.14 Control panel for n machines. (a) Centralized control panel for 1 machine, (b) decentralized control panel for 1 machine, (c) decentralized automation for n-machines, and (d) console for the decentralized automation of n-machines.

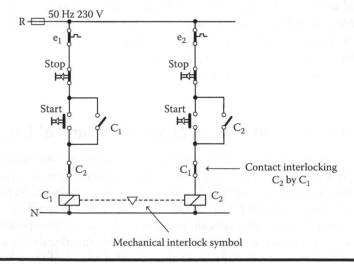

Figure 3.15 Automation circuit with relays under electric and mechanic latching.

In cases of expensive and large motors with inversion, as presented in Section 3.4.1, the power circuit contains two relays for the change of phases (change of rotation), the electrical latch is not sufficient. Due to the catastrophic results that would result from an unwanted potential mechanical energizing of those two relays, a "mechanical latch" is added in the automation. The mechanical latch is a mechanism adapted on two relays and the activation of one relay is not allowed in any way when the other relay has been energized. The operation is called mutual exclusion.

A different kind of latch could happen when energizing a relay before another one which should have already been energized. It is straightforward that this latching is achieved through the utilization of an NO contact of one relay in the activation branch of the other relay. The combined operation of this form, for two or more machines, is generated from the way that the physical process is being carried through the utilized machines. Such an application will be examined subsequently.

3.3.1 Sequential Start—Latch of Machines (Chain Latch)

In industrial production lines where multiple machines are lined up one after another and are functionally cooperative for the completion of a physical production procedure, it is common to cancel the operation of a machine just in case the previous machine has not been set in operation. With this exception, it is ideal that the machines should start progressively (like a chain), with only one operational command. The desired automation circuit for four machines is presented in Figure 3.16, and the number of utilized machines can be generalized independently.

The circuit of Figure 3.16 presents a special functionality, which is desirable in many cases. Let's assume that during the operation of the four machines, a fault takes place and the thermal relay e_2 is energized, and thus machines C_2, C_3, and C_4 stop operating, which is likely to happen (the breaking of the operational chain). If machine C_1 continues to operate, the thermal relay is de-energized, and then the three machines will start operating again, a situation that could be forbidden for some specific physical processes. In these cases, after the de-energizing of the thermal relay, we don't want the machines to restart, instead we need a new command (the press of a START button). This functionality can also be considered as an indirect way of forcing the operator to notice the break in operation in the machines. The characteristics of this desired automation are achieved through the automation circuit shown in Figure 3.17.

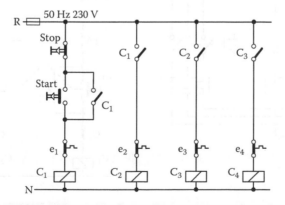

Figure 3.16   Sequential start-up of four machines with chain electric latching.

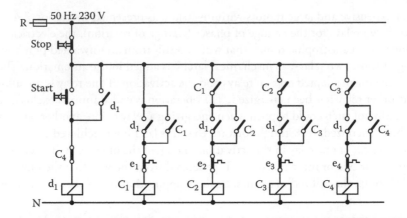

Figure 3.17 **Sequential start-up of four machines with chain electric latching without reoperation in case of overload relay trip-reset.**

3.3.2 Motor Operation with Power Supply from Two Different Networks

In large industrial units it is possible, for safety and cost reasons, to have the ability to select the supply from a different power network, instead of the one commonly provided. In this case, switching to a different power network is ideal. The automation circuit and power circuit needed for this application and the safe transition from one network to the other is presented in Figure 3.18. It should be highlighted that the utilization of the double latch is achieved through the contacts of the relays and from the utilization of the buttons. The utilized buttons have double contacts to allow initially one relay to be de-energized and subsequently the second relay to be energized during the change in the supply.

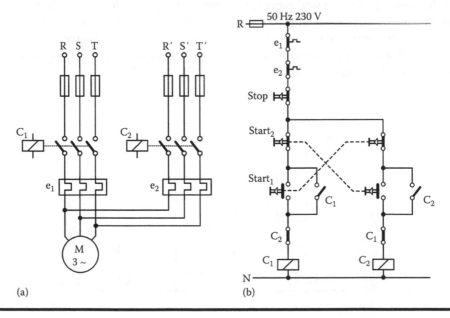

Figure 3.18 **(a) Power and (b) automation circuits of a motor connected in two different supply networks.**

3.4 Automation Circuits for Motors

3.4.1 Motor with Inversion in Rotation

For every kind of motor, the theory of electrical machines provides the schematics for both the power circuit and the functional operation of the machine, or the operation sequence of the relays that supply the electrical motor with power. For a motor with two rotation directions (with inversion), the power circuit is presented in Figure 3.19. When the relay C_1 is energized, the coils of the motor are connected to the R, S, and T phases; while when the relay C_2 is energized, the coils are connected to the T, S, and R phases. In the first case, the motor is rotating clockwise, while in the second case the motor is rotating counterclockwise. Additionally, in the presented case, the motor is being protected by the thermal overload protection (e), while three buttons are needed for the manual operation of the motor; one for the clockwise rotation, one for the counterclockwise rotation, and one for pausing the motor operation (STOP). The energizing of one of the power relays should have a mutual exclusion with the other relay, since synchronal energizing of the two relays, even for a very short time, will cause a direct short circuit. All these desired operations are satisfied through the automation circuit shown in Figure 3.20a, which also includes a light indication for every direction of rotation and a mechanical latch. With this circuit, it is not possible to change the motor's direction of rotation, while it is in operation, since it is needed initially to stop the motor and afterwards to press the button corresponding to the desired direction of rotation. Moreover, the circuit provides the ability for constant operation. In cases where an instant operation is needed (as long as the buttons are being pressed) e.g., in the case of the cranes, a simpler automation circuit can be utilized, as shown in Figure 3.20b. The buttons with double actuation contacts for latching during start could also be utilized for both presented automation circuits.

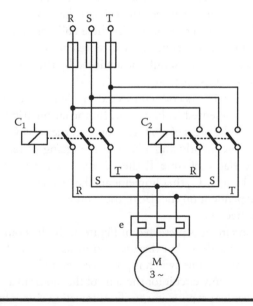

Figure 3.19   Power circuit of a motor with two directions of rotation.

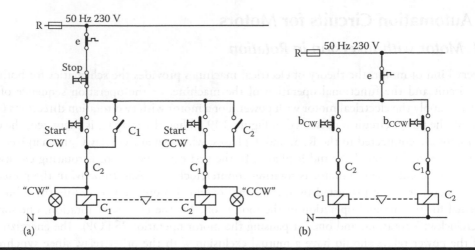

Figure 3.20   Automation circuits for (a) permanent or (b) instant operation of a motor with two directions of rotation.

3.4.2 Motor with a Star-Delta (Y-Δ) Start

In order to reduce the starting current that deteriorates the effective life of certain types of motors (especially in the larger, more powerful ones), a special starting procedure is usually followed, which is called START in a Star-Delta (Y-Δ) configuration. This configuration refers to the connections between the motor's internal coils and the power supply phases that are able to achieve a specific reduction of the startup currents in the Y configuration. However, since this Y coil configuration also represents a lower torque for the motor, a proper connection back to the Δ (higher nominal currents) is needed in order to allow the motor to handle bigger loads.

In the case that no automation is utilized, the coils of the motor can be directly wired in a Y or a Δ configuration, indicated in Figure 3.21b. In these coil configurations, it should be noted that the metallic bridges (gray dashed connections) represent constant connections, while in an automatic operation, these connections should be achieved by the proper utilization of two relays C_2 and C_3, as will be presented subsequently.

For every motor that starts with its coils connected in a Star (Y) and subsequently its coils are connected in a Delta (Δ), as presented in Figure 3.21a, the automation circuit should satisfy the following requirements. With the press of a START button, the relay C_3 that creates the star junction should be energized. Subsequently and after the energizing of relay C_3, the relay C_1 should also be energized. After the elapse of time T, the relay C_3 should be de-energized, and without de-energizing the relay C_1, the relay C_2 that implements the Delta connection should be energized. The time duration of T is dependent on the size of the motor, which should be adjustable, thus the utilization of a time relay is necessary.

The requested automation circuit is presented in Figure 3.21c. It should be noted that in this circuit, there is a contact with a time delay, but there is no time relay as an electrical device. This means (as in all similar situations) that the time switch is pneumatic and placed in relay C_1. The automation circuit in Figure 3.21c is not the only circuit for the start of the motor in a Y/Δ configuration. On the contrary, every automation vendor proposes its specific automation circuit for the same operation that, in principle, has more contacts than the one indicated in Figure 3.21c, which is the minimum contact implementation. Such a circuit with more contacts will be presented in Section 3.6.

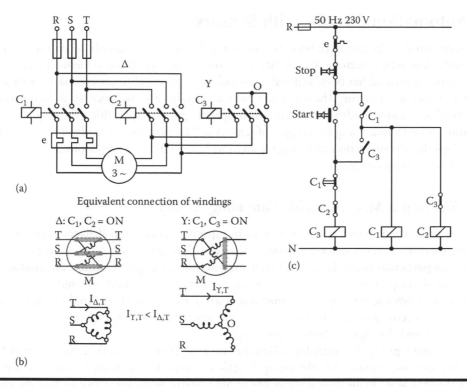

(a)

(b)

(c)

Equivalent connection of windings

Δ: C_1, C_2 = ON   Y: C_1, C_3 = ON

$I_{Y,T} < I_{\Delta,T}$

Figure 3.21 **(a) Power circuit, (b) windings connection, and (c) automation circuit of a motor with Y/Δ start-up.**

3.4.3 Automation of Various Motor Types

In Sections 3.4.1 and 3.4.2, the automation circuits for the most basic types of electrical motors have been presented in order to highlight the characteristics of these types of automations. In general, there are a lot of types of electrical motors, while every type of motor has its specifically designed automation circuit. These automation circuits are typical and standardized with respect to the manufacturing vendors, thus no more references to these circuits will be provided. Subsequently, the most common types of electrical motor automations will be presented and, when needed, the engineer can directly refer to the standardized circuits for automation and operation of these electrical motors:

1. With inversion and Y/Δ
2. With start through autotransformers
3. With start through resistors
4. With one direction of rotation and two speeds
5. With two directions of rotation and two speeds
6. Case (4) with an additional Y/Δ start
7. Case (5) with an additional Y/Δ start
8. With one direction of rotation, two speeds of rotation, and two internal coils
9. With one direction of rotation and three speeds of rotation (one internal coil in a Dahlander connection and an internal simple coil)
10. With two directions of rotation and three speeds
11. With one direction of rotation and four speeds (two internal coils in a Dahlander connection)

3.5 Automation Circuits with Sensors

The automation circuits that have been examined until now contained only relays (power or auxiliary ones), time relays, buttons or switches, and contacts as logical components. In cases where an automatic command has been utilized, this took place in an abstract approach, with a reference to a hypothetical sensor. These circuits did not contain sensors, since we were not addressing physical operations needing the detection of specific physical quantities. Subsequently, more automation circuits containing basic types of sensors will be examined. Finally, it should be noted that the logic for the utilization and integration of additional different types of sensors than those addressed here would be similar.

3.5.1 Starting a Machine with Canceling Ability

Every machine, when in operation, is performing specific actions and delivers a specific result. However, there are cases of machines where their operation, after their start, are dependent specifically on a particular result. In cases of malfunction, where no specific results are produced, the machine should stop, while in the normal operation the machine should continue to operate. The involved time from the starting of the machine until the creation of the final result varies with respect to the specific application. Subsequently, the operation of such a machine with canceling ability is presented through the following example.

Let's assume a pump that transfers a liquid from Tank A into Tank B, as indicated in Figure 3.22. Moreover, we assume that the pump is self-lubricated by the same liquid that is passing through the tank (e.g., in the case of water, the pump is hydro-lubricated). In case that the pump is in operation without being lubricated, which is equivalent to the situation that no liquid is passing through the pump, it is more than likely that the pump will be destroyed (burnt). For this reason, it is desirable to design an automation circuit for putting the pump in operation; if there is a flow of the liquid in the pipe, the pump will continue to operate, while if there is no flow, the overall operation will be canceled, which means that the operation of the pump should be stopped automatically. Since the liquid needs some necessary time (T) sec to reach from Tank A to the pump, the automation circuit should cancel the operation of the pump only after this specific time of T plus some small-time tolerance (specifically T = distance from tank A to flow switch/flow velocity). The existence of a liquid flow in the pipe is detected through the utilization of a flow switch, which is placed closed to the pump. The case of no flow might be caused by various factors, including the

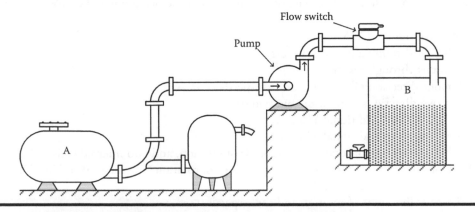

Figure 3.22 Example of a pump which stops if there is no fluid flow.

fault of the pump itself, the absence of liquid, the blocking of pipe, etc. The necessary automation circuit is presented in Figure 3.23a. The time responses of the most characteristic components are presented in the same figure and correspond to the two potential cases of presence or absence of flow in the pipes, after the operation of the pump. It should be additionally noted that this circuit covers also the case where the flow will stop in an unforeseen time after a proper operation of the pump.

In this specific case, the pump will stop operating after a time T from the time instance that the flow has stopped. Thus, the operator would like to put the pump in operation again, without being able to recognize the reason for the pump having stopped in order to resolve the previous fault situation. For this reason, a more complete automation circuit should energize and retain an indication of operation canceling due to a flow termination case, in order to be further distinguished from the case of stopping due to a thermal protection error. The automation circuit shown in Figure 3.24 is fully equipped with all the necessary indications (operation, thermal fault,

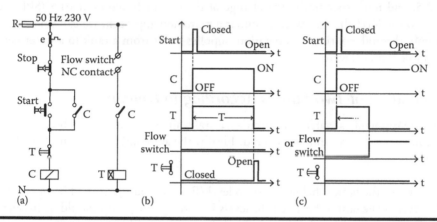

Figure 3.23 **(a) Automation circuit for pump operation with cancellation possibility and time graphs of basic components, without flow (b), and with flow (c).**

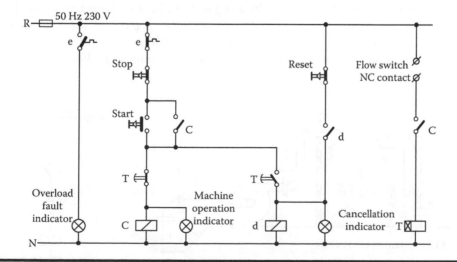

Figure 3.24 **Automation circuit for pump operation with cancellation possibility and complete signaling.**

canceling of operation) for a proper operation of the discussed application. In the case of cancellation, the automation circuit will lock, while not allowing further operation of the pump. In order to allow the further pump operation, the operator should have to press the "Reset" button, which forces a notification of the reason that caused the cancellation of the operation.

3.5.2 Pump Operation Based on Level Control

One pump is utilized to transfer water from a watering channel to a tank A, which should be kept continuously full, as presented in Figure 3.25. In this example, the operation of the pump should be fully automatic. With the help of an electronic level switch (ELS) as a controller, it is possible to keep the level of the tank between the levels E_1 and E_2 constantly without any human intervention. This is achieved through the simple automation circuit also displayed in Figure 3.25.

For a full understanding of this operation, recall the operation of the ELS as presented in Section 2.2.5, and more specifically the change of the output function of ELS (SPDT contact) with respect to the level change. With a similar automation approach, we can understand a case where a pump, through continuous operation, supplies water from a tank to a water network, as long as the tank contains water.

3.5.3 Operation of Two Pumps According to Demand

Two pumps are supplying a water tank from a natural source of water, and subsequently, the tank supplies a consumer with a random demand. Ideally, the automatic operation of the two pumps is as follows:

1. If the level of the tank is below a certain level (E_1), both pumps should be in operation.
2. If the level of the tank is between the levels E_1 and E_2, one pump should be in operation.
3. If the level is greater than level E_2, none of the pumps should be in operation.

As indicated in Figure 3.26, two level controllers are utilized for detecting the levels E_1 and E_2. The automation circuit is presented in Figure 3.26, along with the indicated contact status of ELS_1 and ELS_2, for the case of an empty tank. Let's assume that, in the beginning, the tank has a certain amount of water. A big demand will indicate that the tank will get emptied quite fast, and thus the level will drop below E_1, and with respect for the desired automation, both pumps will be in

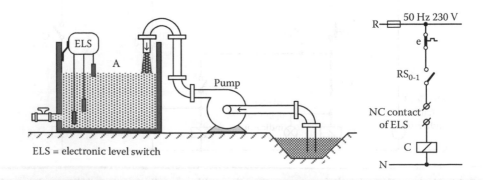

Figure 3.25 Automated filling of a tank.

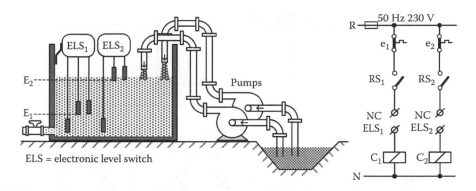

Figure 3.26   Automatic operation of two pumps according to the demand.

operation, while assuming that the water supply of those pumps covers the requested demand. With the operation of the two pumps and a low demand, the tank will continue to fill up, thus if the level becomes bigger than the E_1 level, one pump will operate (relay C_2). If the level of the tank continuously increases and extends to level E_2, none of the pumps will be in operation. If the water level starts to drop again, then one pump will start operating again as soon as the level drops below E_2. Throughout the time where the level remains in between E_1 and E_2, the demand will be equal to the provided supply, while only one pump will continue to operate. In this automation approach, attention should be paid to the configuration of the level sensors' electrodes, which in this case, have been put approximately on the same level.

3.5.4 *Automation of a Garage Door*

In complex machines, where a recurrent operation takes place, this is commonly achieved through the utilization of a motor with the capability of rotation inversion, as has been indicated in Section 3.4.1. In the current application, the recurrent component is a garage door, which is equipped with a motor with reverse capability. As indicated in Figure 3.27, a combined operation of the door is necessary through the utilization of two limit switches (b_1, b_2) and one photocell-sensor (u). The limit switches stop the movement of the door automatically (safety switches) when it reaches the corresponding motion limit. The photocell is utilized for the detection of humans, in order to instantly terminate the movement of the door upon human detection. The operation of this

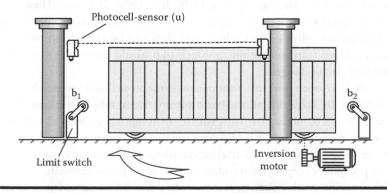

Figure 3.27   An electrically driven gate with sensors for automatic operation.

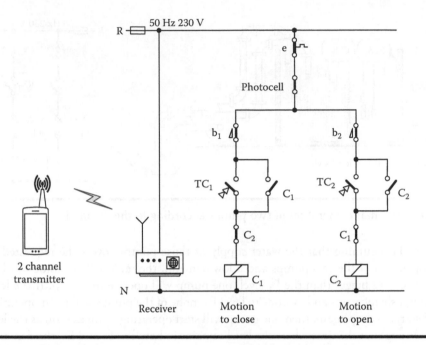

Figure 3.28   Automation circuit of the electrically driven gate of Figure 3.27.

automation is achieved by the utilization of a remote control of two channels (two independent signals), one for every direction of movement. The automation circuit is presented in Figure 3.28. As long as a button from the remote control transmitter is being pressed, a corresponding contact at the receiver (TC_1 or TC_2) remains closed, thus retaining the movement (more automatic operation) of the latch principle that needs to be utilized. A similar operation of the door can also be achieved through a remote control transmitter of one channel. In this case, the automation circuit is different and the utilization of a pulse relay is needed.

3.6 Automation Circuit Design Regulations

For the design of the automation circuits, multiple standards have been defined from national and international regulations, such as NEMA, DIN, BS, ANSI, IEC, etc. These standardization approaches from one point of view provide us with the freedom to select the best automation symbols or design regulations, while on the other hand they can be considered problematic, in that since there is not a unified standardization approach that would enable the direct comprehension of various automated circuit designs.

Until now, there have been multiple automation symbols utilized progressively in the various presented applications. In Table 3.1, all the basic automation symbols that have been utilized are presented, while it should be noted that with this set of symbols, the majority of all of the industrial automation could be designed. It is also obvious that these symbols are not the complete set of symbols that could be utilized in industrial automations and there are many more that could be referred to in the previously mentioned standardization automation manuals.

Except for the specific symbol that represents each device in an automation circuit, an Arabic letter is additionally utilized. The different devices and the corresponding letters are presented in

Table 3.1 Symbols of Switching Contacts and Automation Devices

a) NC contact of a relay b) NO contact of a relay		Any kind of electromagnetic contactor, relay and electrically operated switch
a) Button with NC contact b) Button with NO contact		On-delay timer
Contact of a sensor e.g., limit switch a) NC contact b) NO contact		Off-delay timer
Hand-operated switch a) SPDT b) SPST		Pulse type relay
Contacts of an On-delay timer a) NO contact b) NC contact		Abbreviation symbol for a Star-Delta automation circuit
Contacts of an Off-delay timer a) NO contact b) NC contact		Lamp indicator
Contacts of an overload thermal relay a) NO contact b) NC contact		Siren alarm or horn
Terminals for connecting a remote device (Klemens)		Solenoid valve coil
a) Low-voltage melt-type fuse b) Three melt-type fuses of a 3-phase network		Measuring instrument
Single-line symbol a) 3-pole power relay b) 3-pole power switch		Main terminals of a thermal overload relay
Multi-line symbol of a 3-pole power relay		3-phase motor

(Continued)

Table 3.1 (Continued)   Symbols of Switching Contacts and Automation Devices

R ——— S ——— T ———	a) Multi-line symbol of 3-phase power supply b) Single-line symbol of three conductors		3-pole switch and fuses combination

Table 3.2. In Chapters 3 and 4, and based on these two tables, the industrial automation design will take place.

Another issue that arises with the design of large industrial automations is that of the relay's contact location definition, especially when distributed in a size automation circuit. As an example, the automation circuit in Figure 3.29 is considered, which represents the starting automation based on a Y/Δ configuration, which is still different from the one examined in Section 3.4.2. Based on this automation circuit example, and in order to define the exact number and location of the required relay contacts, the following procedure should be followed:

1. We number all the vertical branches of the automation circuit.
2. Under each relay, we create a small table with two columns, with the notations "O" and "C" representing the number of open and closed contacts, respectively.

Table 3.2   Meaning of Letters in Automation Circuits

Device	Letter	Example
Power switch	a	Load isolation switch, motor protection switch, etc.
Auxiliary switch	b rs	Selector switch, button, limit switch Rotary switch
Power relay	c	Power relay for motor supply
Auxiliary relay	d	Relay used as logical component
Protection or safety devices	e	Melt fuses, automatic fuses, thermal overload relay, etc.
Measurement transformers	f	Current and voltage transformers
Measuring instruments	g	Ampere meter, voltage meter, power meter, etc.
Indication devices	h	Lamp, siren, horn, etc.
Capacitors	k	Capacitor for power factor correction
Machines and power transformers	m	Motors of any type, power transformers
Rectifiers and batteries	n	Bridge rectifiers, batteries 24 V DC for automation circuits supply
Resistors	r	Heat resistors, induction motor resistors, etc.
Special type devices	u	Any device not belonging to the above categories

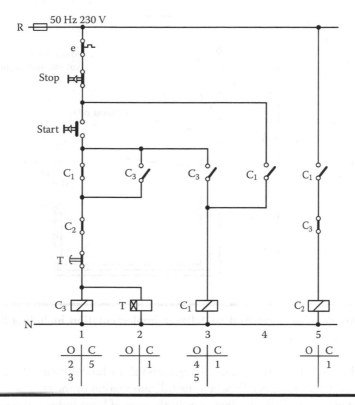

Figure 3.29 **An alternative automation circuit for the Y/Δ start-up of a motor, and an example of the localization manner of switching contacts and relays.**

3. In every column, we denote the number of the branch where a closed or an open contact exists, respectively, and this is happening for all the existing contacts of the relay.

4. We repeat step (3) for all of the remaining relays.

As an example, in the table of the C_3 relay and in the column O, there are the numbers 2 and 3, which indicate that one open contact of C_3 exists in branch 2 and one in branch 3. With this approach, the numbering of all utilized NO and NC contacts that are needed for the corresponding relay takes place. This procedure is extremely important, especially in the case of selecting the necessary relays and implementing a specific automation circuit.

3.7 Implementation of Automation Circuits

After the initial stage of designing an automation circuit for satisfying specific operations or functionalities for a machine, independently of their size and complexity, the next stage concerns the implementation of this automation circuit and, in general, the appropriate manufacturing of the whole automation system. In general, the implementation of the automation circuit cannot be considered a different task from the implementation of the power circuit. Both circuits are implemented and operate at the same time and inside (in most cases) the same industrial panel or rack. In Figure 3.30, a representative implementation of the simple automation circuit for a motor of direct start and with a START-STOP button (such as the one in Figure 3.3) is presented. Such

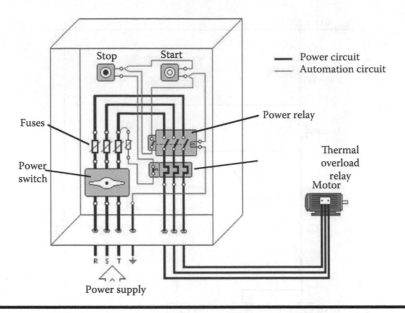

Figure 3.30 Example of an industrial switchgear implementation including both power and automation circuits.

real implemented circuits present some challenges, and show how difficult the checking of their proper operation (or logic) is, especially when the full automation in the rack or a cabinet is not in the electrical drawings. This difficulty increases in the case of large industrial circuits that can be geographically spread over large areas, while containing a large number of buttons and switches on the same rack or cabinet, or in a remote control panel. After manufacturing the industrial switchgear, the installation of the automation system in the production area takes place. If the industrial automation design involves sensors, then the installation should also contain the placement of sensors and their wiring with the switchgear or control panel.

Overall, it should be mentioned that the development of an industrial automation system involves the following stages:

■ Collection of the operational requirements for the complex machine or production line
■ Design of the automation circuit based on the gathered regulations
■ Calculation of the circuit components (e.g., power relays, thermal overload relays, auxiliary relays, wires, power supplies, operating voltages, etc.)
■ Manufacturing of the electrical industrial switchgear
■ Installation of the industrial switchgear, which includes:
  – Placement of the switchgear
  – Power wirings (e.g., motors)
  – Low-power wirings (e.g., sensors, actuators, signaling devices, control buttons and switches in control panels, etc.)
  – Testing of the whole system and tuning of sensors (e.g., position tuning of the limit switches, photo cells, etc.)
■ Implementing and programming, where the automation or parts of the automation in PLCs are needed and perform the necessary wiring connections from and to the PLC and the rest of the automation equipment, as described previously.

3.8 Applications

Here, some practical real-life automation examples are considered in order to increase the knowledge of designing automation circuits and applying the same principles in similar electromechanical projects. Up to this point it has been assumed that the reader has a sufficient background in designing automation circuits, and that we will only provide a general description of the automation design. In contrast, the automation problem will be briefly described with respect to the operational requirements, and the full automation circuit will be provided.

3.8.1 Machine Operation Control from Multiple Positions

Let's consider the problem of controlling a machine from multiple different locations, which means that it is required to have the ability to start or stop the machine from every control position. Let's also consider that for this automation, there are only two available wires and there is no possibility for installing more. As presented in Section 3.2.9, the classical automation circuit for the operation of a machine from multiple positions requires three wires, without including the wirings for the operation indication in every control position.

A solution to this problem is the utilization of a pulse relay and to apply the automation circuit of Figure 3.31. It is obvious that the operation start and pause are achieved by pressing the same button. This generates some problems, especially in the case of an operator who is not aware of the double operation of this button, and tries to stop the machine, e.g., in an emergency. Moreover, it should be noted that the automation circuit of Figure 3.31 has the characteristic of reoperation. If such an operation is not desirable, then the automation circuit should be altered as is presented in Figure 3.32 with the help of an additional auxiliary relay.

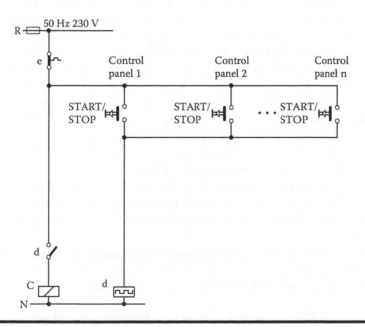

Figure 3.31   Use of pulse relay for the ON-OFF manual control of a machine from multiple control panels with a two-wire interconnection.

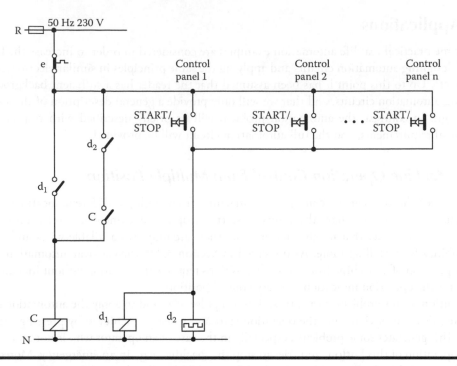

Figure 3.32   The automation circuit of Figure 3.31 in order not to present the reoperation feature.

3.8.2 *Operation Control of a Power Transformer*

Every three-phase power transformer contains a device for the level control of oil, which is denoted as a Buchholz device, and a temperature sensor for controlling the temperature of the oil. Both of these devices have two output contacts SPST configured in two levels, one warning and one final level. This means that if the level of the oil is below a specific level, the SPST contact will be energized, which subsequently will be properly utilized for signaling an alarm. If the level of oil continuously decreases, or a bubble flow is being created, then the second SPST contact is energized, which will be utilized for achieving an operation STOP. A similar functionality can be considered for the other two contacts of the thermometer. As presented in Figure 3.33, the power transformer is connected from one side to the high voltage network through the utilization of a medium voltage disconnector switch and the automatic switch (oil or air based). From the lower voltage side, the power transformer is connected to the low voltage main switch. The medium voltage automatic switch, as well as the low voltage main switch is equipped with a trip coil through which the transformer can be set out of operation, which means that each one breaks the corresponding circuit. The trip coil is activated and causes the breaking when the voltage is applied on its terminals.

At this point, we would like to design a proper automation circuit suitable for the indication of operational faults as well as for the breaking operation in critical errors. More specifically, the whole automation is desired to satisfy the following specifications:

1. For the four kinds of error types, a separate light indication should exist.
2. For the two alert signaling errors (SPST contacts) of the Buchholz device and the thermometer, a sound signaling (horn) should exist.

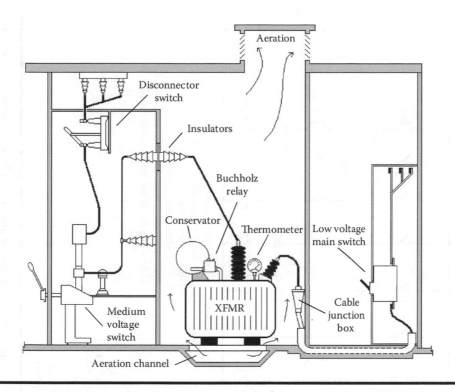

Figure 3.33 General view of a transformer power station.

3. The second error signal from the Buchholz device should create a breaking of the medium power automatic switch.
4. The second error signal from the temperature sensor should create a breaking of the main low voltage switch.
5. The horn should be able to be muted, while the fault continues to exist.
6. A testing capability of the proper operation (without fault) of the light indications and the horn should be provided.
7. When the fault is created by any kind of operation breaking, the light indications should exist even after the event of breaking.

The desired automation circuit is presented in Figure 3.34a. The nominal value for the operational voltage of the circuit is 24 V DC in order to be compatible with the voltage of a battery. To satisfy the seventh demand, after the event of the breaking, a proper transition in the supplying voltage should take place. As presented in Figure 3.34b, this is achieved by the utilization of an additional auxiliary relay, denoted by d_1, which when de-energized connects the battery to the automation circuit.

3.8.3 Operation of Two Pumps with a Cyclic Alteration, Based on a Low-High Demand

Let's consider a small watering network being supplied from two similar pumps according to the demand. This demand is measured directly through the utilization of two pressure sensors, due

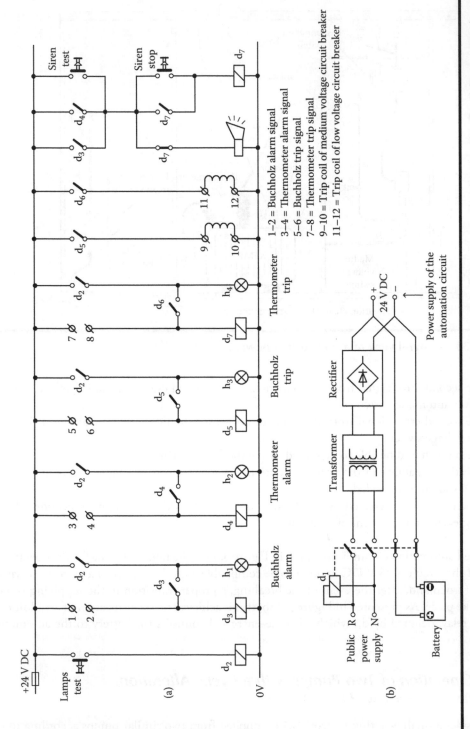

1–2 = Buchholz alarm signal
3–4 = Thermometer alarm signal
5–6 = Buchholz trip signal
7–8 = Thermometer trip signal
9–10 = Trip coil of medium voltage circuit breaker
11–12 = Trip coil of low voltage circuit breaker

Figure 3.34   (a) Automation circuits for a transformer station safety operation, and (b) change over of power supply of the automation circuit.

to the simple fact that a high demand will create a pressure drop. These two pressure sensors are tuned into two different nominal pressures, which correspond to a low and a high demand. Since the low demand usually lasts longer than the high one, within an operational cycle (e.g., 24 hrs), the pump that covers the low demand is called a base pump. Based on a similar approach, the pump that covers the high demand is called a peak pump and it is able to operate (in parallel to the base pump) for smaller time durations and only for a few times within the operational cycle. If the base and peak pumps were always the same, then after a large time window of operation (e.g., one year), the base pump would have completed a very large number of operating hours, especially when compared to the peak pump. This is a situation that might result in a fault in the base pump and thus should be properly avoided through cyclic operation of the base and peak pumps.

The operation circuit for these two pumps is desired to have the following functionalities:

1. The automation circuit should provide the ability to choose between the manual and the automatic operation of each pump separately.
2. If the base pressure sensor is energized, the based pump should be activated.
3. If the base pump is in operation and the peak pressure sensor is energized, then the peak pump should be set into operation. If the peak pressure sensor is de-activated, then the operation of the peak pump should also be stopped.
4. If the base sensor pressure is de-activated, the based pump should stop. However, as soon as the base pressure sensor is activated again, the peak pump should be put in operation and considered as the new base pump, and the previously considered base pump should be considered now as the new peak pump (cyclic operation). The same procedure of changing the consideration among the current base and peak pumps should be followed for every operational cycle, until the end of the overall operation where we would desire the complete operation stop for both pumps.

The overall automation circuit is presented in Figure 3.35, and contains four auxiliary relays, with the relay d_3 responsible for the cyclic operation of the two pumps.

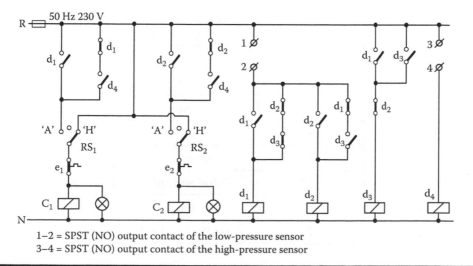

1–2 = SPST (NO) output contact of the low-pressure sensor
3–4 = SPST (NO) output contact of the high-pressure sensor

Figure 3.35   Automation circuit for a two pump operation with cyclic alternation.

3.8.4 *Operation of Three Air Compressors with Predefined Combinations*

Air compressors are devices that consume large amounts of current, especially during their startup. The motors of these large air compressors are usually motors with a Y/Δ start, while it is obvious that significantly more current is consumed, especially in the case of having two or more compressors starting up at the same time. In such cases, and in order to avoid significant voltage drop (that could lead to the generation of faults for electrical devices), a time delay is inserted in the start of their operation. If the demand for compressed air is varying, it is generally useful to have the ability to operate one, two, or more air compressors depending on the general demand. We would like to have a semi-automatic operation starting ability for the air compressors in more detail, based on the following specifications:

1. With the help of a rotational selector switch of eight positions, it is desired to pre-select the operation of the air compressors, based on Table 3.3.
2. When more than one air compressors are started, the second one should start with a constant time delay with respect to the first one (positions four, five, six, and seven). The same feature of time delay should also exist in the starting of the third motor with respect to the second one (position seven).
3. The motors of the three air compressors should all start in a Y/Δ configuration.

In Figure 3.36, the necessary automation circuit for the described automation is presented. In the position of the blocks with the Y/Δ indication, a full automation circuit for the Y/Δ starting of the motors is considered, with two wire commands (without buttons), that have been omitted from this design in order to simplify the presentation and the comprehension of this automation.

Table 3.3   Air Compressor Operation
According to Selector Switch Position

Selector Switch Position	Air Compressors in Operation
0	None
1	1st
2	2nd
3	3rd
4	1st and 2nd
5	2nd and 3rd
6	1st and 3rd
7	1st, 2nd, and 3rd

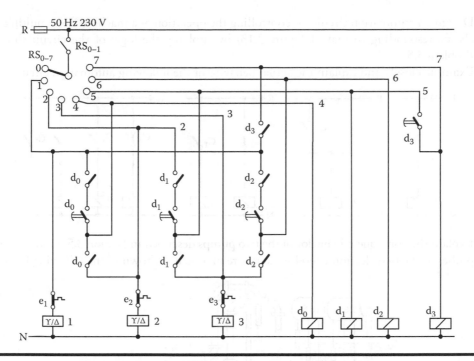

Figure 3.36 Automation circuit for operation of three air-compressors with selectable combinations.

Problems

3.1. Design two automation circuits corresponding to Figures 3.6 and 3.7a by utilizing pneumatic ON-Delay timers.

3.2. Design an automation circuit corresponding to Figure 3.7a by utilizing the pneumatic OFF-Delay timer.

3.3. Design two automation circuits corresponding to Figure 3.8 by utilizing a pneumatic timer for an ON-Delay and OFF-Delay type.

3.4. Redesign the automation circuit of Figure 3.13 in order to add an indicator of the machine's operation at each control panel.

3.5. Examine the operational difference of the following circuit in comparison with the circuit of Figure 3.1d.

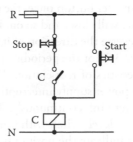

3.6. Design an automation circuit for controlling the operation of a machine from multiple panels (corresponding to that of Figure 3.13) by applying the logic of the circuit shown in Problem 3.5.

3.7. Examine closely and explain the actual behavior of the following automation circuits:

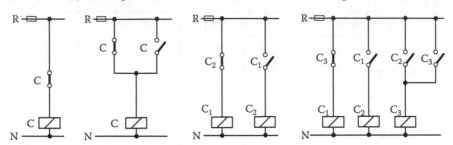

3.8. Explain the consequent behavior of the two pumps described in Section 3.5.3 if the electrode probes of the two electronic level switches are mounted as shown in the following figure:

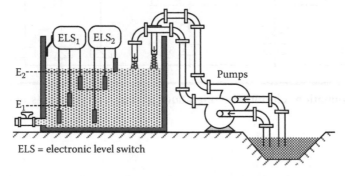

ELS = electronic level switch

3.9. The sliding electric gate, described in Section 3.5.4, is controlled by the circuit of Figure 3.28. Examine if it is possible for the moving gate to be stopped in an intermediate location by normal mode and by an unconventional action. Design an automation circuit that will offer to the user the possibility to stop the gate in any intermediate location, via an additive tele-control signal (i.e., a 3-channel transmitter).

3.10. Design an automation circuit for a motor with two directions of rotation, which will permit the direct change of rotation without the need to previously press the STOP button. Please note that the circuit of Figure 3.19 does not have this feature.

3.11. Design an automation circuit so that the sliding electric gate, described in Section 3.5.4, operates with only one tele-control signal (i.e., a 1-channel transmitter). Specifically, we would like by pressing the transmitter button once, the gate will open. By pressing the same button for a second time, the gate will close, and so on. During the gate movement, pressing of the transmitter button will reverse the direction of its motion.

3.12. The automation circuit of Figure 3.9, for the periodic operation of a machine with two time constants, presents the characteristics of reoperation. Design a similar automation circuit, which will not reoperate after a power supply interruption or restoration.

3.13. A pump is going to empty a water tank continuously. The desired level of the water in this tank is controlled by an electronic level switch with immersed electrode probes. Design the required automation circuit and indicate the positioning of the electrodes into the tank for the differential operation of the pump.

3.14. Design an automation circuit for starting up a Star-Delta motor via an SPST output contact of a sensor (2-wire command).

3.15. Some air-compressors have two electrovalves in their compression chamber as safety exhaust outlets. In order for the air compression in the chamber to be feasible, the two electrovalves must be closed. Furthermore, the two electrovalves are used for a step-by-step start-up of the air compressor, in order to avoid percussive loading of either the compressor or the power supply network. Design an automation circuit so that after the Star-Delta starts the motor, the first valve closes with time delay T_1 and the second one closes with time delay $T_2 > T_1$. Both time constants T_1 and T_2 are measured from the changing time instant from the Star-Delta connection.

3.16. Redesign the automation circuit of Problem 3.15 with only one timer (except the timer needed for the Y/Δ transition) and with the following time constants $T_1 = T$ and $T_2 = 2T$.

3.17. Which of the following circuits are operationally correct or not and for what reason? All the relays and lamps have nominal operating voltage +24 V DC.

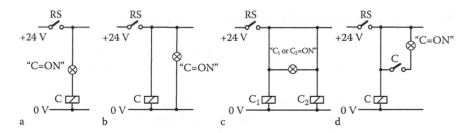

3.18. In an industrial process where a general shutdown of the power supply network took place, is it possible to get some form of electrical signal 3 minutes after an interruption while the shutdown occurs? If yes, explain how to achieve it and design the required automation circuit.

3.19. Although we accept that the indicator lamp of the circuit informs us if the machine M operates or not, this is not strictly true. Describe three cases of fault due to which the machine M does not rotate while the lamp is on. Show how to make the indication literal, i.e., for the lamp to show whether the machine M really rotates or not.

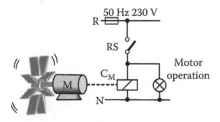

3.20. After an instant START signal from an NO button, the heating resistor R is connected to a nominal supply voltage. After a period of 10 minutes required to heat the viscous fluid, the pump starts to operate, and simultaneously the electrovalve V_1 opens in order to supply the pipe network with the fluid. After an instant STOP signal from an NC button, the heating resistor is disconnected, the electrovalve V_1 closes, the electrovalve V_2 opens, and the pump operates for a period of 5 minutes. During this period, the pump supplies the pipe network with water for its cleaning. At the end of this period, the pump stops and the electrovalve V_2 closes automatically. Design the required automation circuit.

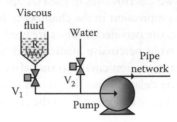

3.21. In a special machine tool, the carriage is equipped with a reversible motor, as shown in the figure. Thus, the carriage can be moved between the two limit positions A and B, which are detected by the two proximity sensors PS_1 and PS_2, correspondingly. The carriage lies initially at the left position (A) and we want it to be moved according to the following specifications:

a. By pressing instantly button b1 the carriage moves right to position B.

b. When the carriage reaches point B (with a signal from PS_2), it stops moving.

c. With the carriage at position B, the instant pressing of button b1 causes it to move left. When the carriage reaches position A (with a signal from PS_1), it stops. The steps (a) and (c) can be repeated as many times as we want.

d. If the button b2 is pressed during the carriage's movement in any direction, then the carriage stops at its current location. By pressing button b_1 again, the carriage continues moving in the same direction before it had been stopped.

Design an automation circuit to satisfy the described specifications.

Chapter 4

Logical Design of Automation Circuits

4.1 Introduction to Logical Design of Automation Circuits

In Chapter 3, we investigated the design of automation circuits based on empirical methods, relying mainly on acquiring knowledge from simple to more complicated circuits that are commonly utilized in the industry. In Chapter 4, the focus will be on a systematic method for the design of automation circuits, especially for the cases of complicated automated demands, where principles from Boolean logic can be adopted to simplify the overall design methodology.

In the Boolean logic approach, everything that is being utilized has two states, such as an ON or an OFF state (for example, an electrical switching contact can be either open or closed) or a relay can be energized or de-energized. In Section 1.1, where the industrial system was defined, it has also been mentioned that various components participating in such a system have two states of operation. Thus, since automation circuits are mainly considering devices with two states, they are directly compatible with the Boolean logic. Furthermore, methods from Boolean logic design can be adopted and applied in the design of automation circuits. In general, there are multiple, specific methodologies for designing Boolean logic diagrams, e.g., the methods based on sequential step diagrams, the cascade method, the Huffman method, and others that mainly consist of variations of these methods. Each one of these methods has its own specific advantages and disadvantages and can be applied to problems of a specific size, while they all have their own level of complexity during application.

From all these methods, the most general approach which is not dependent on the specific kind of automation and is based on the state diagrams, will be presented. The specific aim of this chapter is to highlight these general methodologies that are directly applicable in industrial automation, and not to confuse the reader by presenting numerous design approaches and further extend theoretical analyses of them that will not be directly applicable in real-life industrial problems. From this point of view, the presented state diagram approaches will act as a generally valuable tool in designing automations for complex and large industrial processes.

4.2 Boolean Logic Components

The binary system of numbers has only two values, 0 and 1, while it is utilized for the mathematical description of various physical systems characterized from a binary logic of two states. In electrical systems, the condition "in operation" (ON) is indicated with a "1" and the condition "not in operation" (OFF) is indicated with a "0". Similarly, the values 0 and 1 represent correspondingly the open and the closed state of an electrical switching contact (e.g., relay, hand switch, button, etc.) as depicted in Figure 4.1a. Every switching component, with two possible values, is represented with a capital letter and constitutes a binary system variable, while the necessary tool for the mathematical foundation of these principles is Boolean algebra.

Boolean algebra, first introduced by the English mathematician Boole, is an algebra of "logic", harmonized with a human-based way of thinking. Due to the fact that the variables of Boolean algebra can have only two values (0 and 1), this type of algebra is ideal for the binary system, especially in the way that the switches are operating. The Boolean values of 0 and 1 are not necessarily the arithmetical values of an arithmetic system, but in this case, these can represent symbols of a certain state. As an example, the values of a Boolean variable could be "white" or "black", "low" or "high", and "true" or "false".

In its general form, Boolean algebra is defined as the set of the elements a, b, c, ..., or B={a, b, c, ...} in which the equality, as well as the following operations, are valid:

1. The operation of the logical OR, which is represented by the (+) operator
2. The operation of the logical AND, which is represented by the (·) operator
3. The operation of the inversion or the complement (NOT), which is represented by the ($^-$) operator

In the previous definition, the elements a, b, c, ... are not specifically defined, but in the case of digital systems and digital logic, set B is the set of the utilized switches (switching algebra).

The application of Boolean algebra in digital systems initially took place in 1938 by C. Shannon, who in "A Symbolic Analysis of Relay and Switching Circuits" introduced a way of representing telecommunication circuits through mathematical expressions. With the help of the mathematical modeling of these circuits, containing switches and relays, their methodological design and calculation has been achieved.

In Figure 4.1b, two electrical switching contacts, one open and one closed, with different symbols are displayed. If "A" is denoted as the open contact (A=0), then the complement $\overline{A}$ is a closed contact, or $\overline{A} = 1$. There are three other possible assignments of the Boolean variable A to the open and closed contacts, shown in Figures 4.1c–e. Subsequently, the symbols and the corresponding assignment of Figure 4.1c will be utilized as symbols for contacts, since these are closer in the inversion operation from the design point of view, and because they are similar to the switching symbols that are utilized in the programming of the PLCs.

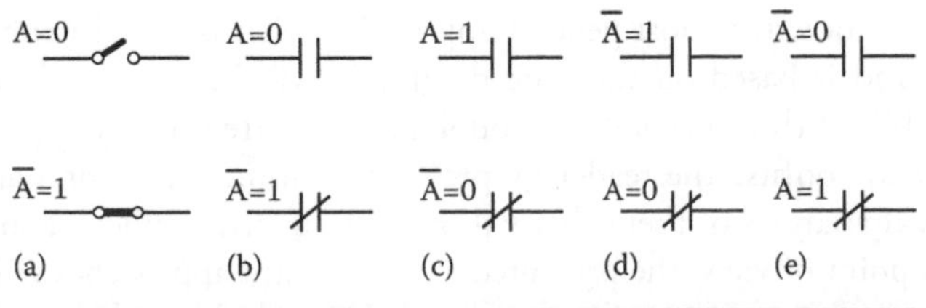

Figure 4.1 Switching elements and possible assignments of binary variables: (a) binary representation of NO and NC relay contacts, (b) open contact as A = 0, (c) open contact as A = 1, (d) closed contact as A = 0, and (e) closed contact as A = 1.

In general, a Boolean variable represents the level of voltage in a wire of a digital electronic circuit. Most commonly, the state with the high energy level is denoted as "1" or "high level" and the corresponding one with the low energy level as "0" or as "low level". In the typical digital electronic circuits (transistor–transistor logic or TTL circuits), the electrical voltage can take only two values: 0 V and +5 V, which are also presented in Figure 4.2 and, more especially, every electrical voltage below 0.4 V is equivalent to a "logical 0", while every voltage above 2.4 V is equivalent to a "logical 1".

In electrical circuits, the electrical components can be connected only in two ways, which is through series or parallel connections. The series connection of electrical components or switching contacts corresponds to the logical operation AND, while the parallel connection corresponds to the logical operation OR. These two operations, as well as the inversion operation, are implemented with the help of the related logical gates AND, OR, and NOT.

A logical function $Z = f(A, B, C, \ldots)$ is a function where the independent variables A, B, C, … as well as the dependent variable Z all belong to the Boolean algebra. In a logical function, the three operations of Boolean algebra can exist together in a simple or complex form, or with variations of them. As we have seen in Chapter 3, automation circuits contain switching components that are connected in series, parallel, or in mixed connections, as well as through coils and relays. It can easily be concluded that these type of automation circuits can be straightforwardly described from an equivalent logical function. For example, consider the pushbutton shown in Figure 4.3a, where its contact is NO; the left connection terminal is considered to be under a voltage and thus equivalent to a logical 1. The activation or inactivation of the button can be described by the binary variable b, while the variable Z represents the voltage that exists in the right connection terminal. In this case, the following relationships can be extracted:

If b=0 (the button is not pressed) then $\overline{Z}$=0 (there is no voltage).
If b=1 (the button is pressed) then Z=1 (there is voltage).

Thus, these two statements can result in the function Z=b, that can be considered as the describing function for an NO button. In a similar way, for the NC button shown in Figure 4.3b, the corresponding logical function is $Z = \overline{b}$. Thus, all automation circuits can be approached independently of their complexity. As a characteristic example, the circuit in Figure 3.1d, can be described by the logical function:

$$C = \overline{STOP} \cdot (START + C),$$

where the STOP and the START labels can be considered as binary variables of the corresponding buttons.

Voltage

5 V ┆ High level or logical 1

0 V ┆ Low level or logical 0
                                        t

Figure 4.2   Waveform of a digital signal.

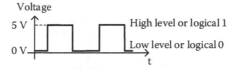

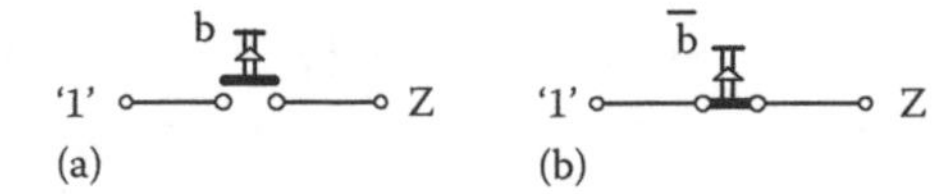

'1' o——o o——o Z   '1' o——o Z
(a)   (b)

Figure 4.3   The pushbutton as a binary variable: (a) NO button and (b) NC button.

4.2.1 Postulates and Theorems of Boolean Algebra

Boolean algebra, as well as classical algebra, were founded on some basic postulates like those of commutative and distributive property of the two logical operations of AND and OR, the existence of the neutral elements 0 and 1, and many more. From these postulates, it is possible to extract a series of theorems that can be utilized in the simplification of the logical functions. In Figure 4.4, some of the most fundamental postulates and theorems are presented in graphical (contact symbol) form as well as in Boolean form. The profound logic behind these theorems is that they are not only useful in their mathematical form in the simplification of the logical functions (i.e., during the logical design that is going to be described subsequently), but they are also very useful in the empirical design of automation circuits.

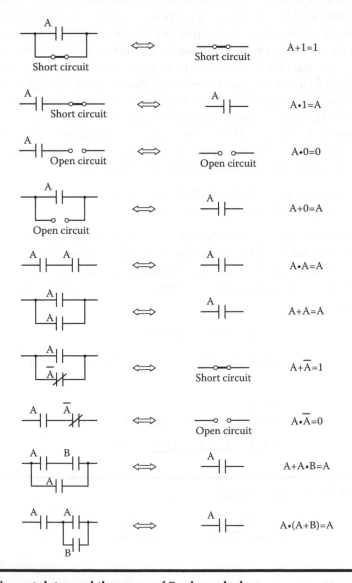

Figure 4.4   Basic postulates and theorems of Boolean algebra.

In such an empirical design procedure, multiple automation drawings are created by erasing or adding additional open or closed contacts or even whole automation branches, until they conclude at the correct automation circuit. In this approach, it is very easy to make design mistakes, which is the case in an automation circuit that contains unnecessary logical elements, like the ones presented in Figure 4.4. For this reason, knowledge of Boolean postulates and theorems will enable the easy spotting of mistakes even in an empirical design procedure.

Finally, the De Morgan Theorem provides the equivalent complement of a complex logical function by replacing each of the containing variables with its complement, and each of the operations with the corresponding dual one. The most simplified applications of the De Morgan Theorem are as follows:

$$\overline{A + B} = \overline{A} \cdot \overline{B} \quad \overline{A \cdot B} = \overline{A} + \overline{B}$$

In general, when the logical function has a complex form, simplification using the algebraic form is quite difficult and time consuming, while in many cases it is not obvious if this function can be further simplified. For this reason, various methods of simplification have been developed, the Karnaugh method being the most popular. In this book, none of these methods will be examined further, since in automation circuits the existence of additional and unnecessary contacts or secondary relays do not dramatically increase the implementation cost, which is happening in digital circuits. However, if for any reason it is absolutely necessary that an automation circuit with minimum realization be implemented, the designer should apply methodologies like the Karnaugh method, which will not be analyzed in the contents of this book.

4.3 State Diagrams

4.3.1 Classical State Diagrams

The design methodology of a state diagram and its logical processing is based on a specific procedure, and is a graphical approach in designing automation circuits, with its main aim to systematically simplify the design procedure, especially in big, complicated industrial automations. In general, the state diagram accurately represents the states of a complex industrial automation system, e.g., a complex machine, as well as the electrical signals that force the automated system to change from one operating state to another. After the proper design of the state diagram, the extraction of Boolean logical expressions that describe each one of these states can be easily derived, and thus the final automation circuit can be further extracted, as will be presented using a more simplified approach. In this methodology, the most important thing is the identification of the exact states and the transition signals, and not the design of the automation itself, as presented in the previous chapters with ad-hoc methodology. Specifically, the methodology in design automation with the state diagram approach involves the following steps:

1. A detailed, extended description of the desired operation of the complex machine or the industrial automation and definition of the operating states
2. Construction of the state diagram based on the design rules that will be described subsequently
3. Extraction of logical Boolean expressions
4. Design of the industrial automation based on the derived logical expressions

The state diagram itself is a direct method of modeling industrial systems that contain a set of logical variables, where their number defines the order of the system. In general, the state of an industrial system or a complex machine can be considered as a set of logical values that contains these logical variables. For example, the state of a complex machine can be the following set of logical values: "The first motor is in operation, the valve is energized, the second motor is not in operation"; where in most cases and for simplification purposes it can be equivalently characterized by verbal terms like "the machine is getting ready". The state diagram is constructed according to the following rules:

1. We denote with circles the different states of the complex machine or industrial system. In every circle, we define the corresponding state.
2. Every possible transition among different states is denoted by an arrow connecting the corresponding states. The arrow's direction denotes the transition's direction.
3. On every directed arrow, we denote the Boolean variable or the logical expression of the Boolean variables that cause the change of state. We denote "Turn OFF" as the logical expression of the variables that cause the change of the dual variable value, characterizing the changed state, from 1 to 0, and "Turn ON" as the logical expression of variables that cause the corresponding change of the same dual variable value from 0 to 1.
4. For all the dual variables of the state diagram we calculate the following expression:

$$X = \overline{\text{Turn OFF}} \cdot (\text{Turn ON} + \text{Present State of X variable}) \tag{1}$$

The "Present State of X variable" is used as a memory type element that updates the current value of the variable. As will be presented subsequently, this term has the same role as the "self-latching contact" principle, which was presented in Section 3.1.1, describing the empirical methodology of circuit design.

5. Based on the extracted and simplified logical expressions from Step 4, we design the industrial automation circuit.

The described procedure for the logical design of the automation circuits, based on the state diagram and logical expression (1), are explained and discussed further through the following application examples.

Example 4.1: START/STOP Operation of a Motor

Let's consider a machine operating with a direct on line starting motor. From the analysis in Chapters 2 and 3, it was clear that the automation circuit will contain a power relay C, which will control the power supply of the motor, and inherently has two states: C=0 where the motor is not in operation and C=1 where the motor is in operation. These states, being the possible operating states of the machine, can be represented by two corresponding circles, indicated in Figure 4.5. It is also defined from the desired industrial application that the transition between these states will take place through the utilization of two buttons b_1 and b_0 for the START and STOP operations, correspondingly. At this point it should be highlighted that the type of the devices causing the transitions are completely dependent on our selections, e.g., it can be an SPDT contact of a sensor, the contact of a button, a relay, etc. In these cases, even if we have selected the type of the transition device, we do not need to define the type of the contacts (e.g., one contact will be NO or NC), since this is something that will be produced from the logical design that we are

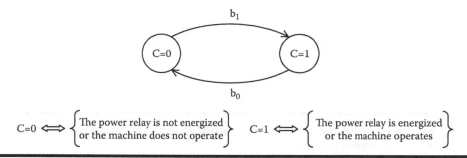

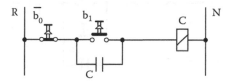

Figure 4.5 State diagram for the direct on line starting motor.

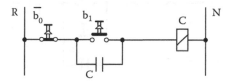

Figure 4.6 The automation circuit extracted from the state diagram of Figure 4.5.

introducing in Chapter 4 directly. Thus, for this example, we can define state C based on Step 3, so that:

$$\text{Turn OFF of } C = b_0 \text{ and Turn On of } C = b_1$$

Subsequently, the logical formula (1) of Step 4 should be calculated, in this case, for the C state as follows:

$$C = \overline{b_0} \cdot (b_1 + C)$$

where for this expression, the digital implementation is the one depicted in Figure 4.6.

Example 4.2: START/STOP Operation of a Motor with Thermal Overload Protection

In Example 4.1, the issue of thermal protection for the motor has not been taken under consideration. If we would like the automation to consider this option (as we already know, the activation of a thermal relay [e] will stop the motor) using the same approach so that the STOP button b_0 would stop the motor, only the logical expression for Turn OFF would change into:

$$\text{Turn OFF of } C = b_0 + e$$

Subsequently, the START-STOP operation of the motor from a secondary control position will need to be controlled through the buttons f_0 (for STOP) and f_1 (for START). In this case, the state diagram will be altered to the following one, presented in Figure 4.7, where we have:

$$\text{Turn OFF of } C = e + b_0 + f_0 \text{ and Turn On of } C = b_1 + f_1$$

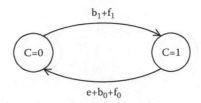

Figure 4.7   State diagram for the direct on line starting motor with thermal overload protection and two START-STOP control positions.

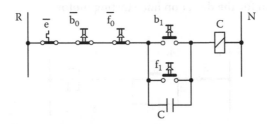

Figure 4.8   The automation circuit extracted from the state diagram of Figure 4.7 for the direct on line starting motor with thermal overload protection and two START-STOP control positions.

By applying the logical expression (1) according to Step 4 for the C variable, we obtain that:

$$C = \overline{e + b_0 + f_0} \cdot (b_1 + f_1 + C)$$

or

$$C = \overline{e} \cdot \overline{b_0} \cdot \overline{f_0} \cdot (b_1 + f_1 + C)$$

where the automation circuit shown in Figure 4.8 can be extracted.

4.3.2 State Diagrams with Sensors

The introduction of the electrical latch and the related operation through sensors (two-wire command), via the state diagrams approach, will be presented in the following example. In general, the electrical latch can be considered as a signal that "stops" the operation of a motor, identically to a STOP button. On the other hand, the operation of a motor through a sensory command means that as long as the corresponding NO contact of the sensor is closed, the motor will also be in operation. Since this command signal is not instantaneous, it is not required to have a memory element to store it, and thus this is equivalent to the fact that in the logical expression (1) of Step 4, the last term of the current state has to be omitted.

Example 4.3: Machine Operation by Pushbuttons and Sensor

To investigate the previous remark further, let's consider the case of a machine that contains two direct on line starting motors being supplied by power through the relays C_1 and C_2. The first motor (C_1) is being controlled manually through the utilization of two pushbuttons, b_0 for STOP and b_1 for START. The second motor (C_2) is being controlled from a sensor s. Moreover, in this automation we would like to introduce the concept of mutual exclusion, which means that the operation of one motor should exclude the operation of the other one. Thus, in this case, it means that for C_1 we will have that 'Turn OFF' = $b_0 + C_2$, as indicated in Figure 4.9. In the same approach, for the C_2 motor, the C_1 operation signal will be equivalent to the stop of the C_2 operation. If we apply the logical expression (1) for both variables C_1 and C_2 we will get that:

$$C_1 = \overline{b_0 + C_2} \cdot (b_1 + C_1) = \overline{b_0} \cdot \overline{C_2} \cdot (b_1 + C_1)$$

and

$$C_2 = \overline{C_1} \cdot (s + C_2)$$

The implementation of these logical expressions is displayed in Figure 4.10, where an error exists, since the energizing of C_2 and the operation of the corresponding motor cannot be altered. This is due to the fact that, from the moment the relay C_2 gets energized, this will be in a continuous latch, independent of the operation of the sensor. Moreover, the energizing of C_1 is not possible due to the corresponding latch in the first branch of the automation circuit, and thus C_2 will always be energized. If the C_2 variable with the logical expression (1) is applied without the term of the "present state of variable", then we will obtain that:

$$C_2 = \overline{C_1} \cdot s$$

which can be implemented from the third correct branch, indicated in the same figure. Thus, as a general design consequence, when the operation signal (or the Turn ON signal) for a state is not

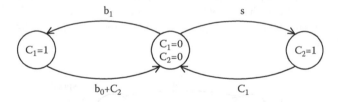

Figure 4.9   A state diagram where the states constitute de-energizing signals (electrical latching).

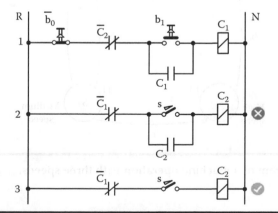

Figure 4.10   Logical design of an automation circuit including electrical latching and sensory command.

instantaneous but is of the general type "operation as long as the signal is ON", then for the formula (1) in Step 4, the last term of "current state of variable" should be omitted.

In the following example, the utilization of the state diagram method by means of a "state machine" technique for the automation of an industrial process machine with multiple states will be presented. In this case, the introduction of complementary auxiliary variables will also be presented, which actually introduces the auxiliary relays in an industrial automation circuit. Finally, it should be noted that the necessity of utilizing auxiliary relays for implementation of an "automation logic" have also been presented in Chapter 3.

Example 4.4: A Three-Speed Machine Automation

One complex industrial machine is operating by utilizing a motor with three speeds. The operation of the machine is achieved through four pushbuttons. The button b_0 is for STOP, while the buttons b_1, b_2, b_3 are for the slow, medium, and high speeds, correspondingly. Moreover, we would like to have the medium speed to be commanded only if the motor is operating in slow speed before, and the high speed to be commanded only if the motor is operating in the medium speed before. The corresponding automation diagram for this problem (states S_0, S_1, S_2, and S_3) is indicated in Figure 4.11. In cases of complex applications or equivalently of state diagrams with a lot of states, the states should be coded through a binary code for the proper functionality of the presented methodology, which will allow for the proper extraction of the logical expressions needed for the design of the automation circuit. As an indicative guide to how many binary variables are needed for state diagrams with n states, the code should contain m binary variables so that the relation $n \leq 2^m$ is valid. Issues such as the best coding selection, the multitude of possible binary assignments, the absence of a systematic method for determining the required binary code, and others can be looked up in the related literature. However, in this book, we will suggest a simple design rule according to which "there should not be a synchronous value change of two Boolean variables as we are moving from one state to another one" and thus the most appropriate coding for this case is Gray coding.

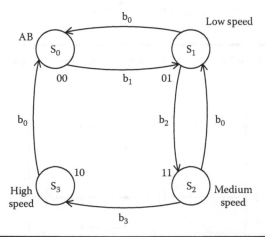

Figure 4.11   State diagram of a machine operation with three speeds.

Thus, the state diagram shown in Figure 4.11 requires coding with two auxiliary variables A and B, while their binary assignment coding for each state is also indicated in the same figure. Subsequently, the procedure for extracting the logical functions remains similar to the one presented before. In the beginning, we apply the logical expression (1) for every one of the auxiliary variables of the selected coding, after defining the logical expressions of the Turn ON and Turn OFF transitions carefully. Moreover, we implement the logic for every state of the diagram, based on the corresponding assigned code of the auxiliary variables. Thus, for this specific example we have that:

$$\text{Turn OFF of A} = \overline{S_2 S_1}(b_0)\Big|_{B=1} + \overrightarrow{S_3 S_0}(b_0)\Big|_{B=0} = b_0 B + b_0 \overline{B} = b_0$$

where $\text{Turn OFF of A} = \overrightarrow{XY}(Z)\Big|_w$ means generally that the Turn OFF change from 1 to 0 of the variable A is caused by the signal Z and through the transition from the X state to the Y state, under the W condition (AND logic). In the same approach, we have that:

$$\text{Turn ON of A} = \overrightarrow{S_1 S_2}(b_2)\Big|_{B=1} = b_2 B$$

$$\text{Turn OFF of B} = \overrightarrow{S_1 S_0}(b_0)\Big|_{A=0} + \overrightarrow{S_2 S_3}(b_3)\Big|_{A=1} = b_0 \overline{A} + b_3 A$$

$$\text{Turn ON of B} = \overrightarrow{S_0 S_1}(b_1)\Big|_{A=0} = b_1 \overline{A}$$

The application of the formula (1) for the A and B variables will result in:

$$A = \overline{b_0}(b_2 B + A)$$

$$B = \overline{b_0 \overline{A} + b_3 A} \ (b_1 \overline{A} + B) = (\overline{b_0} + A)(\overline{b_3} + \overline{A})(b_1 \overline{A} + B)$$

Moreover, for the three power relays C_1, C_2, and C_3 (corresponding to three speeds and hence to three states S_1, S_2, and S_3) we will have, according to the states' coding, that:

$$C_1 = \overline{A}B, \ C_2 = AB, \ C_3 = A\overline{B}$$

The implementation of the previous logical expressions will result in the automation circuit depicted in Figure 4.12.

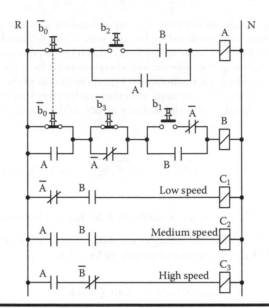

Figure 4.12   The automation circuit of a machine with a three-speed motor as extracted from the state diagram of Figure 4.11.

4.3.3 Step-by-Step Transition due to a Discrete Successive Signal

Attempting to code the states of a state diagram according to the rule that "from one state to another state, no simultaneous variation of two auxiliary variables must exist", usually we prefer an indirect transition of a state to another instead of a direct one. As shown in Figure 4.13, the instant signal "x" causes the transition of a system from state S_0 to state S_2, through the state S_1, which is understood to a human observer as a one-step transition. In fact, the system first passes through state S_1 and then into state S_2, due to an instant signal (an example would be the single actuation of a pushbutton). In such a case, the state S_0 has the code "00", while the state S_2 can have the code "11", which is impossible if a direct transition from S_0 to S_2 has been designed due to the same signal.

On the other hand, there are applications where it is desirable for the same pushbutton to cause different actions through its successive activations. For example, maybe it is desirable that the first activation of a button causes the low speed rotation of a machine, the second activation of the same button causes a medium speed rotation, the third activation of the same button a high speed rotation, and so on. If we follow the Section 4.3.2 approach of state diagram design, the

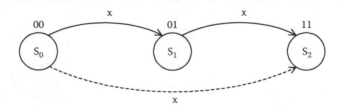

Figure 4.13   Indirect transition from one state to another, for satisfaction of the "one-change each time" rule of state diagram design.

state diagram will lead to a false automation circuit, which will cause the high speed rotation of the machine directly, with only one activation of the pushbutton. Let us demonstrate this situation through an example.

Example 4.5: Combined Operation of Two Machines

Two machines M and N operate through the action of a pushbutton b and two sensors s_1 and s_2, according to the following specifications:

a. By pressing the push button b once, the machine M starts to operate.
b. By pressing button b for a second time, the machine N starts to operate, while the machine M continues to operate.
c. If both machines operate and sensor s_1 is energized, then the machine M stops, while the machine N continues to operate. If instead of sensor S_1, sensor S_2 is energized nothing must happen.
d. If only the machine N operates and sensor S_2 is energized, then the machine N stops also.

The state diagram, shown in Figure 4.14, has been designed based on the assumption that the first activation of the button b leads the system to the state S_1 and it remains there for as long as time is required. Similarly, the second activation of the same button b causes the transition from S_1 to S_2, where both machines are in operation. The introduction of two auxiliary variables and the code assignment, shown in Figure 4.14, give the following equations:

$$A = (\bar{s}_2 + B)(bB + A)$$

$$B = (\bar{s}_1 + \bar{A})(b\bar{A} + B)$$

$$M = \bar{A}B + AB = B$$

$$N = AB + A\bar{B} = A$$

By converting these Boolean expressions into an automation circuit, the circuit shown in Figure 4.15 is obtained.

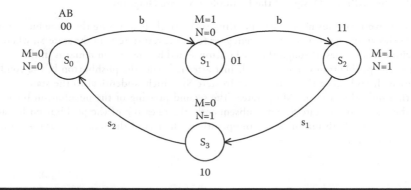

Figure 4.14   State diagram with double transition due to an instant signal b.

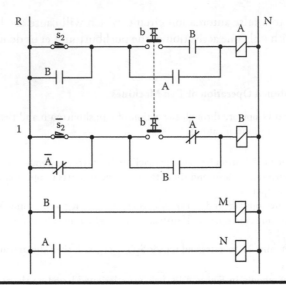

Figure 4.15 Automation circuit confirming the double transition of the state diagram, which means the auxiliary relays A and B are energized both with only one press of the button b.

It is clear by inspection of this circuit logic that by pressing once the button b, both relays A and B are energized (the first relay B is actuated and then immediately relay A is also actuated) leading to the simultaneous start of the operation of both machines M and N. Therefore, the question is "How can we succeed the desired step-by-step activation of different machines through the discrete successive signal coming from the same device?" Such a device may be a sensor, a pushbutton, a switch contact, etc. It must be noted that the two power relays M and N are identical to the auxiliary relays A and B from the logical point of view and hence the auxiliary relays are not necessary. The introduction of the auxiliary variables are in order to follow exactly all the steps of the state diagram design method.

The key action for the step-by-step transition to succeed is to introduce the signal $\bar{b}$ as condition for outputting from a state. This means that the operator who presses a button once to activate a machine has to release the button b first and then press it again for a second time in order to activate another machine. This state diagram designing procedure is applied to Example 4.5, as follows.

Example 4.6: Different Design of the Example 4.5 State Diagram

The pair of two machines must start operating successively by pressing the same button b two times. As shown in Figure 4.16, the system goes from the rest state S_0 to the state S_1, where only the machine M operates, through the pushbutton signal b. The system remains in this state for as long as the pushbutton is not released. Just after releasing the pushbutton b, the condition $\bar{b}$ is satisfied, and hence the system goes to state S_2, which is identical to the state S_1, which means that only the machine M operates. The second pressing of the pushbutton b similarly causes the transition to state S_3, and subsequently the releasing of the pushbutton b leads to the state S_4, where both machines are in operation. The rest of the state diagram is similar to that of Figure 4.14.

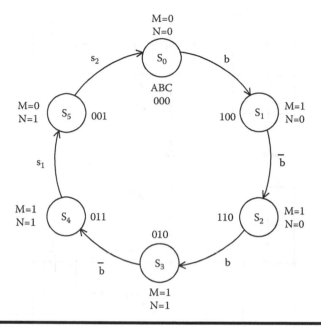

Figure 4.16   State diagram including the complement of a signal ($\bar{b}$) as a condition for separating successive actuations.

The introduction of three auxiliary variables is necessary for the six states of the system. For the code assignment shown in Figure 4.16, we obtain the following equations:

$$\text{Turn ON of A} = b\bar{B}\bar{C}, \quad \text{Turn OFF of A} = b B\bar{C}$$
$$\text{Turn ON of B} = \bar{b}A\bar{C}, \quad \text{Turn OFF of B} = s_1 A C$$
$$\text{Turn ON of C} = \bar{b}AB, \quad \text{Turn OFF of C} = s_2 A\bar{B}$$

Applying the formula (1) to the three auxiliary variables, we obtain,

$$A = (\bar{b} + \bar{B} + C)(b\bar{B}\bar{C} + A)$$

$$B = (\bar{s_1} + A + \bar{C})(\bar{b}A\bar{C} + B)$$

$$C = (\bar{s_2} + A + B)(\bar{b}AB + C)$$

For the operation of the two machines we have,

$$M = A\bar{B}\bar{C} + AB\bar{C} + \bar{A}B\bar{C} + \bar{A}BC = A\bar{C} + \bar{A}B$$

$$N = \bar{A}\bar{B}\bar{C} + \bar{A}BC + \bar{A}\bar{B}C = \bar{A}B + \bar{A}\bar{B}C$$

Converting the above equations to an automation circuit, we obtain the circuit shown in Figure 4.17.

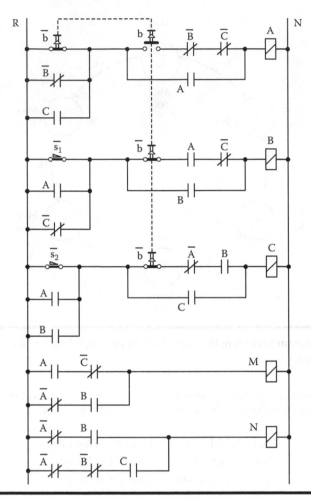

Figure 4.17   The automation circuit of the two machine Example 4.6 extracted from the state diagram of Figure 4.16.

Example 4.7: Power Factor Correction by Manual Insertion of Capacitors

The manual correction of the power factor (cosφ) in an industrial electric AC power station is performed by inserting successively capacitors to the power circuit until it reaches the desired factor value. The insertion of capacitors is achieved by energizing an equal number of relays and by pressing the same button several times. In a similar way, the technician who monitors and corrects the power factor can subtract capacitors by pressing another button. The number of inserting capacitors may be greater than 10 or 12 stages, but in this example, we will examine the case of four capacitors and only the insertion mode due to space limitations. The required state diagram is shown in Figure 4.18, containing ten states and four auxiliary variables. Also, it is obvious its form is extendable and hence this can be a guide for the case of larger number of capacitors. It should be noted that the transitions from state S_8 to S_0 are included for the tutorial scope only in order to have a closed diagram and hence a repeatable procedure. In other words, it is not technically accepted to subtract all the capacitors in one step. In a real system, these transitions are missed and the return of the system from state S_8 to state S_0 will be performed via the above-mentioned subtraction pushbutton.

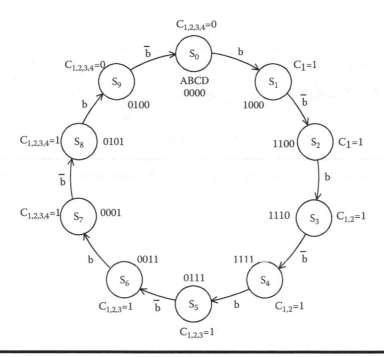

Figure 4.18 State diagram for the power factor correction by manual insertion of capacitors.

4.3.4 State Diagrams with Time Relays

In cases where the automation application contains the parameter of time, as analyzed in Chapter 3, one or more time relays (timers) should be utilized, depending on the application needs. In this case the introduction of a time relay in a state diagram should follow the next additional design principles:

1. Every time relay should be considered as a common relay, and should represent a state of the system.
2. The output contact of the time relay that is going to be energized with a time delay should be considered as the signal that will cause the transition of the system from one state into another one.

Example 4.8: Starting of Two Motors with Time Delay Between Them

Let's consider a machine that contains two motors with direct on line starting and supplied power by two relays, C_1 and C_2. We would like one of the motors to start immediately with the press of a button b_1, while the second motor should start automatically after the elapse of a time T. In every state of operation, the press of a button b_0 should stop the operation of both motors. For this example, the state diagram is the one presented in Figure 4.19.

By applying the same methodology, we have that:

$$\begin{array}{ll} \text{Turn OFF } C_1 = b_0 & \text{Turn ON } C_1 = b_1 \\ \text{Turn OFF } C_2 = b_0 & \text{Turn ON } C_2 = d_T \\ \text{Turn OFF } d = b_0 + C_2 & \text{Turn ON } d = b_1 \end{array}$$

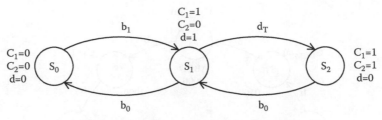

$$d = \text{Timer}, \ d_T = \text{Timer output signal with time delay } T$$

Figure 4.19   State diagram for starting two motors with a time delay between them.

and

$$C_1 = \overline{b_0}\,(b_1 + C_1)$$

$$C_2 = \overline{b_0}\,(d_T + C_2)$$

$$d = \left(\overline{b_0 + C_2}\right)(b_1 + d) = \overline{b_0}\,\overline{C_2}\,(b_1 + d)$$

The implementation of these logical expressions is indicated in Figure 4.20. For the resulting automation circuit, it should be noted that the energizing branch of the time relay d, as derived from the application of the logical expression (1), contains the normal (without delay) contact d, which the time relay might not contain. Therefore, proper care should be paid when selecting the time relay, either by selecting a timer with normal and time-delayed auxiliary contacts, or by utilizing an additional auxiliary relay, as a means to increase the auxiliary contacts of the time relay.

The automation circuit shown in Figure 4.20 does not consist of a minimum circuit realization for this automation application, from a number of contacts point of view. As an example, it is straightforward that the simplification can be achieved with the three contacts of the button b_0 in

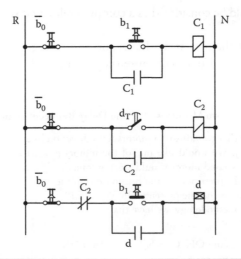

Figure 4.20   Logic design of an automation circuit containing timer.

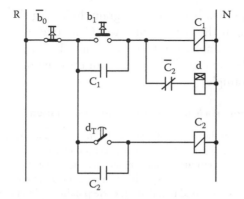

Figure 4.21   The final automation circuit of Figure 4.20, after simplification.

one common for all the three branches. As has been mentioned in Section 4.2.1, we will not address further the problem of minimum realizations, since it is not an issue of further priority in the area of industrial automations. However, this does not mean that after the initial logic design of our circuit we should not perform the straightforward and easy simplifications in the resulting circuit. Thus, the automation circuit shown in Figure 4.20 can be easily simplified to the one depicted in Figure 4.21.

4.3.5 Components' State Diagram Method

Some types of automation circuit design problems may be solved by composing a state diagram different from that presented in Section 4.3.1. Any automated industrial system consists of a number of digital active components (by means of ON-OFF states) or devices, as explained in Section 1.1. Such devices may be relays, coils, solenoids, auxiliary relays, lights, timers, etc. Each one of the system's digital components is represented in the diagram by a corresponding state when it is in the ON state. Since such a state is not a system state but the ON state of the corresponding component, it is denoted as the C_i state of the ith component, instead of the notation S_i. Hence, the diagram contains as many states as there are digital components of the industrial system. Furthermore, the diagram contains one more state, which corresponds to the common rest state of all components, which is also a system state, where all devices are OFF and it is denoted as S_0. The further composition of the diagram is based on the following rules:

1. A pair of transitions must be introduced between S_0 and each state C_i: One for Turn ON and another for Turn OFF of the state C_i.
2. The logic of the system operation, derived from the corresponding specifications, is allocated on the transitions as conditions for firing them.
3. It is permitted to have two or more synchronous transitions. Hence, two or more states C_i may exist simultaneously in an energized situation.
4. The ON or OFF mode of a state C_i may be a condition of a transition.
5. If two or more components are identical from the point of view of the operation's conditions, then they are expressed by one common state.
6. After completion of the diagram, the logical formula (1) is applied for each state C_i. Hence, there is no need for the introduction of auxiliary variables and further coding assignments.

The synthesis of a component-based state diagram, following the above rules, leads to a star form diagram with state S_0 in the center and states C_i around S_0. The system's states are then various sets of states C_i. For example, a system state S_1 may be the set $\{C_1, C_3, C_4\}$, while another system state e.g., S_3 to be the set $\{C_1, C_2\}$. Let's explain the composition of the component's state diagram by the use of an example.

Example 4.9: Automation of a Multi-Conveyor Assembly Station

Consider the assembly station shown in Figure 4.22. From three locations A, B, and C of initial production, three corresponding objects are sent via the central conveyor D to the assembly table. The conveyors A, B, and C operate continuously and are therefore not considered in this automation system. Each of the three workers always puts the same object on the conveyor, for which they are responsible at random times, but only under the condition that the corresponding light H_1, H_2, or H_3 is on. When an object reaches the conveyor D, it is detected by the corresponding proximity switch (PS_i) and then the corresponding light (H_i) turns off, indicating that another object should not be placed on the conveyor. There is no specific order for the placement of the three objects from the workers. When the last object reaches the conveyor D, this means that there are three different objects on the central conveyor D. First, the light H_4 turns on, and second, the central conveyor D can operate if the worker at the assembly table presses the button b_0. After the button pressing, the conveyor D operates for 45 s. At the end of this period the conveyor D stops, the lights H_1, H_2, and H_3 turn on while the light H_4 turns off.

The digital components of the system are four lights, the motor for operation of the central conveyor D (powered by the relay C_M), and a timer T, required to measure the operation time of 45 s. Therefore, the component-based states of the system are: $\{C_1$ for $H_1\}$, $\{C_2$ for $H_2\}$, $\{C_3$ for $H_3\}$, $\{C_4$ for $H_4\}$, and $\{C_5$ for C_M and T$\}$, since the timer T and the relay C_M are identical from the operation point of view. The component state diagram is shown in Figure 4.23, where the transition to state C_i for i=1 to 3 is triggered by the corresponding signal PS_i, and may happen simultaneously. The transition to state C_4, where H_4 is ON, depends on the condition $C_1C_2C_3$, which means that three objects are ready to be placed on the conveyor D. The transition to state C_5 is triggered by the condition b_0C_4, which means the state C_4 is energized and the button b_0 is pressed. Subsequently, we can write the logical equation (1) for all states C_i.

$$\text{Turn ON } C_1 = PS_1 \qquad \text{Turn OFF } C_1 = T$$
$$\text{Turn ON } C_2 = PS_2 \qquad \text{Turn OFF } C_2 = T$$
$$\text{Turn ON } C_3 = PS_3 \qquad \text{Turn OFF } C_3 = T$$
$$\text{Turn ON } C_4 = C_1C_2C_3 \qquad \text{Turn OFF } C_4 = T$$
$$\text{Turn ON } C_5 = b_0C_4 \qquad \text{Turn OFF } C_5 = T$$

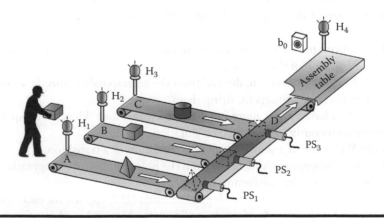

Figure 4.22 A multi-conveyor system feeds an assembly station.

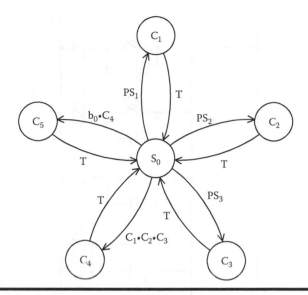

Figure 4.23 The component state diagram of the multi-conveyor system feeding an assembly station.

$$C_1 = \bar{T}(PS_1 + C_1)$$
$$C_2 = \bar{T}(PS_2 + C_2)$$
$$C_3 = \bar{T}(PS_3 + C_3)$$
$$C_4 = \bar{T}(C_1 C_2 C_3 + C_4)$$
$$C_5 = \bar{T}(b_0 C_4 + C_5)$$

For the operation of the active components we have

$$H_1 = \bar{C}_1,\ H_2 = \bar{C}_2,\ H_3 = \bar{C}_3,\ H_4 = C_4,\ C_M = T = C_5.$$

Converting the above equations to an automation circuit and after some obvious simplifications, we obtain the circuit shown in Figure 4.24.

As we ascertain by the inspection of the automation circuit shown in Figure 4.24, the relays C_1, C_2, and C_3 may be characterized as auxiliary relays, but have been introduced in the circuit without the use of auxiliary variables and binary coding that is a basic easiness of the above modified state diagram method applicable to similar problems.

4.3.6 State Diagrams and Minimum Realizations

In most cases, automation problems do not have a unique solution. Different methods of circuit design can produce different automation circuits. As many solutions as there are to an automation problem, only one of them is the minimum implementation, i.e., the solution with the smallest possible number of relay contacts and auxiliary relays. The design of an automation circuit which is not the minimum implementation does not constitute a mistake or an implication problem. The main and important goal is for the circuit to operate correctly, satisfying the specifications for the overall system's operation. Let's clarify these non-identical solutions derived from different design methods by an example.

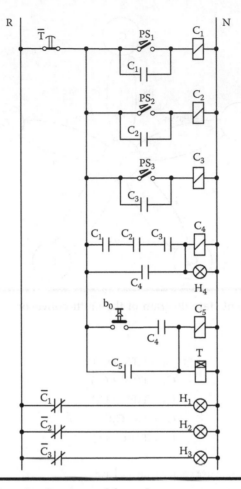

Figure 4.24 The automation circuit extracted from the component state diagram of Figure 4.23.

**Example 4.10: A Drilling and Milling Machine Tool
with Two Different in Duration Processes**

Consider the complex machine shown in Figure 4.25, where a processed object of a production line undergoes two different treatments. The two treatments, which are performed via the operation of two corresponding motors M_1 and M_2, start simultaneously, but one of them lasts 20 s longer than the other. It is desired to design an automation circuit, so that an instant signal START will cause the simultaneous operation of the two motors, while an instant signal STOP will stop one motor immediately and the other motor after 20 s.

FIRST SOLUTION BASED ON THE EMPIRICAL METHOD

For an engineer with relatively little experience in automation projects, it would be easy to design an automation circuit like the one shown in Figure 4.26. The timer contact C_1 is the delay output of an off-delay pneumatic timer mechanically mounted on the relay C_1. For this reason, there is no electronic timer in the automation circuit as an independent device under possible activation.

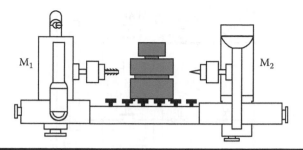

Figure 4.25 **A metal drilling and milling machine tool with two different processes in terms of duration and type.**

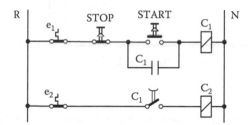

Figure 4.26 **Automation circuit for Example 4.10 based on the empirical designing.**

SECOND SOLUTION BASED ON THE COMPONENT STATE DIAGRAM METHOD

The digital components of the system are the two motors M_1 and M_2, and the timer T for measuring the delay time of 20 s. Therefore, the components states of the system are, {C_1 for M_1}, {C_2 for M_2}, and {C_3 for T}. The component state diagram is shown in Figure 4.27, where the transition to states C_1 and C_2 is triggered simultaneously by the instant start signal "x" from a button. The signal "T" is produced by the timer T, the activation of which happens on state C_3 by the instant stop signal "s" from a button. The deactivation of state C_2, due to signal T, triggers subsequently the deactivation of state C_3, or in other words the turn-off of timer T, since there is no longer a reason to operate.

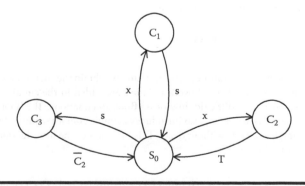

Figure 4.27 **The component state diagram for Example 4.10.**

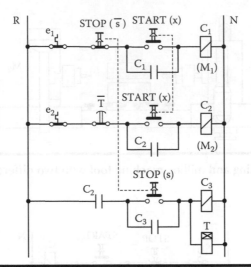

Figure 4.28 Automation circuit for Example 4.10 based on the component state diagram of Figure 4.27.

The next step of circuit design is to write the logical formula (1) for all states C_i, as follows:

$$\text{Turn ON } C_1 = x \qquad \text{Turn OFF } C_1 = s$$
$$\text{Turn ON } C_2 = x \qquad \text{Turn OFF } C_2 = T$$
$$\text{Turn ON } C_3 = s \qquad \text{Turn OFF } C_3 = \bar{C}_2$$

$$C_1 = \bar{s}(x + C_1)$$
$$C_2 = \bar{T}(x + C_2)$$
$$C_3 = \bar{\bar{C}}_2(s + C_3) = C_2(s + C_3)$$

For the operation of the active components we have,

$$M_1 = C_1, \; M_2 = C_2, \; T = C_3$$

By converting the above equations to an automation circuit, we obtain the circuit shown in Figure 4.28.

The contacts e_1 and e_2 of the thermal overload relays are added in the circuit *a posteriori*, since there is no reason to increase the difficulty in writing Boolean equations. The Boolean equation for state C_3, where the timer T is activated, also introduces a simple contact without delay. Usually, the timers do not offer simple contacts without delay. This necessitates the introduction of two distinct devices, the relay C_3 and the timer T.

THIRD SOLUTION BASED ON THE SYSTEM STATE DIAGRAM METHOD

The system of the drilling and milling machine tool has two operation states, except that of the so-called "rest state". The first one corresponds to the operation of both motors M_1 and M_2, while the second one to the operation only of the motor M_2. The timer T is activated simultaneously with M_2.

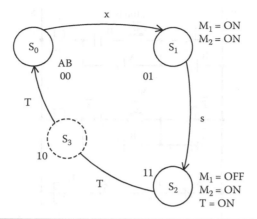

Figure 4.29   The system state diagram for Example 4.10.

The three-state diagram is shown in Figure 4.29, where the signals "x" and "s" correspond to the START and STOP buttons as previously. The code assignment of the auxiliary variables A and B present a simultaneous change of them from the state S_2 to state S_0. The violation of this basic rule, concerning the state diagram synthesis, can be bypassed by introducing the pseudo-state S_3, keeping the same signal T for both its transitions. For the code assignment shown in Figure 4.29, the following equations can be derived:

$$\text{Turn ON } A = sB, \quad \text{Turn OFF } A = TB$$
$$\text{Turn ON } B = x\bar{A}, \quad \text{Turn OFF } B = TA$$

Applying logical formula (1) to the two auxiliary variables we obtain,

$$A = \overline{T\bar{B}}(sB + A) = (\bar{T} + B)(sB + A)$$
$$B = \overline{TA}(x\bar{A} + B) = (\bar{T} + \bar{A})(x\bar{A} + B)$$

For the operation of the two motors and the timer we have,

$$M_1 = \bar{A}B$$
$$M_2 = \bar{A}B + AB = B$$
$$T = AB$$

By converting the above equations to an automation circuit, we obtain the circuit shown in Figure 4.30.

A careful inspection of the circuit can lead to the conclusion that the power relay M_2 is identical to the auxiliary relay B. Therefore, the relay B can be omitted and its contacts may be substituted by corresponding ones of relay M_2.

The comparison of the three automation circuits shown in Figures 4.26, 4.28, and 4.30, derived by three different methods, leads to the following significant comments. First of all, it should be highlighted that all the three automation circuits are different, with each one having a different total number of contacts and auxiliary relays, while all the three circuits are operating correctly, which is also the most important demand when building industrial automations.

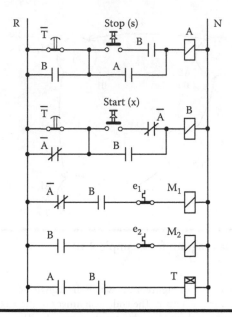

Figure 4.30 Automation circuit for Example 4.10 based on the system state diagram of Figure 4.29.

Second, the minimum realization of the circuit is the one generated from the empirical approach, which is something that might sound awkward. However, we should remember that in these state diagram approaches, state reduction techniques have not been applied and thus, such a result when examining minimum realizations might happen. In many real cases it is possible to have an industrial automation circuit with more switching contacts or a few more auxiliary relays than another one implementing the exact same logic. This happens mainly due to neglect of the state reduction approaches during the design phase, and it is not a serious cost problem, since the automation circuit, in most cases, will only be implemented once for a specific and unique application. This is in contrast to the logical design of integrated circuits (ICs), where the extended application of state reduction techniques is fundamental, mainly due to their production number, which can be equal even to thousands or millions of replications of the same circuit, and thus, in this case, the corresponding cost demands minimum realizations.

Third, the characteristics of the automation circuits shown in Figures 4.28 and 4.30 are worth mentioning. The circuit in Figure 4.28 has two buttons of dual contacts, in contrast with those shown in Figure 4.30 that have single contacts. The timer T in Figure 4.29 has only one delayed contact, while the corresponding one in Figure 4.30 has two delayed contacts. Since timers usually have only one delayed contact or output, the implementation problem can be easily overcome by an additional auxiliary relay.

4.3.7 Sequential Automation Systems

A category of simple automation circuit design problems are the so-called sequential logic automation problems. The simplicity refers to the difficulty of the state diagram synthesis, which offers a standard design procedure. Chapter 7 will present that for these specific types of systems, a specific programming language has been developed for PLCs, which is called the sequential function chart (SFC) and has been adopted by the Standards IEC 61131-3. Let's illustrate this type of application in more detail by the following example.

Example 4.11: Chemical Process Automation in a Reactor

Consider the chemical reactor shown in Figure 4.31, where two different liquids are mixed in order for a new product to be produced under concrete stirring and temperature conditions. The mixing process starts only if the "Operation Cycle On" condition is true and is realized by a simple ON-OFF switch. If during the mixing process the condition is confuted, the mixing process continues until the end of operation cycle, and then the reactor stops to operate. The mixing process must follow the next seven steps:

1. The valve V_1 closes
2. With precondition that the valve V_1 has been closed, the refrigerator (RFG) starts to operate and simultaneously the valve V_2 opens. The valve V_2 remains open until the liquid A level inside the reactor reaches the height L_2 and is detected by a corresponding sensor.
3. With step 2 completed, the stirring impeller (STR) starts to operate. The stirring impeller and the refrigerator stop to operate at the end of the mixing cycle.
4. With the precondition that the stirring impeller operates, the valve V_3 opens and remains open, until the liquid B level reaches the height L_3 detected by a second corresponding sensor.
5. With Step 4 completed, a time interval of 40 min follows for the mixing and the chemical reaction of the two liquids.
6. At the end of this reaction period, the valve V_1 opens and the reactor starts to empty.
7. When the level of the produced liquid drops to height L_1, then the valve V_1 closes and a new operation cycle begins if the corresponding condition exists.

The required state diagram is shown in Figure 4.32, containing six states and three auxiliary variables, where the signal "HS" corresponds to the "Operation Cycle On" condition, which is realized by a hand-operated switch of SPST type.

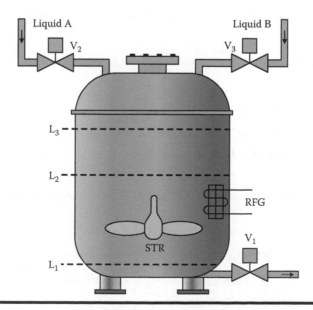

Figure 4.31   A chemical reactor for mixing two liquids and compounding a new product.

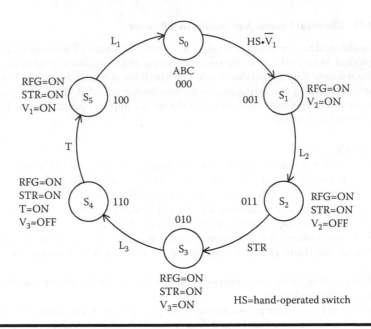

Figure 4.32 State diagram for Example 4.11.

For the code assignment shown in Figure 4.32, we obtain the following equations:

$$\text{Turn ON A} = L_3 B\bar{C}, \qquad \text{Turn OFF A} = L_1 \bar{B} C$$
$$\text{Turn ON B} = L_2 A\bar{C}, \qquad \text{Turn OFF B} = TA\bar{C}$$
$$\text{Turn ON C} = HS \cdot \bar{V}_1 \overline{AB}, \qquad \text{Turn OFF C} = STR \cdot \overline{AB}$$

Applying logical Equation (1) to three auxiliary variables, we obtain:

$$A = \overline{L_1 \bar{B} C}(L_3 B\bar{C} + A) = (\bar{L}_1 + B + C)(L_3 B\bar{C} + A)$$
$$B = \overline{TA\bar{C}}(L_2 A\bar{C} + B) = (\bar{T} + \bar{A} + C)(L_2 A\bar{C} + B)$$
$$C = \overline{STR \cdot \overline{AB}}(HS \cdot \bar{V}_1 \overline{AB} + C) = (\overline{STR} + A + \bar{B})(HS \cdot \bar{V}_1 \overline{AB} + C)$$

For the operation of the two machines (STR and RFG), valves, and the timer, we obtain:

$$\overline{RFG} = \overline{A}\overline{B}\overline{C} \Rightarrow RFG = A + B + C$$

$$STR = \bar{A}BC + \bar{A}B\bar{C} + AB\bar{C} + A\bar{B}\bar{C} = \bar{A}B + A\bar{C}$$

$$V_1 = A\overline{B}\overline{C}, \; V_2 = \overline{A}\overline{B}C, \; V_3 = \overline{A}B\bar{C}, \; T = AB\bar{C}$$

By converting the above equations to an automation circuit, we obtain the automation circuit presented in Figure 4.33.

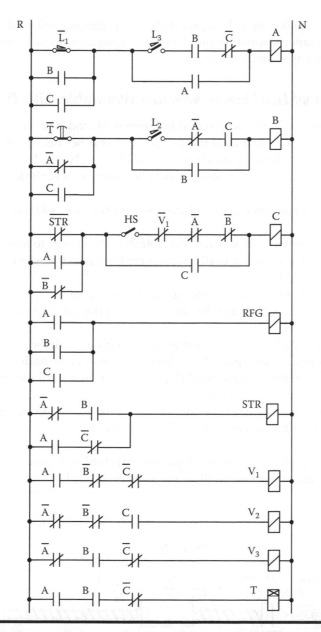

Figure 4.33 Automation circuit for Example 4.11 based on the state diagram of Figure 4.32.

4.4 Applications

In Section 4.4, we will examine some complex setups of industrial automation, where we will investigate how the corresponding logical design method allows for the simplification in the design of solutions for complex problems using a systematic and straightforward approach, especially in cases where the empirical approach will not work. In the following examples, the description of the desired industrial application will be provided at the beginning, followed by the presentation of the logic state diagram construction and the logical expressions derived from equation (1), and finally, the corresponding automation circuit.

4.4.1 Bidirectional Lead Screw Movable Worktable with Two Speeds

The lead screw moving worktable is presented in Figure 4.34, and is operating with the help of a two-direction motor and two rotational speeds. The limit switches that are indicated in the same figure have been placed in those points where we would like to change the speed direction, or the moving speed. Thus, the desired automation should be able to achieve the following:

1. With the press of a button b_1, the lead screw worktable (T) should move to the right with a low speed (R_{LS}).
2. When the limit switch x is energized, the table should continue to move towards the right, but with the high speed (R_{HS}), until it reaches the limit switch y, where it returns to the low speed of motion (R_{LS}).
3. As soon as the limit switch z is energized, the direction of the movement should be inverted, which means that the table should move to the left (L_{LS}) without a change in the speed.
4. The movement to the left should continue in a similar way until the press of the limit switch y, where it continues at low speed (L_{LS}). From the limit switch y until the limit switch x, the movement is happening at high speed (L_{HS}) and from the x until the w at low speed (L_{LS}), where again the motion is reversed.
5. This palindromic movement of the table continues until a button b_0 is pressed, only while the lead screw table moves toward the right at low speed.

For the described problem, the state diagram is displayed in Figure 4.35. Overall, the diagram has five states and thus three auxiliary variables are needed, which are coded in the same figure.

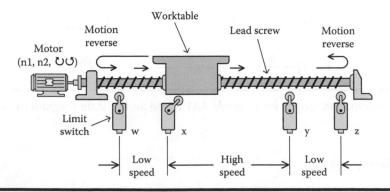

Figure 4.34   Bidirectional and two-speed movable worktable via lead screw.

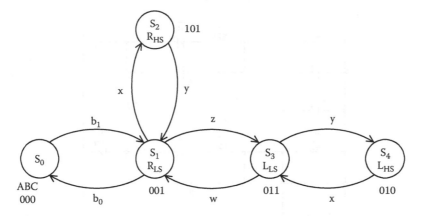

R = motion to the right, L = motion to the left, LS = low speed, HS = high speed

Figure 4.35   State diagram for application in Section 4.3.1, shown in Figure 4.34.

For each of the auxiliary variables, we have:

$$\text{Turn ON A} = \overrightarrow{S_1 S_2}\,(x)\Big|_{B=0,\,C=1} = x\bar{B}C, \qquad \text{Turn OFF A} = \overrightarrow{S_2 S_1}\,(y)\Big|_{B=0,\,C=1} = y\bar{B}C$$

$$\text{Turn ON B} = \overrightarrow{S_1 S_3}\,(z)\Big|_{A=0,\,\bar{C}=1} = z\bar{A}C \qquad \text{Turn OFF B} - \overrightarrow{S_3 S_1}\,(w)\Big|_{A=0,\,C=1} = w\bar{A}C$$

$$\text{Turn ON C} = \overrightarrow{S_0 S_1}\,(b_1)\Big|_{A=0,\,B=0} \qquad\qquad \text{Turn OFF C} = \overrightarrow{S_1 S_0}\,(b_0)\Big|_{A=0,\,B=0}$$

$$+\,\overrightarrow{S_4 S_3}\,(x)\Big|_{A=0,\,B=1} = b_1 \bar{A}B + x\bar{A}B \qquad +\,\overrightarrow{S_3 S_4}\,(y)\Big|_{A=0,\,B=1} = b_0 \bar{A}B + y\bar{A}B$$

Applying the logical Equation (1) to three auxiliary variables, we obtain,

$$A = \overline{y\bar{B}C}\,(x\bar{B}C + A) = (\bar{y} + B + \bar{C})(x\bar{B}C + A)$$

$$B = \overline{w\bar{A}C}\,(z\bar{A}C + B) = (\bar{w} + A + \bar{C})(z\bar{A}C + B)$$

$$C = \left(\overline{b_0 \bar{A}B + y\bar{A}B}\right)(b_1 \bar{A}B + x\bar{A}B + C) = (\bar{b}_0 + A + \bar{B})(\bar{y} + A + \bar{B})(b_1 \bar{A}B + x\bar{A}B + C),$$

and also

$$R_{LS} = \bar{A}\bar{B}C,\; R_{HS} = A\bar{B}C,\; L_{LS} = \bar{A}BC,\; L_{HS} = \bar{A}B\bar{C}$$

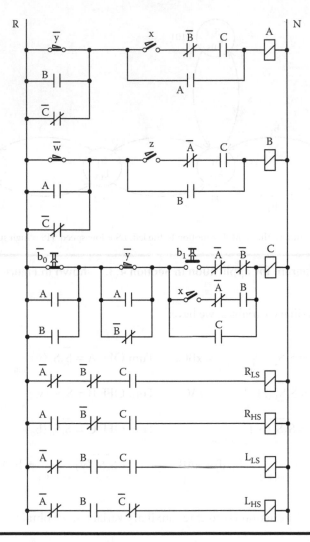

Figure 4.36 Automation circuit for application in Section 4.3.1 based on the state diagram of Figure 4.35.

The implementation of these logical expressions results in the automation circuit shown in Figure 4.36, where it can also be checked that its operation satisfies the desired described automation.

4.4.2 Palindromic Movement of a Worktable with Memory

In Figure 4.37, a simplified form of the carrier (lead screw worktable) of a machine tool is depicted, which is called a "lathe". The worktable of the lathe is desired to be moved in between two limit positions to the left and to the right, according to the following specifications:

1. Initially, we define the movement of the table to the right with "S_R" and the movement to the left with "S_L". In both states, and with the press of a button "s" (STOP), the table stops in its current position.

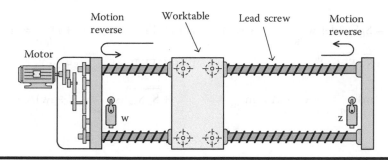

Figure 4.37   The movable carrier (lead screw worktable) of a machine tool.

2. With a press of the button "m" (memory button), the table continues moving in the same manner before it was stopped, due to the press of the button "s".
3. If the table is moving to the right (S_R), then either by the press of a button "a" or when it reaches the end of its movement where the limit switch "z" is energized, the direction of the motion will be inverted, which means that the table should move to the left (S_L).
4. With the same approach, when the table is moving to the left (S_L), either with a press of a button "d", or when it reaches in the end of its movement where the limit switch "w" is energized, the direction of the motion will be inverted, which means that the table should move to the right (S_R).
5. If, during the movement of the table to the left (S_L), the limit switch "w" is energized, while the limit switch "z" remains energized, due to a fault (e.g., the limit switch has been blocked), then the table should stop, like in the case where the button "s" had been pressed.

Overall, we have the following operational buttons and limit switches:

s = STOP button
m = motion continuation button
a = S_L motion button
d = S_R motion button
z = limit switch of the S_R motion
w = limit switch of the S_L motion

The corresponding state diagram is indicated in Figure 4.38, where the STOP state of S_{0L} and S_{0R} is noted, with a previous S_L or S_R motion correspondingly.

Based on this remark, we have the following Turn ON and Turn OFF sets for the auxiliary variables:

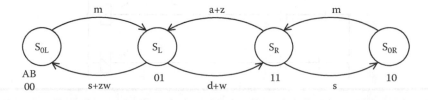

Figure 4.38   State diagram for application in Section 4.4.2, shown in Figure 4.37.

$$\text{Turn ON } A = \overrightarrow{S_L S_R}\,(d+w)\Big|_{B=1} = (d+w)B, \quad \text{Turn OFF } A = \overrightarrow{S_R S_L}\,(a+z)\Big|_{B=1} = (a+z)B$$

$$\text{Turn ON } B = \overrightarrow{S_{0L} S_L}\,(m)\Big|_{A=0} \qquad \text{Turn OFF } B = \overrightarrow{S_L S_{0L}}\,(s+zw)\Big|_{A=0}$$

$$+ \overrightarrow{S_{0R} S_R}\,(m)\Big|_{A=1} = m\overline{A} + mA = m \qquad + \overrightarrow{S_R S_{0R}}\,(s)\Big|_{A=1} = (s+zw)\overline{A} + sA = zw\overline{A} + s$$

Applying the logical Equation (1) to the A and B auxiliary variables, we obtain,

$$A = \overline{(a+z)B}\big((d+w)B + A\big) = \big(\overline{a}\,\overline{z} + \overline{B}\big)\big((d+w)B + A\big)$$

$$B = \big(\overline{zw\overline{A} + s}\big)(m + B) = (\overline{z} + \overline{w} + A)\,\overline{s}(m + B),$$

and also,

$$S_L = \overline{A}B, \quad S_R = AB$$

The implementation of these logical expressions is presented in the automation circuit shown in Figure 4.39.

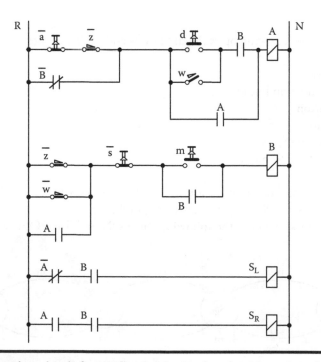

Figure 4.39 Automation circuit for application in Section 4.4.2 based on the state diagram of Figure 4.38.

4.4.3 *Operation of N Machines with Pause under Specific Conditions*

Let's assume that we have a set of n identical machines, which we would like to start and stop manually with a corresponding n number of START-STOP pairs of buttons. For the START operation of the machines no specific condition is needed. However, for the STOP operation, due to some functional specifications, it is important that the STOP action is applied immediately to all the machines, except for the last in operation machine, which should terminate its operation only when a sensor is energized. The operation of the rest of the machines, except the last one, must stop independently of the sensor's state. The difficult part of this problem is the fact that the last machine in operation is not predefined, rather it is randomly selected from the n of total machines. In the case that the sensor is activated, and the STOP button of the final operation machine is not pressed; then the machine continues to operate. To summarize, every machine from the n total can act as the last machine in operation, where we would like to stop it in cooperation with a sensor. In this case, it is assumed that the rest of the machines will have been stopped already.

At this point is should be mentioned that this problem is not an abstract example for tutorial purposes; it occurs frequently in the central autonomous heating systems of multiple apartments. In this case, the machines are replaced from the central electrovalves of the apartments (heating or cooling of the apartment). In these systems, every inhabitant can stop the heating at any time it is desired. In the case that the inhabitant is the last one who switches off the heating, in order not to have hot water trapped in specific areas of the pumps' network, the automation system should prevent the electrovalve of the last apartment to close, even if the inhabitant keeps it open until all the thermal heating of the apartment is reduced to an acceptable level (based on a specific temperature sensor) before closing the final electrovalve.

Since the construction of the state diagram for 1, 2, 3,..., n-1,..., n machines is very complicated, we will only represent the case of three machines, but in a way that the expansion of the diagram to more machines would be straightforward. In Figure 4.40, this state diagram is presented with all the potential operational combinations of the machines; with the indications of the machines (A, B, C) and the three states (S_1, S_2, S_3), where only the last machine is in operation, and from where the transition to S_0 requires the logical condition for the sensor to be energized. In this automation system, the sensor signal, which is also the transition event, is represented by "t". Additionally, we denote with "s_i" and "p_i" the START and STOP buttons of the ith machine, with i=1, 2, 3 for the illustrated example.

Based on this analysis, the Turn ON and Turn OFF logical expressions for the auxiliary variables A, B, and C are defined as:

$$\text{Turn ON A} = \left.\overrightarrow{S_0 S_1}\,(s_1)\right|_{B=0,\,C=0} + \left.\overrightarrow{S_3 S_2}\,(s_1)\right|_{B=1,\,C=0}$$

$$+ \left.\overrightarrow{S_7 S_6}\,(s_1)\right|_{B=1,\,C=1} + \left.\overrightarrow{S_4 S_5}\,(s_1)\right|_{B=0,\,C=1}$$

$$= s_1\overline{B}\overline{C} + s_1 B\overline{C} + s_1 BC + s_1\overline{B}C = s_1\overline{B} + s_1 B = s_1$$

$$\text{Turn OFF A} = \left.\overrightarrow{S_1 S_0}\,(p_1 t)\right|_{B=0,\,C=0} + \left.\overrightarrow{S_5 S_4}\,(p_1)\right|_{B=0,\,C=1}$$

$$+ \left.\overrightarrow{S_2 S_3}\,(p_1)\right|_{B=1,\,C=0} + \left.\overrightarrow{S_6 S_7}\,(p_1)\right|_{B=1,\,C=1}$$

$$= p_1 t\overline{B}\overline{C} + p_1\overline{B}C + p_1 B\overline{C} + p_1 BC = p_1\overline{B}C + p_1\overline{B}t + p_1 B$$

$$= p_1\overline{B}C + p_1 B + p_1 t = p_1 B + p_1 C + p_1 t = p_1(B + C + t)$$

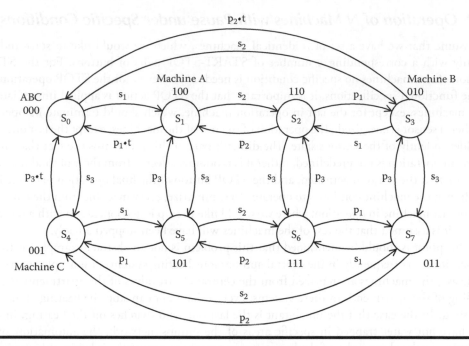

Figure 4.40 State diagram for the operation of three machines with pause under specific conditions (Application 4.4.3).

$$\text{Turn ON B} = \overrightarrow{S_1S_2}\,(s_2)\Big|_{A=1,\,C=0} + \overrightarrow{S_0S_3}\,(s_2)\Big|_{A=0,\,C=0}$$
$$+ \overrightarrow{S_5S_6}\,(s_2)\Big|_{A=1,\,C=1} + \overrightarrow{S_4S_7}\,(s_2)\Big|_{A=0,\,C=1}$$
$$= s_2A\overline{C} + s_2\overline{AC} + s_2AC + s_2\overline{A}C = s_2\overline{C} + s_2C = s_2$$

$$\text{Turn OFF B} = \overrightarrow{S_1S_2}\,(p_2)\Big|_{A=1,\,C=0} + \overrightarrow{S_3S_0}\,(p_2t)\Big|_{A=0,\,C=0}$$
$$+ \overrightarrow{S_7S_4}\,(p_2)\Big|_{A=0,\,C=1} + \overrightarrow{S_6S_5}\,(p_2)\Big|_{A=1,\,C=1}$$
$$= p_2A\overline{C} + p_2t\overline{AC} + p_2\overline{A}C + p_2AC = p_2\overline{C}A + p_2\overline{C}t + p_2C$$
$$= p_2\overline{C}A + p_2C + p_2t = p_2C + p_2A + p_2t = p_2(C + A + t)$$

$$\text{Turn ON C} = \overrightarrow{S_0S_4}\,(s_3)\Big|_{A=0,\,B=0} + \overrightarrow{S_1S_5}\,(s_3)\Big|_{A=1,\,B=0}$$
$$+ \overrightarrow{S_2S_6}\,(s_3)\Big|_{A=1,\,B=1} + \overrightarrow{S_3S_7}\,(s_3)\Big|_{A=0,\,B=1}$$
$$= s_3\overline{AB} + s_3A\overline{B} + s_3AB + s_3\overline{A}B = s_3\overline{B} + s_3B = s_3$$

$$\text{Turn OFF } C = \overrightarrow{S_7 S_3}\,(p_3)\Big|_{A=0,\,B=1} + \overrightarrow{S_6 S_2}\,(p_3)\Big|_{A=1,\,B=1}$$

$$+ \overrightarrow{S_5 S_1}\,(p_3)\Big|_{A=1,\,B=0} + \overrightarrow{S_4 S_0}\,(p_3 t)\Big|_{A=0,\,B=0}$$

$$= p_3\overline{A}B + p_3 AB + p_3 A\overline{B} + p_3 t\overline{A}\overline{B} = p_3\overline{B}A + p_3\overline{B}t + p_3 B$$

$$= p_3\overline{B}A + p_3 B + p_3 t = p_3 B + p_3 A + p_3 t = p_3(B + A + t)$$

Applying logical Equation (1) to the A, B, and C auxiliary variables, we obtain:

$$A = \overline{p_1(B + C + t)}\,(s_1 + A) = \left[\overline{p_1} + (\overline{B}\,\overline{C}\,\overline{t})\right](s_1 + A)$$

$$B = \overline{p_2(C + A + t)}\,(s_2 + B) = \left[\overline{p_2} + (\overline{C}\,\overline{A}\,\overline{t})\right](s_2 + B)$$

$$C = \overline{p_3\left(B + A + t\right)}\,(s_3 + C) = \left[\overline{p_3} + (\overline{B}\,\overline{A}\,\overline{t})\right](s_3 + C)$$

The implementation of the previous logical expressions gives us the automation circuit shown in Figure 4.41 for the case of three machines, where it is obvious that this can be easily expanded and generalized for the n machine case.

The automation circuit shown in Figure 4.41, although it operates well, presents a basic disadvantage. The pressing of pushbutton STOP of the last machine must happen simultaneously

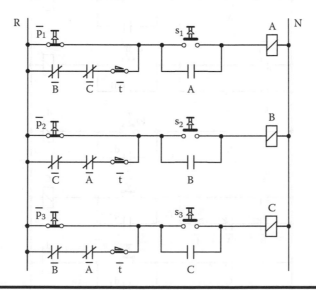

Figure 4.41   Automation circuit for application in Section 4.4.3 based on the state diagram of Figure 4.40.

with the activation of the sensor. This requirement, specifically for the case of central heating, is not acceptable, since each inhabitant can't wait for when the sensor will be energized in order to press the STOP pushbutton. In addition, each inhabitant does not have information about the state of the sensor, if it has already been energized, or when it is going to be energized. In fact, the condition "$p_i \cdot t$" for the STOP of the last machine in the state diagram means that the button "p_i" must be pressed simultaneously with the sensor "t" activation, something that was intentionally stated in order for the reader to establish the improvement of the automation circuit succesively. What is desired in this case is the STOP signal, caused by the press of the button, is saved in the "memory" of the automation circuit and when the sensor stops the last machine is realized. One approach to solve this problem may be the addition of one more state per machine in the state diagram of Figure 4.40. This new state will represent the condition "STOP has been pressed and the sensor awaits" where the electrovalve continues to be open. With three new states in the already complex diagram of Figure 4.40, someone has to write the Turn ON and Turn OFF expressions of the auxiliary variables again, applying the logical formula (1) and extracting the new automation circuit. A second approach, based on empirical design knowledge, is to create the "memory" element for saving an instant signal, based on the self-latching principle described in Section 3.1.1. According to this principle, in the world of automation, one auxiliary relay constitutes a memory with a size of 1 bit, when it is energized by a momentary pushbutton signal, and remains energized through its self-latching contact. Therefore, we need three more auxiliary relays d_i (i=1, 2, 3) than

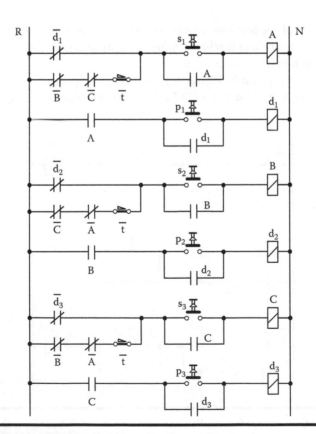

Figure 4.42 The improved automation circuit for application in Section 4.4.3 by adding the possibility to save the stopping commands.

those included in the automation circuit of Figure 4.41, in order to save the three stopping signals p_i (i=1, 2, 3) and also to replace the "p_i" contacts with "d_i" corresponding ones. The new automation circuit, without the above-mentioned disadvantage, is shown in Figure 4.42.

Problems

4.1. In the reciprocating lead screw set-up described in Section 4.4.1, the working table motion stops by pressing the button b_0 only if it is moved right at a low speed or, in other words, it is in state S_1. In any other state of the automation system, pressing the button b_0 does not cause any action. Design an automation circuit which will permit the working table to stop by pressing the same button b_0, also from state S_3 when it moves left at a low speed.

4.2. Based on the description of Problem 4.1, design an automation circuit which will permit the working table to stop from any state (S_1, S_2, S_3, or S_4) of operation.

4.3. Based on the description of Problem 4.1, design an automation circuit which will permit the working table to stop, by pressing the button b_0, at the end w when it is moving left or at the end z when it is moving right, independently of time instant when the button is pressed.

4.4. Explain why the state diagram of Figure 4.20 does not include, as a stop condition, the case where the working table is moving right and the limit switch z is pressed while the limit switch w has been previously energized and remains energized due to a fault.

4.5. For the application described in Section 4.4.3, design an improved automation circuit which will save the instantaneous signal Stop of the last operated machine and will cause its stopping with delay T after the sensor is energized.

4.6. In a processing station, the motor pushes, via the worm screw and the sliding platform, objects from location 1 to locations 2, 3, and 4 in order for these to accept the corresponding treatment. Then the sliding platform returns back to initial place 1. Design an automation circuit only for the movement of objects according to the following specifications:
   a. The platform slides one space at each press of button b.
   b. The sliding is sequential, cyclic, and non-reversible from an intermediate location.
   c. Hence, starting from the initial place 1,
      • with the first pressing of b, the platform slides to location 2
      • with second pressing of b, the platform slides to location 3
      • with third pressing of b, the platform slides to location 4
      • with the fourth pressing of b, the platform returns back to location 1
      • with fifth pressing of b, the platform slides to location 2 and so on.
   d. The detection of the platform's successive positions is achieved via the corresponding proximity sensors.

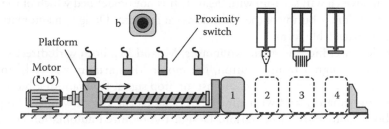

4.7. From three locations A, B, and C of initial production, objects are sent via the central conveyor D to the assembly table. The conveyors A, B, and C operate continuously and therefore are not considered in this automation system. Each of the three workers always puts the same object on the conveyor, for which he is responsible at a random time but only under the condition that the corresponding light H_1, H_2, or H_3 is on. When an object reaches the conveyor D, it is detected by the corresponding proximity switch PS_i. The workers must put two objects A, two objects B, and one object C, and then the corresponding light H_i turns off, which means "do not put another object on the conveyor". For example, two activations of PS_2 cause the turnoff of the light H_2. There is no specific order for the placement of the five objects from the workers. When the last object reaches conveyor D, that means there are five objects on the central conveyor D, then the light H_4 turns on and the central conveyor D can operate if a technician in the assembly table presses the button b_0. After pressing the button, conveyor D operates for 45 s. At the end of this period the conveyor D stops, the lights H_1, H_2, and H_3 turn on while the light H_4 turns off. Design the required automation circuit using the state diagram method.

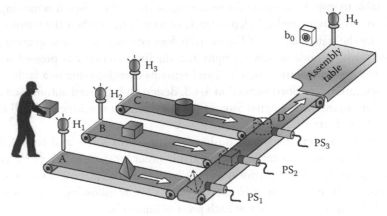

4.8. A processed object in a production line undergoes three different treatments on a complex machine tool. The three treatments, which are performed via the operation of three corresponding motors, start simultaneously but have different durations. Design an automation circuit so that an instant signal Start to cause the simultaneous operation of the three motors, while an instant signal Stop to stop one motor immediately, the second one after 20 s, and the third one after 30 s.

4.9. N machines start to operate or stop by hand via an equal number of button pairs Start-Stop. For the Stop operation of any machine, there are no specific requirements. For the Start operation, however, it is desired to respond immediately to any of the N machines except for the first one, which will start-up with delay T. It is not predefined which of the N machines will start first, it may be any of the N machines at random. Design an automation circuit for N = 3 in an extendable form.

4.10. In a complex machine, there are two motors, M_1 and M_2, both with direct start-up. After an instant Start signal from a button, the motor M_1 starts immediately and with time delay T starts the motor M_2. After an instant Stop signal from another button, the motor M_2 stops immediately and with time delay T stops the motor M_1. The Stop signal acts only if both motors are in operation. Design the required automation circuit using the state diagram method.

4.11. Three machines, M_1, M_2 and M_3 (motors with direct on line starting), start to operate after a common instant Start signal and with null-time latch sequence. This means that if the machine M_1 has been started, only then the machine M_2 starts immediately; and if M_2 operates, then M_3 immediately starts to operate. If any of the three machines stop during operation for any reason (e.g., the thermal relay is energized) then all the machines must stop. Also, the three machines will stop after a common instant Stop signal. Design the required automation circuit.

Note: If a machine M_i starts, and then a machine M_{i+1} operates after passing a time period, we refer to this functionality as a "delay time latch sequence".

4.12. Automate the chemical mixer of the following figure by designing an automation circuit satisfying the following specifications:
   a. The mixing process starts with an instant signal from a button (mixing cycle).
   b. If one of the tanks A, B, and C do not contain the minimum required amount of liquid (dashed lines), the mixing process can't begin.
   c. The filling process includes (in the following respective order) the filling of a central tank T with liquid A up to level L_A, then with liquid B up to level L_B, and finally with liquid C up to level L_C, by opening the corresponding valves V_i, (i=1, 2, 3).
   d. After the end of the filling process, a time period of 10 minutes follows for a chemical reaction to occur.
   e. At the end of this period, the pump M starts to operate until the tank T is empty. This expresses the end of the mixing cycle.
   f. For the detection of various liquid levels, use electronic level switches with immersed electrodes, whose location must be defined in all tanks.

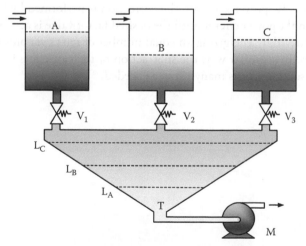

4.13. Centrifugal separators are machines designed especially for liquid-based applications. Using centrifugal force, they separate substances and solids from liquids. They are equally as effective at separating liquid mixtures at the same time as removing solids. Because of the large inertia of the rotated mechanical parts and the high rotation speed, these machines must not vibrate during their operation. Design an automation circuit of a centrifugal separator with two speeds of rotation according to the following specifications:
   a. The separator starts at low speed by pressing the Start button.
   b. After 30 s, the separator goes over the high speed automatically.

c. If the vibration sensor (SPDT output) is activated during the operation at high speed, then the separator must stop immediately, and cannot reoperate except if a reset button is pressed (lock of operation due to vibrations). The operator of the machine is informed about the lock of operation via a light indicator.

d. With an instant signal Stop from a corresponding button, the separator stops to operate independently of the rotation speed.

Note: If the vibration sensor is activated during the rotation at a low speed, which is entirely possible, then this fact should not have any effect on the operation of the separator.

4.14. The automation system of the assembly station described in Problem 4.7 can be redesigned in order to make better utilization of processing time and also to include the remaining parts of the set-up that have deliberately not been encountered. Specifically, the control and operation of the three conveyors A, B, and C (Figure 4.7) must be incorporated in the automation circuit in order to permit the placement of the second two objects even if the corresponding light is turned off. This is reasonable when a worker or object production delays the formation of the objects' triple on conveyor D. Then, the other two workers can put second objects on their corresponding conveyors, but these objects must stop just before conveyor D, otherwise there will be duplicate objects on conveyor D. In order to achieve this, additive sensors are required to be placed on conveyors A, B, and C and possibly additive signaling on initial production places. It is clear that this problem becomes complex and should be faced as a project of an automation system study where you can improvise choosing additional equipment and not as a simple tutorial problem. Therefore, design the automation of the assembly station so that its operation is as efficient and flexible as possible.

4.15. Design an automation circuit for the manual correction of the power factor ($\cos\varphi$) in an industrial electric AC power station. The correction is performed by inserting successively five capacitors in the power circuit until the desired factor value is achieved. The insertion of capacitors is achieved by energizing an equal number of relays via pressing the same button b_{in} several times. In a similar way, the subtraction of the capacitors is achieved by pressing the button b_{out} successively as many times as needed.

Chapter 5

Elements of Electro-Pneumatic Components

5.1 Introduction to Electro-Pneumatic Components

The term "pneumatic automations" usually refers to the automation systems that base their operation on the utilization of pressurized air. Pressurized air, even if it is a carrier of low energy, has a number of significant properties that make its utilization extremely popular in multiple industrial applications. For the use of pressurized air in industrial automation, the most commonly operated devices are pressurized air flow regulators, widely known as pneumatic valves and pneumatic actuators, which are devices that transform the energy of pressurized air in motion. In cases where the energizing of air flow valves is performed by the utilization of electrical solenoids, these pneumatic systems are called electro-pneumatic automations.

The general area of pneumatic automations is very extensive, and thus the aim of this chapter is to cover the case of electro-pneumatic automations in detail, with the main emphasis on the final automation circuit.

5.2 Pressurized Air

Air has been utilized as a means of movement creation since ancient times, as can be seen from numerous related literature on devices and machines that were operating by utilizing pressurized air. It should also be highlighted that the English term "pneumatics" originates from the Greek word "πνεύμα", which meant "breathing" or "light blowing", and today means "spirit". Subsequently, the physical properties of air are considered well known *a priori* from classical physics courses, and thus no depth of analysis will be provided here.

Although it had not been utilized to a large extent in the past, pressurized air was adopted by industrial automations rather late in the 1960s. The real utilization of pneumatics in industry happened after the creation of the need to automate the largest portion of industrial operations. The rapid growth in this field that followed, as well as the establishment of pneumatic automations, was mainly due to the fact that there is a specific category of industrial automation problems that can only be solved efficiently and at low cost, by the utilization of different means, if not pressurized air. Nowadays, there is no industrial automated production that is not utilizing pneumatic automations with pressurized air. Pressurized air has some significant advantages, such as:

- Air exists everywhere and in unlimited supply, and thus the cost of the raw material (air) for pneumatic automations is zero.
- Pressurized air can be very easily transferred through the utilization of elastic or metallic pipes over long distances. Furthermore, pressurized air does not need to be returned to the source of its production (pressurized tank) which happens in hydraulic automation circuits.
- The compressor for generating the pressurized air does not need to be under continuous operation, since pressurized air can be easily stored and in large quantities.
- The variation of environmental temperature does not usually affect the operation of the pneumatic circuit.
- The utilization of pressurized air involves no danger of ignition or explosion, as in the case of utilizing an electrical actuator in an explosive environment.
- Pressurized air is a clean source of energy, since every leakage in the pipes or in the pneumatic devices does not create dust and dirt, a fact for example that is of paramount importance in the food, wood, medical, and high-tech industries, as in the case of integrated circuit production.
- Pressurized air is a very fast actuation factor. It allows for high speeds of operation (movement), as in the case of the pneumatic cylinders that can reach actuation speeds of 1–2 m/s.
- Factors that influence the operation of a pneumatic automation, such as the speed and actuation force, are regulating factors for the adaptation of the pneumatic system to the needs of an industrial application, a situation that can be easily performed in the case of pneumatics.

In general, a pneumatic automation system consists of a pressurized air production unit, a pneumatic circuit of pressurized air distribution, and pneumatic devices that enable the automatic operation of an industrial process. The first two parts, the generation and the distribution of the pressurized air, will not be addressed further in depth, since they do not contribute to the design of the overall automation scheme. More focus will be provided on the operation of different types of pneumatic devices, as well as on the way that these devices are controlled from an automation circuit.

Similar to pneumatic automations, hydraulic automations are ruled by the same philosophy and principles of operation and automation, and thus will not be further analyzed here. However, the industrial terminology, the methods for automation, and the corresponding hydraulic devices are similar to the ones in pneumatic automation, which means that an engineer with a very good knowledge of pneumatic automations can easily comprehend and design hydraulic automations as well.

5.3 Production of Pressurized Air

Pressurized air is produced from compressors, which are electromechanical machines that have the ability to pressurize the air into a metallic container (tank) under the desired operational pressure. In most of the industrial automation cases, it is suggested to have a central station for the production of the pressurized air, instead of having independent and distributed local compressors. Thus, in this approach, each new pneumatic device is supplied from the existing pneumatic network and its installation in the general automation is faster and easier. In general, it should be kept in mind that small and mobile compressors are rarely utilized in industrial automations.

Pneumatic compressors come in multiple types; however, there are two main categories. The first category includes compressors whose operation is based on the principle of movement, where the air is trapped in a chamber, and subsequently the volume of the chamber is reduced with the help of a pressing piston. The compressors of this category are called "piston based compressors". The second category includes compressors where the pressurization is achieved from the acceleration of the air mass and thus are called "turbine compressors". In all cases, the most common operational pressure in the pressurized air circuits is 6–7 bars (1 bar = 14.5 psi).

5.4 Distribution of Pressurized Air

Every pneumatic automation setup is supplied by pressurized air from a pressurized container of the compressor, commonly known as a compression tank, through the distribution pipes. The diameter of this pipe should be such that the pressure drops among the compressor and the pneumatic installation should be less than 0.1 bar. A bigger pressure drop creates a problem in the operation of the pneumatic automation, a situation that is equivalent to the drop of the electrical voltage. The pipes carrying the pressurized air can be made out of copper, steel, or plastic.

The selection of the pipe's diameter is made based on the utilization of two nomograms. The parameters that are inserted in these calculations are as follows:

1. The volumetric flow of the pressurized air
2. The length of the pipe
3. The operational pressure of the pressurized air
4. The allowable drop of pressure

Additionally, the diameter of the distribution pipes is affected by the number and the type of various equipment that are placed in the pneumatic installation, such as the pneumatic two-way valves, the regulator valves, the pneumatic T-connectors (or curve connectors), the pneumatic switches, etc. This equipment is utilized in the calculations through their "equivalent length". This equivalence refers to their resistance during the flowing of the pressurized air, while the "equivalent length" expresses the length of the pipe of the same diameter that is characterized by the same resistance to the equipment. The way of calculating the diameter of the pipe is presented subsequently by the utilization of an example, which it should be noted is the same procedure for all different types of materials.

Example 5.1: Internal Diameter Calculation of a Pressurized Air Distribution Pipe.

The consumption of pressurized air in a factory is at a level of 800 m³/hr. The distribution pipe will have a length of 300 m and it will contain six T-connectors, five 90°-curve connectors and one two-way valve. For this application, the allowable drop of pressure will be 0.04 bar, while the operation pressure should be 9 bar.

In the nomogram of Figure 5.1, the first vertical line indicates the length of the pipe, while the second line refers to the flow of the pressurized air. The point L = 300 m in the first line and the point Q = 800 m³/hr in the second line defines a straight line that intercepts Axis 1 at Z_1. The fourth line represents the operational pressure and the fifth line represents the allowable pressure drop. Point P = 9 bar in the fourth line and point ΔP = 0.04 bar in the fifth line defines a line that intercepts Axis 2 at the point Z_2. The previously defined points Z_1 and Z_2 in Axes 1 and 2 correspondingly define a line that intersects the third line at one point that expresses the internal diameter of the required pipe. Based on all these calculations, the resulting internal diameter of the pressurized air distributed pipe is D = 90 mm.

Subsequently, the pressure drop that is created from the various equipment of having a diameter equal to the calculated one (D = 90 mm), should be taken into consideration. The nomogram of Figure 5.2 provides an equivalent length for the five types of equipment. Each one has an equivalent resistance to the one that of a pipe of the same diameter and a length equal to the calculated length. The horizontal axis of the nomogram indicates the diameter of the pipe. Thus, for D = 90 mm, we have the following:

$$\text{Equivalent length of T-connector} = 10.5 \text{ m}$$

$$\text{Equivalent two-way valve length} = 32 \text{ m}$$

$$\text{Equivalent length of } 90°\text{-curve connector} = 1.2 \text{ m}$$

$$\text{Total equivalent length} = 6 \times 10.5 + 32 + 5 \times 1.2 = 101 \text{ m}$$

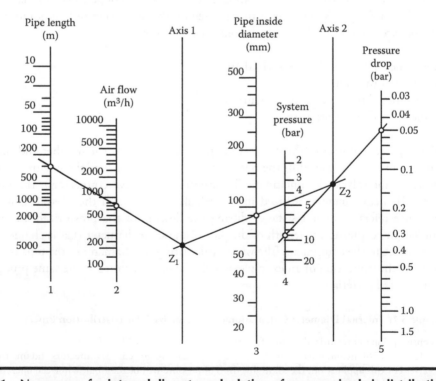

Figure 5.1   Nomogram for internal diameter calculation of a pressurized air distribution pipe.

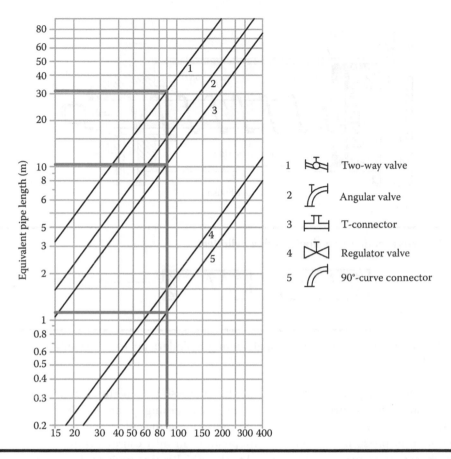

Figure 5.2   Nomogram for the equivalent length calculation of some types of equipment.

Thus, the overall length of the pressure pipe can be considered as 300 m + 101 m = 401 m. With respect to the total length of the pipe, we could repeat the calculations for the correction of the initial calculations. By following the same steps as before, we can calculate the final diameter of the pipe, which is D = 96 mm. Finally, at this point it should be mentioned, as in the case of electrical motors, the aim of this section is not to provide a full and extended analysis on the selection of the pneumatic equipment; instead, it is focused on the way that the pneumatic equipment should be utilized and integrated into fully functioning industrial automations.

5.5 Pneumatic Devices

Pneumatic devices can be split into two categories. The first contains the devices that are able to transform the energy of the pressurized air in a directional or rotational motion, such as pneumatic cylinders and motors, which are also called "pneumatic actuators". The second category includes the devices that control or regulate the operation of pneumatic cylinders and motors, which mainly contains the various valve types.

5.5.1 Single-Acting Cylinders

In the case of single-acting cylinders, pressurized air is applied only on one of the sides of the cylinder and mechanical work is produced only in one direction during cylinder extension. In Figure 5.3,

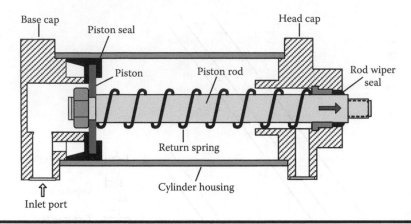

Figure 5.3 Cross section of a single-acting cylinder.

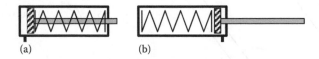

Figure 5.4 Pneumatic symbols of a single-acting cylinder: (a) rod normally in, and (b) rod normally out.

a section of a single-acting cylinder is presented, where the basic parts of the cylinder are also depicted, such as the piston, the pressurized air inlet, the rod, the returning spring, and the elastic O-rings for the sealing of chambers, with the piston and the rod acting as the only moving parts of the cylinder. The application of pressurized air in the corresponding air inlet causes a corresponding extension of the piston rod. The return of the piston rod to the initial resting position takes place either based on the embedded return spring or with the application of an external force. Thus, the force applied from the return spring should be adequate in order to allow for a retraction of the cylinder with an increased speed. In the case of single-acting cylinders, the movement of the rod is restricted by the physical length of the spring, which is usually 10 cm in length. Some of the most common applications for single-acting cylinders include the cases of actuating materials, such as tightening, tossing, lifting, promoting, feeding, and vibrating. The single-acting cylinders are produced in two basic forms, presented in Figure 5.4. These types are contracted and elongated cylinders, and if no pressurized air is energized to it, they are similar to the NO and NC types of electric contacts. The designs depicted in Figure 5.4 also indicate the symbols in the simple graphical representation of single-acting cylinders that are used during the design of automations with pneumatics.

5.5.2 Double-Acting Cylinders

In the case of the double-acting cylinders, the pressurized air is inserted from both inlets of the cylinder and thus the piston has the ability to move in both directions, with the mechanical work produced in both directions of the movement. This type of a cylinder is mainly utilized in applications where it is necessary for the piston to perform a controllable task during extension or retraction, and has the ability to apply a force at the same time. The rod for double-acting cylinders can be significantly longer than those of single-acting cylinders. In Figure 5.5, the longitudinal section of a double-acting cylinder is presented, where except from the general structure, the two positions

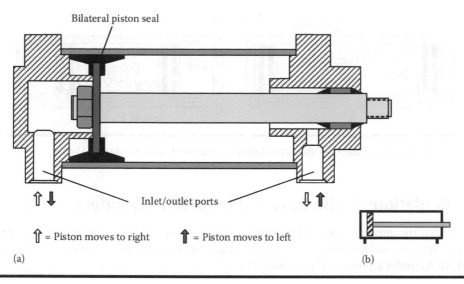

(a)

⇑ = Piston moves to right   ⬆ = Piston moves to left

(b)

Figure 5.5 **Cross section of a double-acting cylinder (a) and a pneumatic symbol (b).**

of the pressurized air connections, the moving piston, the rod with thread at the end for connection to the load, and the O-rings for the sealing of the chambers are visible.

Except for the previous two main types of pneumatic cylinders, as shown in Figure 5.6, there are several additional types of cylinders for specific applications, such as:

- Cylinders with a motion dampener at the end of the movement, or double-acting cylinders with adjustable cushioning at both ends
- Double-acting cylinders with a double rod end
- Non-rotating rod cylinders with twin rods
- Series connection of cylinders and cylinders of multiple positions
- Cylinders with an impact velocity

Furthermore, a general representation of the final concept of cylinders that can transform the translational movement into a rotational one are described in Figure 5.7.

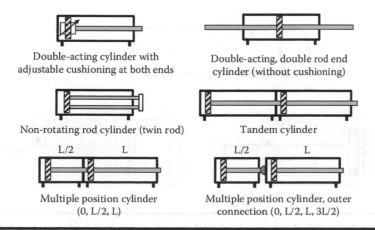

Double-acting cylinder with adjustable cushioning at both ends

Double-acting, double rod end cylinder (without cushioning)

Non-rotating rod cylinder (twin rod)

Tandem cylinder

Multiple position cylinder (0, L/2, L)

Multiple position cylinder, outer connection (0, L/2, L, 3L/2)

Figure 5.6 **Various types of pneumatic cylinders for specific operations.**

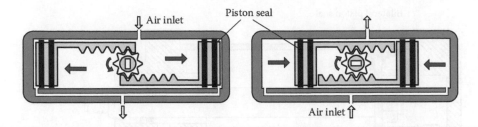

Figure 5.7   Cylinder for altering the translation movement into a rotational one.

5.6 Calculations in the Case of Pneumatic Cylinders

In Figure 5.8a, the internal state of a pneumatic cylinder is depicted; the pressurized air of pressure P has been applied, and the moving piston has an area of A. In general, the force that is generated in a surface A, under a pressure P is provided by:

$$F = AP$$

where F is the force in N, A is the area in m², and P is the pressure in N/m² (1 bar = 10^5 N/m²).

If the area A is representing the area of the piston in a pneumatic cylinder, the friction that is generated on the walls of the cylindrical surface due to the movement of the piston and the force of the spring (if it exists) should be taken into consideration. The friction is in general calculated as a percentage (%) of the nominal force F and varies from 5–20%, depending on the type of fitting for the piston in the internal surface of the cylinder. Thus, we have:

Single-acting cylinders: $F_{real} = F - F_{friction} - F_{spring}$

where $F_{friction} = 5 - 20\% \, F$

F_{spring} = the spring force

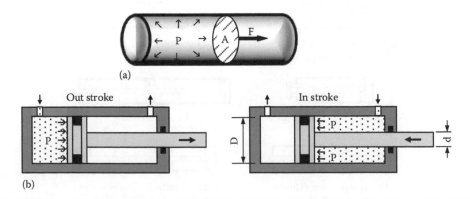

Figure 5.8   Calculation of the force applied by a pneumatic cylinder (a), and the difference between in-stroke and out-stroke applied force of a double-acting cylinder (b).

Double-acting cylinders: $F_{real} = AP - F_{friction}$ (Out stroke force)

or $F_{real} = A'P - F_{friction}$ (In stroke force)

where A is the area of the piston from the left side and A′ is the useful surface area from the right side, as indicated in Figure 5.8b, with the following definitions: $A = \pi D^2/4$ and $A' = \pi(D^2-d^2)/4$.

5.6.1 Length of Piston Stroke

For a fixed diameter of a pneumatic cylinder, the length of the piston's stroke cannot be as long as we want. This is because a large length of stroke will create a large mechanical fatigue on the piston's rod, through which the force is applied on the load (F_{real}). To avoid this fatigue when it is desirable to have a greater stroke, we should select a bigger diameter for the cylinder, and consequently a bigger diameter for the rod. With the utilization of a corresponding nomogram, as in the previous case of a pipe diameter calculation, the allowable stroke with respect to the rod diameter and the applied force can be calculated.

5.6.2 Speed of Piston's Translation

The speed that the piston is translating in a pneumatic cylinder is a function of multiple parameters, resulting from the manufacture of the cylinder itself, or the control unit (valve). The mean speed value for classical cylinders varies from 0.1–1.5 m/s. For a fixed load (force F), a fixed diameter of the cylinder and a fixed inlet valve, the maximum speed of the piston is also defined.

5.7 Pressurized Air Flow Control Valves

The operation of pneumatic cylinders requires the utilization of specific control equipment called valves. With the term "valves", we refer to all control devices that could be used for selecting the flow direction of pressurized air, to apply control signals like START and STOP, to regulate the pressure and volumetric flow, etc. Valves in general can be categorized based on their functionality as follows:

1. Directional valves
2. Non-return or check valves
3. Pressure control valves
4. Flow control valves (volumetric)
5. Isolation valves

In the following subsections of Section 5.7, the operation of the directional valves will be further analyzed, since they affect directly the logic of an electro-pneumatic automation system design.

5.7.1 Directional Valves

Directional valves are devices that have the ability to control the direction of pressurized air in pneumatic installations. The control of the flow direction is mainly utilized for controlling the behavior of the pneumatic cylinders. Each directional valve is characterized by the number of switching positions and by the number of pressurized airways that could be externally connected. Due to the manufacturing complexity of these valves, based on international standards and regulations, the following simplified symbolic representations have been adopted:

- The switching positions of a valve are denoted by rectangles.
- The number of rectangles in a row denotes the number of switching positions that the valve has.
- The route of the flow for the pressurized air is denoted by directional lines within the rectangles.
- The isolation points (stops of air flow) are denoted by a tau (T).
- The external connections of the pressurized air lines are denoted by external lines to the rectangle that represent the rest state of the valve.
- The other operational positions of the valve are created from the translation of the block (left or right) until the point that the flow directions are aligned to the external connections.
- The exhausting of the pressurized air can be achieved either freely toward the environment or in a network of pipes for the collection of low pressure air.

Table 5.1 summarizes all the previous notations and illustrates some symbolic details. For better comprehension of the symbols utilized for the valve representation, refer to Figure 5.9, where the real operation of the two most common valve symbols is indicated for the flow and non-flow of pressurized air as it is actually happening inside the valve, with respect to the valve's symbol as well.

In Figure 5.10 the combined operation of a directional valve with three connection lines, two switching states, and a single-acting cylinder is presented. In one switching state (left) of the valve, the pressurized air is inserted from the line A and it moves the piston toward the right direction, while contracting the cylinder's spring. In the second switching position (right) of the valve, line A is blocked, while the pressurized air in the cylinder is being released through the exhaust line B. Thus, in this position the piston is returning to its initial position, through the action of the returning spring.

In Figure 5.11, the operation of a valve with four connection lines and two switching states, combined with a double-acting cylinder is displayed. As has already become apparent, the valves can be coded directly from the number of connecting lines and, afterwards, the number of switching positions. For example, the valve shown in Figure 5.11 is coded as a four way, two-position valve. The normal switching position for a valve, or rest position, is the position where any actuation mechanism of the valve is inactive.

In Table 5.2, the valve type, the initial position status, and the symbol for the most typical directional valves are presented. It is obvious that for a directional valve to operate in the various switching positions, a kind of actuation is needed, either from a human or from an automatic system. This actuation can be manual, mechanical, pneumatic, or a combination of all these.

5.7.2 Manual Actuation of a Valve

The manual actuation of a valve takes place throughout a mechanism that can be adapted on a valve and can be realized in the form of a button, a lever, a pedal, or any other form of actuation presented in Figure 5.12a. These mechanisms constitute the pneumatic equivalent of electrical buttons and switches. For example, the flow of current throughout an electrical contact is equivalent to the pressurized air flow throughout the airways of the valve. The symbols in Figure 5.12a

Table 5.1   Symbolic Representations of Directional Valves and Their Operations

The switching positions of a valve are represented by a rectangle	
The number of adjacent rectangles expresses the equal number of switching positions of the valve	
The air flow path and its direction at each switching position inside the valve, is represented by an arrowed line	
The air flow shut-off is indicated by a 'T' symbol	
The outer connections of air flow, usually called inlet or outlet ports, are indicated by lines drawn on the outside of the rectangle representing the normal or initial position	
The other positions of the valve operation are obtained by shifting the rectangles until the flow paths inside the valve correspond to the outer connections	
(a) The air exhaust flow path without pipe connection (free exhaust in the environment), (b) air exhaust flow path with pipe connection (closed pressurized-air network), (c) symbol of a pressurized-air source	(a) (b) (c)

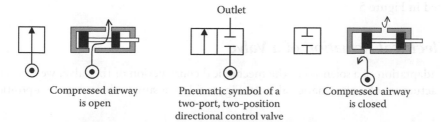

Outlet

Compressed airway is open

Pneumatic symbol of a two-port, two-position directional control valve

Compressed airway is closed

Figure 5.9   The two operation states of a two-port, two-position directional control valve and the corresponding pneumatic symbols.

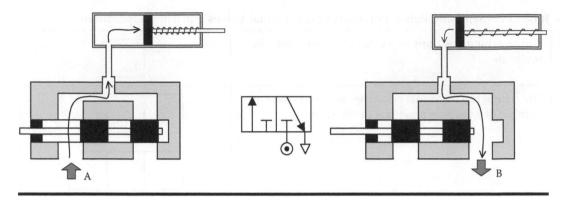

Figure 5.10 The two-state operation of a three-port, two-position directional control valve controlling a single-acting cylinder.

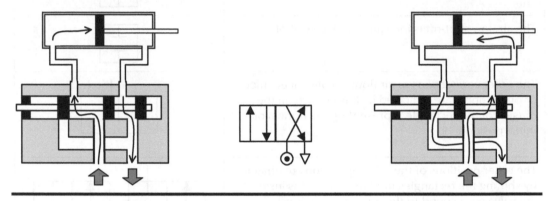

Figure 5.11 The two-state operation of a four-port, two-position directional control valve controlling a double-acting cylinder.

are open in their bottom part, since these symbols are adapted to the general symbols of the valve indicated in Figure 5.12b.

5.7.3 Mechanical Actuation of a Valve

The mechanical actuation of a valve is achieved also with various mechanisms that can be adapted on these valves. In general, these mechanisms require the application of a force (or pressure, based on the same concept of pressing a button), through a mechanical setup and not through a human, as presented in Figure 5.13.

5.7.4 Electrical Actuation of a Valve

With the adaptation of a solenoid in the mechanical construction of the valve, we can achieve an electrical actuation of a pneumatic valve. In Figure 5.14a, a simple three way, two-position valve

Table 5.2 Classical, Basic Directional Valves

Valve Type	Initial Position	Valve Symbol
2/2 Way	Closed	
2/2 Way	Open	
3/2 Way	Closed	
3/2 Way	Open	
3/3 Way	Closed	
4/2 Way	1 Air Line 1 Exhaust Line	
4/3 Way	Middle Position Closed	
4/3 Way	2 Exhaust Lines	
5/2 Way	2 Exhaust Ports	
6/3 Way	3 Air Lines	

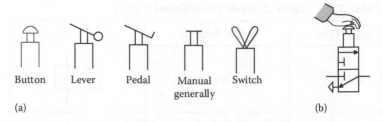

Figure 5.12 Symbols of various actuation mechanisms for manual control of a direction valve (a), and a hand-actuated, 3/2 directional control valve (b).

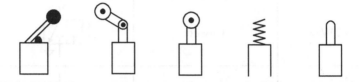

Figure 5.13 Symbols of various actuation mechanisms for the mechanical control of a directional control valve.

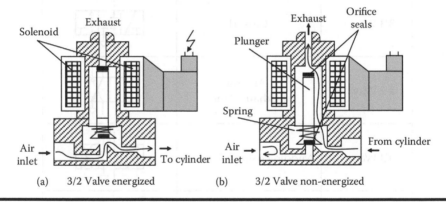

Figure 5.14 A 3/2 directional control valve with electrical actuation in two operation states (a) energized and (b) non-energized.

is indicated, that carries a solenoid and a corresponding plunger. The application of electrical voltage on the solenoid creates a magnetic field that attracts the plunger inside the valve toward the upper position, thus blocking the flow of the pressurized air toward the exhaust point. The pressurized air is applied in the air inlet and is directed toward the only potential exit to the cylinder, since the electromagnetic field of the solenoid is still keeping the plunger in the upper position inside the valve. Thus, if in the air exit a single-acting cylinder has been connected, this operation would cause the extension of the cylinder's rod. If the voltage of the solenoid is interrupted, the plunger returns to its initial bottom position due to the return spring and thus blocks the flow of the pressurized air toward the exhaust and exit points. In this situation, the pressurized air exists in the chamber of the single-acting cylinder, and can find an exit (relief) through the exhaust point, which means that the single-acting cylinder will return to its normal position. As an overview of the described operation, the application of a voltage on the valve's solenoid moves the piston rod in one direction, while the termination of the voltage application causes the return of the piston rod to its rest position. As has been mentioned, the return of the rod to its normal position is achieved by a return spring; however, this could also be accomplished by the utilization of a second solenoid that could actuate in the opposite direction. In Figure 5.15a, the symbol of an electrically actuated three-way, two-position valve with return spring is presented, while Figure 5.15b shows the same valve with two solenoids and without return spring. The direction valves with three switching positions and electrical actuation are carrying two solenoids, one in each side of the valve. Due to the numerous types of valves, solenoids, and their corresponding functionalities, a generalized form of symbolic representation has been adopted for the electro-pneumatic valves, which is as follows:

- In a valve with two switching positions, the "operation" that is indicated by the rectangle located next to the spring symbol, takes place when the valve is not energized.
- In a valve with three switching positions, the "operation" of the central rectangle takes place when the valve is not energized.
- When a solenoid is energized (in any kind of a valve), then the operation of the rectangle that is next to the solenoid takes place.

The previous rules, as well as the operation of two classical directional valves, are presented in Figure 5.16.

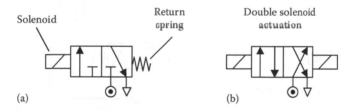

(a)   (b)

Figure 5.15 **Symbol of a 3/2 direction control valve with the electrical actuation of one solenoid and return spring (a), and with two solenoids and a return spring (b).**

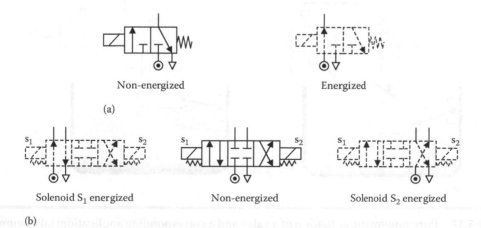

Figure 5.16 **Operation of two basic direction control valves: (a) a 3/2 direction valve, (b) a 4/3 directional valve, and the air flow paths in energized or non-energized state.**

5.7.5 Pneumatic Actuation of a Valve

As mentioned in Section 5.1, a pneumatic automation system can be fully pneumatic. In this case, the pneumatic equivalents of the electrical buttons, the limit switches, the selecting switches, etc., as well as the logical units (based on Boolean algebra) such as AND, OR, etc., exist, thus allowing the implementation of simple or complex functions with the physical magnitude to be pressurized air and not an electrical current. This enables the synthesis of a complete pneumatic industrial automation system, without the utilization of electrical components.

The pressurized air, the flow direction of which is being controlled by a valve, can be utilized also for valve actuation. In Figure 5.17a, a pure pneumatic circuit for the extension and retraction of a single-acting cylinder is presented. Furthermore, in Figure 5.17b the automatic execution of a tank filling with only pneumatic devices is indicated. In all these cases, either in a full pneumatic or electro-pneumatic circuit, the logic that describes the operation of the automation is exactly the same. The only differing factor is the physical variable, pressurized air, or electrical current.

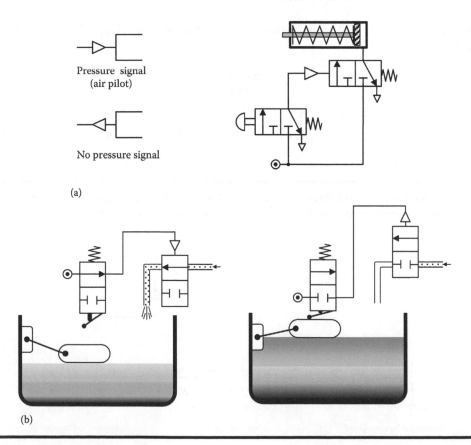

Figure 5.17   Pure pneumatic actuation of a valve and a corresponding application: (a) Pneumatic actuation of a 3/2 valve by another button-type 3/2 valve and (b) pure pneumatic circuit for tank automatic filling.

5.8 Circuits for Electro-Pneumatic Automation

In every electro-pneumatic application of automation, initially the pneumatic operation diagram should be designed. This diagram includes the pneumatic circuit for the flow of the pressurized air, pneumatic cylinders, valves, control devices, and the corresponding connection of all of them. In general, the pneumatic circuit expresses the logical flow of the pressurized air, as well as the logic behind the operation of the pneumatic cylinders, due to the action of the pressurized air.

If the pneumatic automation circuit also includes valves with solenoids, then a separate electrical automation circuit should also be designed for proper completion of the industrial automation, a circuit that will solely control the operation of the electro-valves. For example, in Figure 5.18 a simple pneumatic and electrical circuit is presented for controlling the operation of a single-acting cylinder with an automatic return. When the cylinder is connected to the pressurized air source, the piston rod is extracted, while when the cylinder is connected with the air exhaust line, the rod returns to its initial position due to the return spring. The electromagnetic valve supplies pressurized air to the cylinder when it is energized, while it connects the cylinder to the exhaust line when it is not energized. This operation is presented in the pneumatic circuit of Figure 5.18a. In this case, the extension of the piston rod activates a limit switch, whose contact should be utilized for the automatic return of the piston to its initial position. The implementation solution for this problem is presented in Figure 5.18b, which consists of the electrical circuit of the pneumatic automation. If the extension of the piston rod lasts a long time, this movement can be interrupted through a STOP button.

In Figure 5.19, the pneumatic and electrical automation circuit for the control of a lifting ramp of light objects is presented. We would also like the ramp to stop in any position of this lifting stroke. For this automation problem, the pneumatic equipment contains, as shown in Figure 5.19a, a double-acting cylinder, a four-way, three-position electro-valve, and an air compressor with the pressurized air tank. In Figure 5.19b, the corresponding electrical circuit of the automation is also presented. Except from the classical START-STOP operation of the air compressor, the rest of the

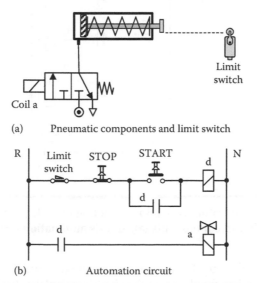

(a)   Pneumatic components and limit switch

(b)   Automation circuit

Figure 5.18   **A single-acting cylinder with automatic return pneumatic circuit and components (a), and an electric automation circuit (b).**

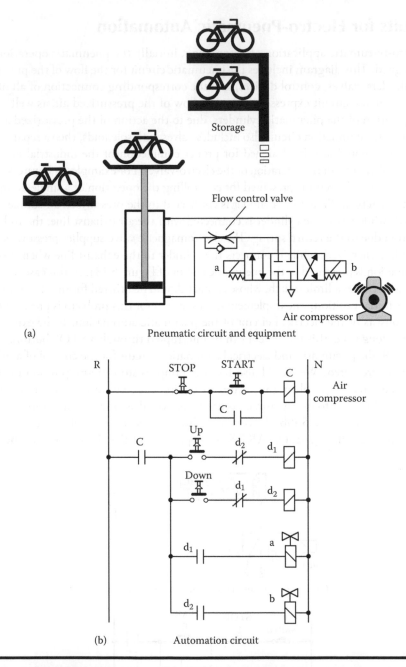

(a)   Pneumatic circuit and equipment

(b)   Automation circuit

Figure 5.19 A light-object elevator with a double-acting cylinder, manual control of its piston position pneumatic circuit and components (a), and its automation circuit (b).

automation circuit is equivalent to the one for the operation of an electric motor with a reverse option. Instead of the two relays for changing the phase connections, in this case, we have the two equivalent solenoids of the electro-valve that also should never be activated at the same time.

When an electro-pneumatic automation includes a large number (greater than three) of cylinders, as well as the desired operation (which involves a complex combination of extension and retraction of the piston rods), then the description of the automation (especially the steps that every cylinder should execute for completing the overall automation task) is problematic. Thus, in these cases, it is suggested to utilize diagrams that can represent the movements for every cylinder in a simplified way, including possible combinations for combined tasks. In general, there exist two types of such diagrams: the diagram of "position-steps" and the diagram "position-time". The word "position" is referring to the position that the piston has (in or out), the word "steps" is referring to the application steps that concerns the operation of the cylinders, while the word "time" is referring to the time that a step lasts in the application. In the following example, the usefulness of these diagrams in complex industrial automations is presented, as well as the need for using them together with the pneumatic and electric circuit.

In Figure 5.20, a change in the translation level of goods is achieved from the conveyor belt M_A to the conveyor belt M_B, with the help of the corresponding cylinders A and B. For the proper operation of this process, the A cylinder should be retained in extension, while the piston B will push the object toward the conveyor belt M_B. In a different case, the object will most likely either fall down or collapse by the piston and the fixed parts of the conveyor belt M_B. This condition can be very easily expressed during the design stage of the automation with the diagram "position-steps" (shown in Figure 5.21) and more especially from Step 2 to Step 3. All the desired movements of the two cylinders that are described in this diagram in a very simple way also helps the design of the overall automation circuit. Another important desire in this automation is the fact that the conveyor belt M_A should not send objects while the piston A is still in extension, because the object will fall to

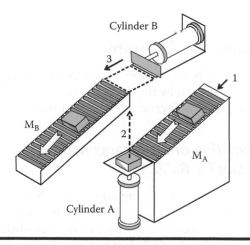

Figure 5.20   A two-cylinder arrangement for level changing in a production line.

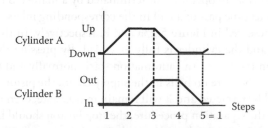

Figure 5.21   Position-step diagram of the two-cylinder arrangement shown in Figure 5.20.

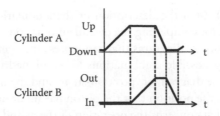

Figure 5.22   Position-time diagram of the two-cylinder arrangement shown in Figure 5.20.

the ground. However, the operation of the M_A is not included in the diagram "position-steps" (it is relevant only for cylinders) and thus this condition is not expressed elsewhere in this diagram.

In many applications, even if the extension of a cylinder lasts for quite a long time (a few seconds), its retraction should be faster in order to keep production cycles to a minimum. As an example, the extension of cylinder B in Figure 5.20 should last for a sufficient time, since the objects might be heavy and fragile. Thus the piston rod, in this case, should move at quite a slow speed, pushing the object gently to the conveyor belt M_B. However, during the return of the piston, a fast movement could be achieved, since there is no reason for delaying this return. In the case that this difference in the time execution of the various tasks makes sense for the proper implementation of an industrial automation, the diagram's "position-time" can be utilized, such as the one presented in Figure 5.22, related to the application in Figure 5.20.

5.9 Electro-Pneumatic Applications

In Section 5.9, we present a set of more complex electro-pneumatic automations where initially the problem will be described, followed by the desired operation, the diagram "position-steps" if needed, the pneumatic circuit, and the electric automation circuit.

5.9.1 *Industrial Automation of an Arrangement for Separating Similar Balls*

In Figure 5.23, a simple mechanism for separating similar balls coming from a feeder is presented. A double-acting cylinder is moving the carrier of the balls left and right, at such a speed that in each translation, extension, or retraction, the mechanism receives a ball and then, due to gravity, releases it in the corresponding inlets 1 and 2. It should also be noted that the regulation of the speed for this translation mechanism is performed by a special valve in the pneumatic circuit.

The overall installation is set in operation or terminated by a manual START-STOP command. The proximity switches detect the pass of a ball in the corresponding inlets. The required pneumatic and electric circuits are presented in Figure 5.23a and b, respectively. In the rest state of operation, the valve is de-energized and the cylinder is fully extended. By pressing the START button, after checking that balls exist in the feeder, the automation starts normally and the overall setup will be able to separate the balls. If there are no balls in the supply chain, the piston rod will be translated to the right, the cylinder will be retracted and it will remain in this state, even if balls are supplied later. In this case, for repeating the separation procedure, the stop button should be pressed first, and then the START button again. It is obvious that a better design of the overall automation can be achieved in order to perform the same operations in a more autonomous approach.

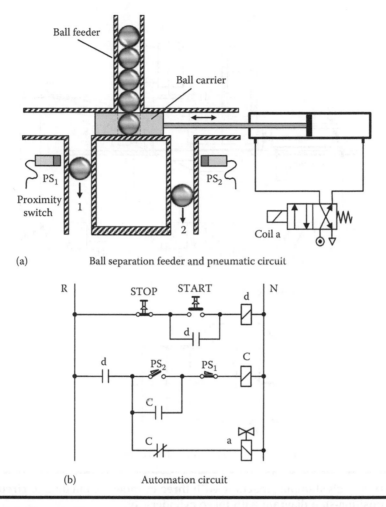

(a)   Ball separation feeder and pneumatic circuit

(b)   Automation circuit

Figure 5.23   A ball separation arrangement with pneumatic circuit and components (a), and an automation circuit (b).

5.9.2 *Industrial Automation of an Object Stamping Machine*

In Figure 5.24a, an object stamping machine is presented that it is operating with the help of three pneumatic cylinders. The objects are received from the supply feeder, with the help of gravity and without a special mechanism. From the cylinders A, B, and C, A places and tightens the object at the stamping position. The B cylinder performs the stamping, while the C cylinder pushes the object onto a conveyor belt. A number of limit switches are necessary in order to detect the various movements of cylinders, particularly limit switches {b, c} for the A cylinder, {d, e} for the B cylinder, and {f, g} for the C cylinder. Furthermore, the limit switch "a" detects the existence of an object for stamping. In general, it is not necessary that every single- or double-action cylinder be accompanied by the two sensors. However, the utilization of two sensors per cylinder simplifies the implementation of the sequential logic of the physical process. In any similar application, we have to initially select the types of cylinders and valves, as well as design the circuit of the pressurized air. These selections are already completed in Figure 5.24a. All the cylinders were selected as double actuation types. Alternatively, cylinders B and C could be of a single actuation type.

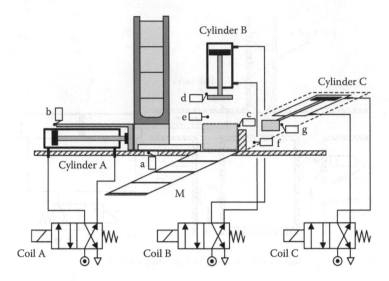

(a)   Stamping machine and pneumatic circuit of three cylinders

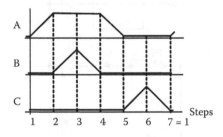

(b)   Displacement-step diagram of three cylinders

Figure 5.24   An objects stamping machine with three cylinders, a pneumatic circuit, and components (a); a position-step diagram with three cylinders (b).

The valves were selected to be of a four-way, two-position type, each with one solenoid. Thus, the automation circuit will be simplified (fewer solenoids, three instead of six, in the case of four-way, three-position valves).

The automation circuit of the stamping machine must be designed in order to obtain the following functionalities:

1. To fulfill the diagram of positions or steps shown in Figure 5.24b.
2. The overall operation of the machine must be initiated by pressing a START button and continue without interruption
3. The operation of the machine should stop if a STOP button is pressed from the operator, or if the objects for stamping are over in the feeder. In both cases, if the STOP function takes place in the middle of the stamping cycle, the stamping operation should be integrated, and then the machine should be returned to its rest position.

According to the previous considerations, the specific application automation circuit is displayed in Figure 5.25. In cases of complicated applications with multiple cylinders and a large

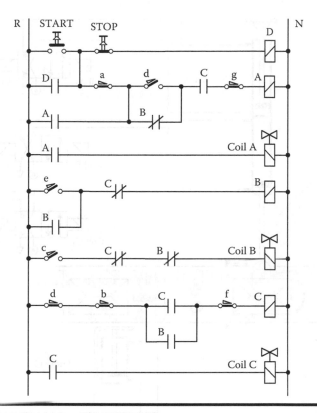

Figure 5.25 Automation circuit of a stamping machine.

number of limit switches, the utilization of a position-step diagram (where the position of the limit switches is also indicated with respect to the position of the piston rods) simplifies significantly the overall design of the automation circuit without logic errors.

5.9.3 Industrial Automation of a Conveyor Arrangement for Objects Shorting

In Figure 5.26, the concept of a very common industrial application is presented, which is the problem of objects shorting and storing that are translated on a conveyor belt. In this figure, the central conveyor belt translates the objects, while the double cylinder A-B (multi-position cylinder) is able to separate the objects that have reached the conveyor belt's junction into three different directions. The selection of the object's type can be achieved in various ways, depending on the overall application, e.g., based on color, size, weight, etc. In this specific example, we assume that the type of object has its unique code that can be recognized from the sensors "a" and "b". The combination of the energized sensors defines the route of the objects, based on the following logic:

- A (ON) and B (OFF) = object in conveyor belt 1
- A (ON) and B (ON) = object in conveyor belt 2
- A (OFF) and B (ON) = object in conveyor belt 3

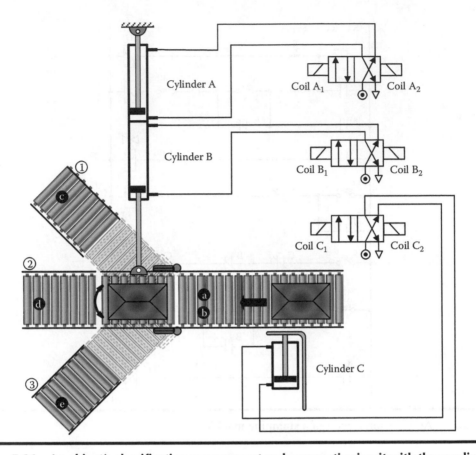

Figure 5.26 An object's classification arrangement and pneumatic circuit with three cylinders.

For the double multi-position cylinder A-B, we also define that,

Position of the movable conveyor part	Cylinder A	Cylinder B
1	In	In
2	In	Out
3	Out	Out

while the cylinder C stops by tightening or inserting obstacles in the continuation of the objects' translation, during the time that the routing of each object lasts. The sensors {c, d, e} are placed correspondingly in the three conveyor belts {1, 2, 3} for detecting the pass of an object and in order to initiate the release from C cylinder of the next one. The individual operational specifications are as follows:

1. When an object reaches the code reading point, the piston C is extended in order to stop the translation of the incoming objects.
2. At the same time as (1), the cylinders A and B are taking the positions corresponding to the recognized code.

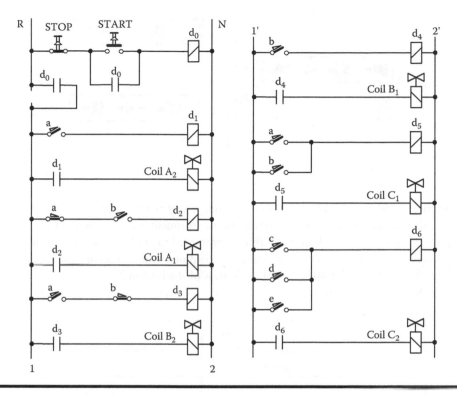

Figure 5.27 **Automation circuit of the object's classification arrangement shown in Figure 5.26.**

3. When an object passes from the corresponding sensor in one of the three final conveyor belts, this sensor signal is utilized for energizing the retraction of the C cylinder, in order to allow the feeding of new objects, and thus a new cycle starts again.

For the pneumatic circuit, four-way, two-position valves have been selected, each with two operational solenoids or coils, one for each of the switching states. The most common type of valve is with one coil and return spring; however, we are selecting two coils in order to demonstrate how easily an industrial automation circuit becomes quite complex. In Figure 5.27, the automation circuit is presented, which is set into operation by classical START-STOP push buttons.

Problems

5.1. Design an arrangement for the separation of similar objects in two parallel transportation lines as shown in the figure. The objects' separation is based on their simple alternative placement through the moving part of the conveyor. In particular, select and design:
  a. The required electro-pneumatic equipment (cylinder, directional valve, sensors, etc.)
  b. The required pneumatic circuit
  c. The required electric automation circuit

Note: The frequency of objects' arrival is such that one object can arrive in the separation point before the previous one reaches one of the two parallel conveyors. This means that in such a case, the motor M must stop as long as needed.

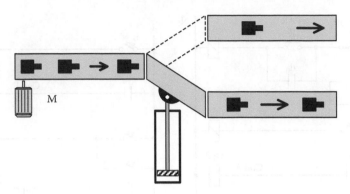

5.2. In a process station, a cylinder pushes objects (one object at a time) from position 1 to positions 2, 3, and 4, in order to undergo the corresponding treatment. After the end of the treatment, the cylinder returns to its initial position. You are prompted to resolve only this part of the automation relating to the movement of the cylinder, according to the following specifications:

a. The piston rod slides one place at each press of the button b.

b. The sliding is sequential, cyclic, and non-reversible and therefore from the rest position 1,

with the	1st	pressing of the button 'b' the piston rod	slides to position	'2'
-//-	2nd	-//-	-//-	'3'
-//-	3rd	-//-	-//-	'4'
-//-	4th	-//-	returns back to	'1'
-//-	5th	-//-	slides to	'2' and so on.

c. The position sensing of the piston rod is achieved via the magnet M situated inside the cylinder and the reed switches mounted outside of the cylinder. When the magnet is in front of a reed switch, its NO contact closes.

Select a suitable pneumatic directional valve and design the required automation circuit.

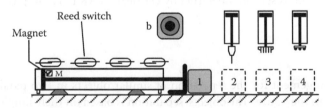

5.3. The figure shows an assembly station. The devices to be assembled are forwarded to technicians through the conveyor belt M and the pneumatic cylinders C_1, C_2, and C_3. On the side of the first (second, third) technician there is the push-button X (Y, Z), in which the technician can request a device, when he has finished the assembly of the previous one. Then, the conveyor operates until a device comes in front of sensor a (b, c). Subsequently, the conveyor stops and the corresponding cylinder is extended. The cylinder pushes the device to the work location until the point 1 (2, 3) where a suitable proximity switch triggers the retraction of the cylinder.

Then the system is on stand-by to accept a new request by another technician for the forwarding of another device. Design the required automation circuit using the state diagram method.

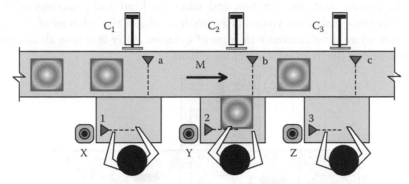

Note: In possible questions such as "How would the automation circuit have to act in a case where a new device request is triggered while simultaneously another one is forwarded?" or any similar one, state whichever answer you consider to be technically correct in order for the station to present maximum effectiveness.

5.4. Objects differing in height are separated in the conveyor system shown in the figure. There are three kinds of objects, two of which will be forwarded to conveyors M_2 and M_3 with the help of cylinders C_1 and C_2, while the third one will continue to the main conveyor M_1. Each cylinder forwards objects of only one kind. The number and order of arrival of objects are random. The desired mode of operation of the system must satisfy the following specifications:

    a. The conveyor system takes effect in standby mode and hence the automation circuit is in operation through a two-position rotary switch RS_{0-1}.

    b. Then, the three conveyors M_1, M_2, and M_3 start to operate continuously, but conveyor M1 must stop every time a cylinder pushes an object or retracts.

    c. Define the placement height of the photoelectric switches PEC_1 and PEC_2 depending on which object you want to detect. In other words, select which cylinder will forward which object.

After selecting the required additive electro-pneumatic equipment, such as the kind of cylinders, directional valves, and additive sensors for detecting the motion of piston rods, design the electric automation circuit.

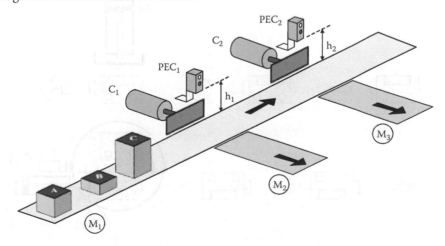

5.5. Objects are moving in opposite directions from two conveyor belts, M_1 and M_2, toward a central point where a cylinder pushes them onto the conveyor belt M_3. In your opinion, define the correct mode of operation and select the kind and placement of the required sensors. The most basic issue you will have to determine is the behavior of the system when two objects arrive simultaneously in front of cylinder. After assessing all the conditions of operation, design the required automation circuit.

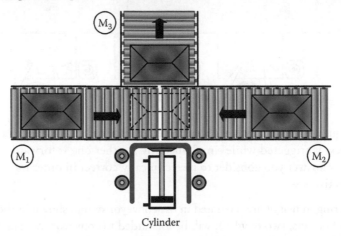

Cylinder

5.6. Non-symmetrical devices are transported on a conveyor belt system where their direction of motion and orientation change through a rotating table. The two cylinders C_1 and C_2 push in and push out devices on the rotating table correspondingly. The conveyor system starts to operate through a two-position rotary switch RS_{0-1} and the two conveyor belts M_1 and M_2 start to operate simultaneously. When a device reaches in front of proximity switch PS_1, then C_1, M_0, and C_2 must perform the movements defined in the position-step diagram of the figure. The proximity switch PS_2 is energized in each half-rotation of the table M_0. The frequency of the device's arrival is such that a device has been transferred from M_1 to M_2 before another one reaches in front of PS_1.

After selecting the required electro-pneumatic equipment such as kind of cylinders, directional valves and additive sensors for detecting the motion of piston rods, design the electric automation circuit.

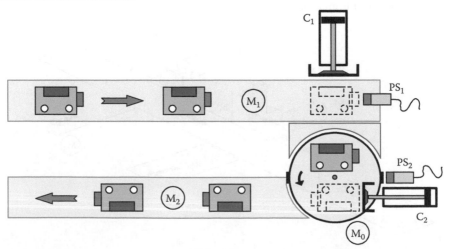

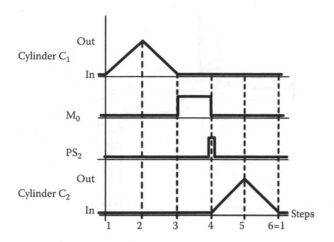

5.7. The double-acting cylinder in the figure has a stroke equal to the distance between limit switches a and c. With the help of the limit switch b, it is desired an intermediate position of the piston rod. Design the pneumatic and automation circuits so that the piston rod follows the position-step diagram.

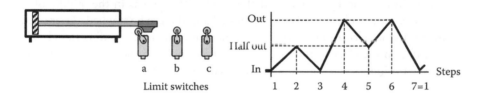

5.8. The figure shows a conveyor setup for the separation of similar objects into two parallel transportation lines, based on the simple alternative placement of objects through the moving part of the conveyor. The cylinder A pushes objects on the movable part of the conveyor, the placement of which is altered by cylinder B. When the sensor S_1 or S_2 detects the passing of an object, then the change of the movable part placement can be performed according to the position-step diagram. The operation of the conveyor system starts with an instant START signal from a button. Similarly, the operation stops with an instant STOP signal from a button but only if the procedure is in the pushing phase (cylinder A in the extension stage). After selecting the required electro-pneumatic equipment (cylinders kind, directional valves, sensors, etc.), design the electric automation circuit by applying the state diagram method.

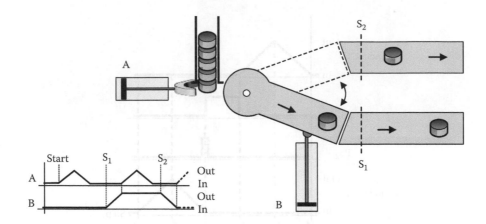

5.9. A simple robotic mechanism consists of two cylinders C_1 and C_2 transferring objects on a horizontal surface from position P_1 to position P_3 through position P_2 according to the position-step diagram. The proximity switch PS detects the presence of an object in position P_1 and then the transportation of the object starts. The robotic mechanism starts to operate through a two-position rotary switch RS_{0-1} and transfers objects continuously as long as they are placed in position P_1. The introduction of objects occurs in position P_1 and they are exported in position P_3 manually. If the switch RS_{0-1} opens in the middle of an object transfer phase, the last object has to be integrated and then to be stopped. After selecting the required additional electro-pneumatic equipment (cylinders kind, directional valves, sensors, etc.), design the electric automation circuit by applying the state diagram method in order for the position-step diagram to be realized iteratively.

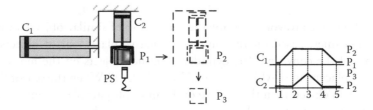

5.10. Two double-acting cylinders C_1 and C_2 are arranged opposite each other as shown in part (a) of the figure, and it is desired to perform a collaborative movement according to the position-step diagram shown in part (b) and to transfer an object between the two piston rods. At the rest state, the object is in the terminal left position. By pressing a button (move) instantly, the object comes to the terminal right position and remains there. By pressing the button move again, the object returns to the terminal left position, and so on. The pneumatic circuit is shown in part (c) of the figure, where S_1 and S_2 are the solenoids of the directional valves. The proximity sensors a_1, a_2, b_1, and b_2 detect the terminal positions of the two cylinders. Design the required electric automation circuit.

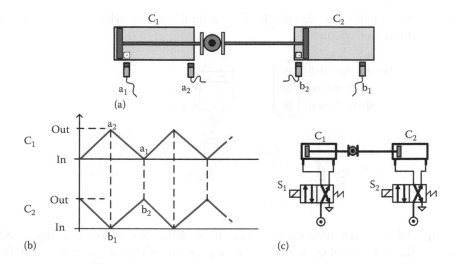

(a)

(b)

(c)

5.11. A painting station includes two conveyor belts, the cylinder A for pushing objects on the painting table, the painting head B which operates with pressurized air, and the cylinder C for removing the painted objects. The conveyor belts operate continuously and transfer objects to and from the painting station. Therefore, they don't affect the automation logic. If an object is detected along the dashed line S from a corresponding proximity switch, the automation system must realize the position-step diagram shown in the figure. After selecting the required additional electro-pneumatic equipment (cylinders kind, directional valves, sensors, etc.), design the electric automation circuit.

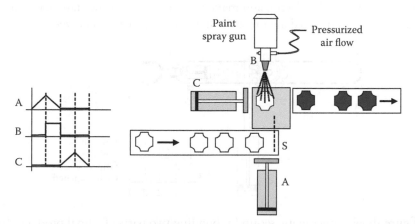

5.12. The figure shows an object stamping station which is switched in/out using a standby operation through two corresponding buttons. The standby operation state is indicated in the control panel and means that the station can automatically stamp an object if it is placed in position 1. In particular, if the station is in standby mode and an object is placed in position 1 (signal S_1), then the object is transferred to position 2 (signal S_2) through the operation of the conveyor belt M. Then, the extraction of the cylinder is energized, the stamping is realized (signal S_5), and subsequently the retraction of the cylinder begins. Just as the piston rod returns (signal S_4), the object is transferred to position 3 (signal S_3) and remains there until a technician removes it. Suppose that only one object exists at the station each time. After selecting the suitable directional valve for the double-acting cylinder, design the required electric automation circuit.

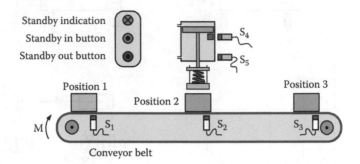

5.13. A simple robotic mechanism operates with the assistance of two double-acting cylinders C_1 and C_2. The extension of cylinder C_1 causes the rotation of the robotic arm by 180° from the right position P_1 to the left position P_2 while transferring an object. The extension of cylinder C_2 causes the closing of the griper and hence the tightening of an object. The sufficient tightening of the object is detected by the pressure sensor PR and only then can the arm rotation start. The two terminal positions of the arm are detected by the corresponding proximity switches PS_1 and PS_2. The robotic mechanism starts to operate through a two-position rotary switch RS_{0-1} and transfers objects continuously, as long as they are placed into the griper in position P_1. The import and export of objects in the griper is done automatically by another machine and hence it will ignore what the robotic mechanism does if there is no object. After selecting the required additional electro-pneumatic equipment (directional control valves, sensors, etc.), design the electric automation circuit applying the state diagram method, in order for the position-step diagram of the figure to be realized iteratively.

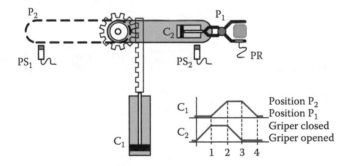

5.14. The figure shows a pneumatic set-up for bonding two parts of a final product. The cylinder C_1, after its extension, applies the required liquid epoxy glue during the retraction phase with the simultaneous operation of the small glue pump M. Then, the cylinder C_2 compresses the two pieces in order to be bonded. A technician puts two pieces for bonding that are detected by the corresponding proximity switches PS_1 and PS_2. With this fact as a condition, the bonding procedure can be triggered by pressing a push-button instantly and then the position-step diagram of actions must take effect. Note that when the cylinder C_2 pushes the left piece, the switch PS_1 stops to be energized. After selecting the required additional electro-pneumatic equipment (cylinders kind, directional control valves, sensors, etc.), design the electric automation circuit applying the state diagram method.

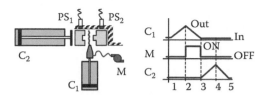

5.15. A machine M_1 processes objects which arrive via a conveyor belt M_2. The last enters within the machine M_1 and exits on the other side, while the processing is performed in motion.

Sometimes the processing of an object fails. In such a case, the faulty object is detected by the specific photoelectric window PE and then the following actions must take effect:

a. The conveyor belt M_2 stops, and the rejected object is transferred from the point K to the point L through the sequential activation of the three cylinders C_1, C_2, and C_3 in order to be re-processed according to the position-step diagram of the figure.

b. With the rejected object at point L, the conveyor belt starts to operate again. Suppose that there is no possibility for collision with another non-processed object.

c. The machine M_1 continues to operate during restoration of the rejected object in order for the raw materials to be at a suitable temperature.

The whole system starts by an instant signal from the START button. Then, a time period of 80 sec must follow for warming-up the machine M_1 (power supply of existing resistors). At end of this period, M_1 and M_2 operate automatically. The system stops after an instant signal from a STOP button, but only if there is no evolution of a faulty object restoration.

After selecting the required additional electro-pneumatic equipment (cylinders kind, directional control valves, sensors, etc.), design the electric automation circuit applying the state diagram method.

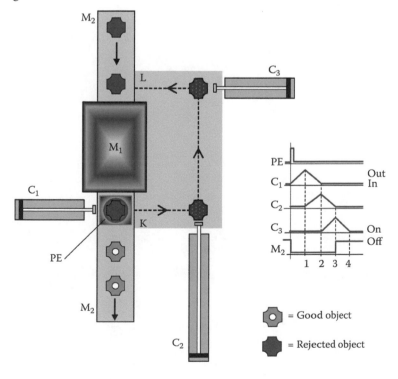

Chapter 6

Basic Operating Principles of PLCs

6.1 Introduction to PLCs

Programmable logic controllers (PLCs) initially appeared in the industry during the 1960s and had a completely different form than those implemented today, since they were built out of logical components that only replaced the operation of the auxiliary relays. Even primitive PLCs were very reliable for a long time when compared to classical electromechanical relays, demanding much less space in the overall automation. Subsequently, their evolution passed through multiple stages, while the most important ones were inclusion of digital components for timing, synchronization and counting, and use of microprocessors. The microprocessors had already started to be a fundamental part of the personal computers (PCs). Nowadays, PLCs can be either simple or complex, come in a variety of sizes, and are equipped with a wide variety of extensions and interfaces that fulfill all the type of needs found at factory level, including the need to communicate with other devices and computers. It should also be mentioned that there are multiple programming languages for tuning the behavior of PLCs so that they can match the different programming skills of the end users. All these issues will be analytically covered in this chapter.

Every PLC, independently of its type and size, can be characterized as a digital device with a microcontroller and a programmable memory that can store and execute user instructions expressing Boolean logic, sequential logic, timing, counting, and mathematical processing, in order to control the operation of a complex machine or an overall industrial process through the utilization of digital and/or analog inputs and outputs (I/Os).

PLCs have the basic structure of a personal computer, with two significant differences. The first is related to the available hardware for the I/Os of the PLC, while the second is related to the microcontroller operation manner and its interaction with the rest of the electronic components of the PLC. A PC's main objective is to communicate with the end user for the successful execution of various arithmetic and algebraic calculations, graphical editing and representation, communication tasks, etc. Thus, in these cases, the end user provides the corresponding commands through a proper interface, such as a keyboard or mouse, while the outcome of these actions is either displayed on the monitor of the PC or printed. The PLC's main task is to communicate with the industrial environment and, more specifically, with either the input devices that are providing the sensorial measurements or with the actuators that interact with the process. For example, typical input devices are sensors, buttons, and switches, while typical output devices are power relays, valve coils, and indicating lights. Since these devices are operating at a different power level than the one that PLCs are usually operating at, it is necessary for PLCs to have the proper I/O hardware to adjust and adapt the power levels accordingly. In Figure 6.1 the basic parts of a PLC are presented: the CPU, the I/O modules, the RAM, and the power supply.

A programming device is a peripheral device that is used only for the programming stage of PLCs, and is not necessary for its operation, therefore it is removed afterwards. In some specific types of small PLCs, the programming device is embedded in its main body. In general, the programming device may be either a specially manufactured digital device (usually portable and specific to a PLC) or a classical PC equipped with the software that the PLC's manufacturer is developing for PLC programming. Before proceeding in analyzing the operation and the interaction of the PLC components, it is very important to define which hardware devices and tasks of the classical industrial automation the PLC is replacing. As has been mentioned in Chapters 2 through 4, a classical industrial automation system needs the following:

1. Auxiliary devices (such as time relays, hour meters, counters, auxiliary relays, etc.) that constitute the basic electrical components of the automation and are mounted in an electrical enclosure.
2. Design of the overall automation electrical circuit that has to achieve the desired operation of the controlled process.
3. Wiring that is needed inside the electrical enclosure for connecting the auxiliary devices between them and also with the I/O devices existing in the enclosure.
4. Wiring that is needed for connecting the electrical enclosure with the I/O devices as a whole, existing far from the enclosure. Input devices may be photoelectric switches, proximity switches, selector switches, etc., while output devices may be motors, electrovalve coils, other actuators, indication lights, etc.

As indicated in Figure 6.2, the first three cases are now embedded in the operation and programming of the PLC, while the last case remains the same, as in classical industrial automations.

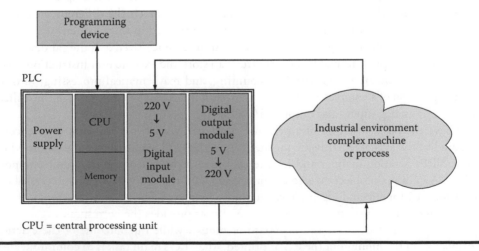

Figure 6.1 Internal structure of a programmable logic controller.

More analytically, a PLC contains several dozens of all the necessary classical industrial automation components (such as auxiliary relays, timers, counters, etc.) due to its digital form. Thus, the implementation of an industrial automation system does not require the purchase and integration of any kind of auxiliary devices. The design of the classical automation circuit, in most cases, is replaced by direct PLC programming. The effectiveness of the overall operation is dependent on the overall complexity of the code (software) for industrial automation, versus the complexity of the wiring needed to embed the automation logic in the electrical circuits. Thus, the role of a PLC nowadays is to transform the hardwiring into flexible software, and to serve as an expert tool for the industrial engineer to solve hard and demanding problems. At this point is should also be highlighted that the PLC is not replacing all the components of an industrial automation, since the power units still remain unchanged (e.g., power relays). As illustrated in Figure 6.2, all the corresponding I/Os remain unchanged, and are used to interact through the software that is running in the PLC.

In Sections 6.3 through 6.8 we will analyze the characteristics of all the components that construct a functional PLC in detail; however, for the proper understanding of this concept, we should initially emphasize the fundamental operational differences between PLCs and PCs. As has already been mentioned, a PLC contains a microprocessor that executes all the internal functionalities of the needed automation, as indicated in Figures 6.1 and 6.2. Furthermore, the processor is responsible for the execution of the user's programmed instructions; the utilization of the memory that stores the automation programs; as well as various types of data that concern the operation of the internal digital components; such as timers, counters, input components that transform high power signals into low power ones that are compatible with the digital logic of the PLC for their usage in the automation program, and output components that are transforming the low power commands from the PLC to the automation devices to proper and compatible high power signals. On the PLC side, there is a specific sequence in which the previous actions are executed. This sequence is cyclic and continuously repeated during the operation of the PLC in the RUN mode.

In Figure 6.3, the cyclic operation of the PLC, as well as the corresponding sequential actions in a more simplified approach, are depicted. Let's consider the fundamental circuit presented in Figure 6.3a. The corresponding logic is simple, and indicates that in the case that the rotary switch RS is closed, then the relay C is energized. If we want to implement the logic of this simple circuit in a PLC, the previous circuit is translated in proper software that it is stored in a specific place in the memory. Regarding the memory itself, there are two additional memory units, where one is dedicated to the storage of the output state and is called "Output Image Table" and the second is dedicated to the storage of the input states and is called "Input Image Table". Since the switch RS is an input device and is connected with the PLC through the input component of the PLC, let's assume the third input. Power relay C is an output device and is connected to the output component of the PLC, so let's assume the fourth output. The components of the program ⊣├·⊣ ├ are instructions that are stored in the PLC memory and refer to the corresponding variables that in our case are input 3 and output 4. The switch RS in the beginning is closed. Let's assume that we would like the PLC to be placed in RUN mode, and that we would like to monitor all the initial steps, which the microprocessor executes based on the corresponding operating system. The input unit, controlled by the microprocessor, is sampling all the inputs, including input 3. This means that the PLC is detecting if there is a voltage or not in every input. Since the switch RS is closed, there is voltage in input 3, as indicated in Figure 6.3b. This voltage subsequently is converted and properly adjusted from the input component in a low power TTL signal. The existence of this TTL signal is stored as a logical 1 in the memory of the input image table and at the position that corresponds to the third input. In the inputs where there is no application of voltage, a

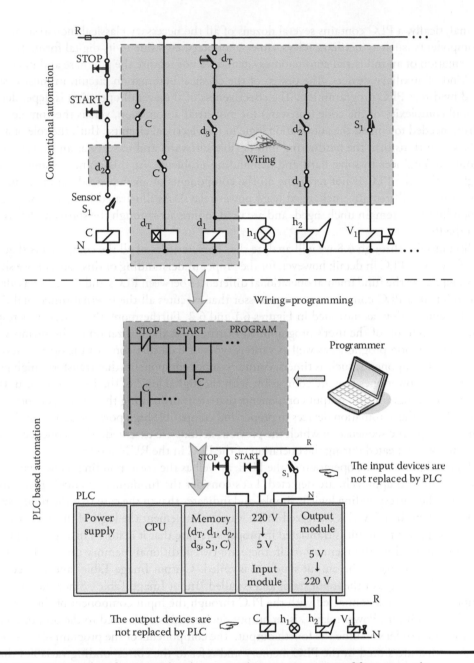

Figure 6.2 **Conventional automation in comparison to programmable automation.**

logical "0" is stored. After sampling of all the inputs, the microprocessor starts with the execution of the program. The instruction ⊣⊢ input 3, by definition means that the point A is at a logical "1" if input 3 is activated, and at a logical "0" if it is deactivated. Thus, for the microprocessor to execute this instruction, it is necessary to sample the status of input 3 through the input image table. Subsequently, the microprocessor executes the following in the list instruction ⊣ ⊢ output 4. By definition, this instruction also means that if point A is in logical "1" then output 4 should be energized, while if it is in a logical "0", it should be deactivated. The activation of an output or not,

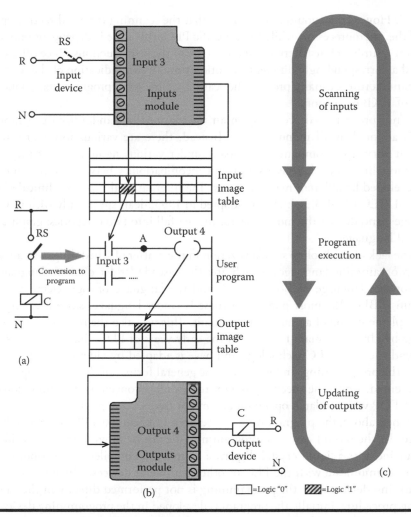

Figure 6.3 **Cyclic operation (or scanning) of a PLC means continuous and repeated reading of inputs, user program execution, and updating of outputs: conventional circuit (a), I/O and user program memory in relation to I/O cards (b), and scanning cycle (c).**

as a direct result of a command execution, means the corresponding writing of a logical "1" or "0" at the output image table. In our example, the output 4 should be energized by the registration of a logical "1" in the corresponding memory position. Subsequently, the rest of the program's branches are executed, if more exist. When the whole program is executed, then all the states of the outputs have been stored in the output image table, digital "1" or "0". Subsequently, the microprocessor transfers the output image table at the PLC's output component. Thus, in output 4 of the output component, a logical "1" TTL signal is transformed to a power signal that can energize a switching component, through which an output device is energized, or the relay C in our case. Subsequently, the microprocessor repeats the sampling of the inputs, executes the program from the beginning, and updates the outputs and repeats again the same cycle, as indicated in Figure 6.1c. This continuous cyclic operation of the PLC is known as the "scanning" mode. The time for a full scanning indicates the operational speed of the PLC, and should vary from a few milliseconds or less. If the variation of an input's state is faster that the scanning time, then these variations are not detectable from the PLC. However, it should also be noted that the scanning time is directly dependent on the speed of the microprocessor, while for a specific PLC, this time is dependent only on the size of the program (number of instructions) and the type of these instructions, since different instructions demand a correspondingly different execution time. As an indication of the scanning speed of a PLC, manufacturers usually provide the scanning time for a program that contains a set of instructions of 1 KB of memory.

PLCs are not programmed according to an internationally standardized programming language that is adopted by all manufacturers. Instead, there are various forms of programming languages that vary from company to company under various names, even if they are similar in their functionality. Also, there is a significant incompatibility between similar programming languages developed by different manufacturers. The International Electrotechnical Commission standard 61131 (IEC 61131-3), which was adopted in 1993, deals specifically with PLC programming languages, and defines the most basic forms that fall into two categories: graphic languages and text-based languages.

Despite the lack of an absolute standardization in the matter of scheduling, we can distinguish three main programming languages, which are: the cascade ladder diagram language (ladder), the instruction list language or Boolean (IL), and the language of logic elements or function block diagram (FBD). The most popular of these is ladder language, since it is very similar to the classic implementation of an automation circuit. This was also the reason for the adoption of this language by almost all manufacturers in the early years of PLCs, because in this way it was easier to spread the novel PLC technology and have it adapted by older engineers who were not familiar with the programming. In Figure 6.4, the general format of the three programming languages is presented, where the specific program represents a conventional automation circuit for the START-STOP with self-latch operation of a motor (see Section 3.1.1).

As mentioned above, the programming device for a PLC can be a PC or a specially designed digital device. In the second case, the programming device or programmer is not just a simple keyboard, but includes a liquid crystal display, a memory, a compiler, and various communication ports to communicate with the PLC or other peripheral devices. These units are necessary in a programming device, since the programming is not performed directly in the PLC. For the programming procedure, initially the program is developed in the programming device and afterwards it is translated into a machine language that is stored initially in the programmer's memory and then transferred to the PLC's memory, provided that there is a communication link between the programmer and the PLC. From the programmer, it is usually feasible to monitor the operation of the PLC in order to detect the status of the PLC's internal elements and to perform various diagnostic tests.

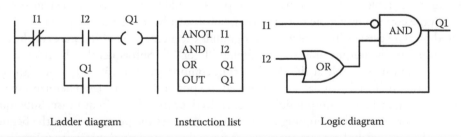

| Ladder diagram | Instruction list | Logic diagram |

Figure 6.4 The conventional START-STOP automation circuit, translated in program for the PLC in three basic languages.

The power supply unit of a PLC is of secondary importance, since it simply provides the various voltages required in each section of the PLC. In certain types of PLCs, it is also possible to supply the circuits of the input devices, but in general the power supply unit never feeds the actuating circuits of the output devices, since the power required to activate the output devices is always supplied from external sources.

The development and spread in utilization of PLCs has been very rapid in recent years, while there is a continuous development of new models with more and more features, smaller sizes, and more affordable costs. Today, PLCs are used in any type of manufacturing process or complex machine, as well as in smaller applications (such as car washes, traffic lights, pumping stations, etc.), since PLCs are one of the most reliable automation solutions. The widespread use of PLCs in industrial automation is attributed to their numerous important advantages, which include:

Adequacy of the contacts. When developing a conventional automation system, during the design of the corresponding automation circuit, we should always evaluate the efficiency of the auxiliary switching contacts of the power relays. When the required auxiliary contacts are numerous and are not available in the utilized power relay, then the engineer has the option to add these additional auxiliary contact blocks, or implement a parallel connection with a second relay in order to use these contacts as auxiliary ones. However, in the case of a PLC, there is no such issue, since the adequacy of the contacts is unlimited, as each internal memory bit location of a PLC can take the role of an auxiliary relay, which could be utilized as many times as we would like in a corresponding automation program. In reality, there is a limit that is dependent on the size of the PLC's memory.

Time saving. For the development of a programmable automation system with a PLC, the writing of the program (design of the automation circuit) can be done in parallel with the installation of the PLC and its connections to the I/O devices, since the program is written in the programming device. In the case of conventional automation (classical automation wirings) this is not possible, since initially the automation circuit should be designed, then the industrial electrical enclosure should be constructed according to the designed automation circuit to perform the installation and its connections to the input and output devices.

Reduced need for space. Since the PLCs are digital devices, they have a comparatively small volume as well as dozens of timers, counters, and hundreds of auxiliary relays, thus their volume is incomparably less than that of a conventional industrial automation enclosure with an equivalent number of auxiliary equipment.

Easy automation modification. The alteration or simple modification of a conventional automation circuit can be performed only by means of removing cables, adding new changes of equipment and, in the worst case scenario, by stopping the operation of the control system for some time. However, in the case of PLCs, all the above modifications are simply equivalent to the direct alteration of the corresponding program that, after the required amendments, can be directly downloaded onto the PLC online or with a pause of the overall operation that lasts for a few seconds.

Easy fault detection. With the help of the PLC's programming device, the status of the PLC's internal elements and the corresponding execution of the loaded program can be directly monitored. In addition, the ON or OFF state of all input and output devices can be further monitored through the utilization of indicative LEDs. Furthermore, the possibility of "forced" (virtual) or simulated notional state changes of an input device, for observing the reaction of the automation system and the overall control logic, can be directly performed in the PLC, mainly due to its digital structure that allows the performance of various tests that assist in the troubleshooting.

Modern and working tools. PLCs have significantly contributed in altering the working environment of engineers, since they have transferred them from the field of cables, auxiliary relays, screwdrivers, etc., to working in an environment similar to one that the PCs have. Engineers have to work with a keyboard or mouse and a program in Windows or in another environment and simply print the automation program instead of designing the automation circuit and applying their knowledge of digital systems. All these concepts created a different and modern operating environment, especially when compared to the corresponding one some decades ago.

Subsequently, in the next sections of Chapter 6 and in Chapter 7, the hardware and the software of PLCs are presented in detail, with a specific focus on the following aspects in the development of an industrial automation system:

1. *Selecting the PLC.* After studying the desired automation control system, the engineer should be able to decide and select what is the most preferable PLC device, what its computational characteristics are, the number of the required I/Os, the number of required switching capabilities, the range of the power supply, etc.
2. *PLC programming.* To understand the operational logic of the automation system, the engineer must write the required automation program for the PLC and evaluate the proper functionality of the developed software.
3. *Installation of the PLC.* After determining the installation specifications of the PLC and its corresponding connections with the I/O devices, the engineer has to operate the overall industrial automation system, as well as to complete the proper initialization of the PLC and modification or adaptation of system parameters during its real online operation successfully.

6.2 Modular Construction of a PLC

PLC manufacturing companies have adopted two basic types of PLC constructions, which are the compact form and the modular shape. As presented in Figure 6.5a, PLCs of the first category are solid structures, usually non-expandable, with a specific number of inputs and outputs. They form an integrated structure that includes, in addition to the I/Os, the power supply, microprocessor, interface with a programming device, STOP-RUN mode switch, LED indicators, etc. PLCs of this category are suitable mainly for small-size automation applications, while medium or large PLCs follow the philosophy of modular construction. This means that in this case, PLCs are composed of independent modular subunits mounted on a common base. As presented in Figure 6.5b, PLCs of this class are comprised of a base with specific dimensions and, therefore, they have specific space capacity for equipment, such as a power supply module, microprocessor unit, and various I/O modules. Furthermore, the modular base has an electrical interface bus (printed circuit conductors) with appropriate plug-in connectors for the modular units, which achieve functional cooperation of all of the PLC's components. Some manufacturers, instead of the universal bus concept, prefer special connecting outlets where each newly-added module is capable of connecting properly with the previous one while, at the same time, providing connections to the next one. The major advantage of the modular assembly of a PLC is that the engineer has the ability

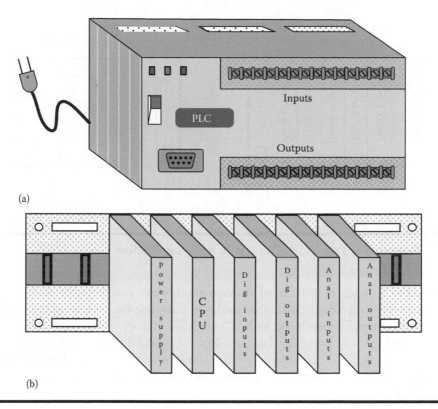

(a)

(b)

Figure 6.5   Compact (a) and modular (b) structural form of a PLC.

to synthesize a PLC that fully corresponds to the requirements of a particular application. Due to the fact that, except for the power supply and the microprocessor unit, there are a wide variety of I/O modules (e.g., discrete or analog, with a number of available I/O channels, nominal operating voltage, etc.), the engineer should consider the type of equipment required for each individual application and synthesize the PLC accordingly. In addition, except for the various types of I/O modules, there are a large number of special purpose modules that make the task of setting up a modular PLC more complex. At this point, it should be mentioned that the minimum setting of a modular PLC includes a power unit, a processor unit, and an input and an output module, while in Figure 6.6, two typical types of commercial compact and modular PLCs are represented.

(a)   (b)

Figure 6.6   Two commercial PLCs in compact form (a) (Schneider Electric) and modular form (b) (Siemens).

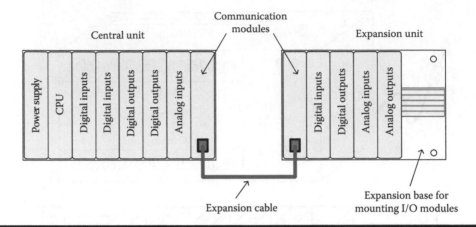

Figure 6.7   Extension of a central PLC unit with modular structure.

Oversized PLCs usually have a central processing unit which can support a much larger number of I/Os than those that may be contained in the modular one-base version of the PLC, even in the case that the latter is fully equipped with I/O modules. In such a situation, a relevant question is "how to expand the initial modular base of a PLC to a second base in order to utilize a large number of I/O modules?" This is achieved by the utilization of two communication units and an interconnection cable that enables communication between the CPU and the expansion base, as presented in Figure 6.7.

6.3 PLC I/O Components

For most PLCs, I/O modules are one of the most common and simple parts of equipment, but also the most important after the microprocessor unit. These I/O modules constitute the interface between the various I/O devices of the industrial automation system and the microprocessor that executes the desired program, and hence perform the control logic. Although the form of the I/O modules differs from manufacturer to manufacturer, without exception, all of them require a power source, carry the necessary electronic circuits for converting the I/O signals and to communicate with the microprocessor, have a terminal block for the cable connection and, finally, carry the necessary mechanism for mounting in the case of a modular structure. The standard and most common I/O modules are the following:

1. Digital input modules
2. Digital output modules
3. Analog input modules
4. Analog output modules

In addition to these modules, there are many others for special purposes. A module can contain only inputs, only outputs, or a combination of inputs and outputs. Digital I/O modules are also called discrete I/O modules by many manufacturers, while both terms refer to the inputs and outputs of ON and OFF states. Many manufacturers, instead of the English term "I/O module" are utilizing the equivalent terms "I/O board", "I/O card", and "I/O block". In Figure 6.8, a common type of a module with digital outputs is presented. In this case, the external view and the general construction is similar for other types of I/O modules that also perform their function independently. Inside the plastic protective cover are the electronic circuits for the I/Os, while the terminal block is usually associated with plug-in contacts on the electronic board. The removability of the terminal block is a significant property in cases of module damage, since it enables the rapid replacement of the module without requiring time-consuming rewiring. Furthermore, PLCs have the ability to connect different groups of I/O devices in a very short time at the same I/O module, provided that the devices are already connected to the corresponding terminal blocks. In general, the cable connection terminals accept conductors up to 2.5 mm^2, while in Figure 6.9, a module used for the control of stepper motors from an Allen-Bradley industrial PLC is presented.

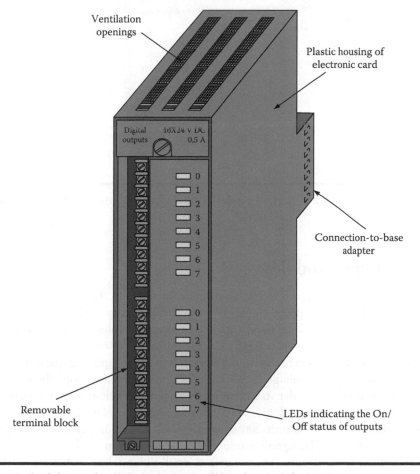

Figure 6.8   Typical form of a digital output module of a modular PLC.

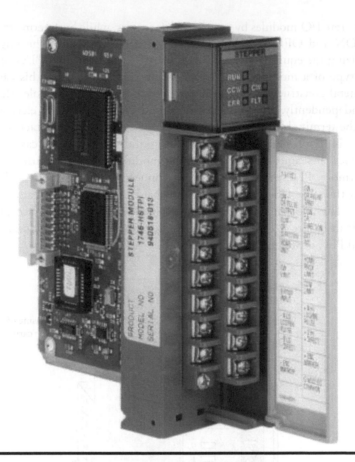

Figure 6.9   An electronic card (PLC module) with digital outputs for stepper motor driving (Allen Bradley).

6.4 Digital Input Modules

Digital inputs modules are designed to receive discrete signals from the input devices and convert them appropriately for further processing, while electrically isolating them before being transferred to the memory of the PLC. These signals are electrical voltages that can have different values, and can be either continuous (DC) or alternating (AC). For this reason, the input module performs a conversion of the voltage level and a modulation of the input signals, so that it becomes compatible with the operating voltage (5 V DC) of the microprocessor and the other electronic components. Subsequently, in order to isolate the high power levels that characterize industrial input devices, from the low power (logic) signals at the microprocessor, the modulated signals are directed to the electrical isolation unit. Subsequently, each shaped and electrically isolated signal is multiplexed with other similar signals produced by the same input module in order to be transferred serially to the microprocessor. For the case of small PLCs, when we have a parallel transfer of the inputs' status, no multiplexing unit is utilized. All the processing steps for the input signals are presented in Figure 6.10. Each digital input module also features a simple visual indication of the corresponding circuit status for each of the inputs (ON or OFF), so that the user is aware of the operational mode of the input devices.

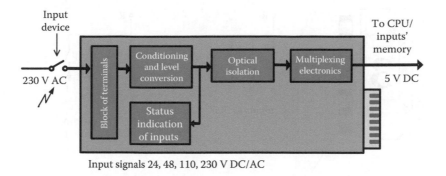

Input signals 24, 48, 110, 230 V DC/AC

Figure 6.10   Block diagram of operations inside a digital input module.

As was mentioned previously, each input signal connected to a digital input module is a discrete signal of two ON and OFF states. More specifically, the ON state corresponds to the nominal input voltage and the OFF state to zero. Because the input voltages may be of different range or type, the digital input modules are offered by the manufacturers in various AC and DC nominal voltages that are compatible with the corresponding standard operating voltages of the input devices. These are the 24, 48, 110, and 230 V DC or AC, the most popular being 24 V DC, since it is selected by the majority of engineers when designing new automation systems, mainly because of the protection from electrical shock that it offers. Each digital input module is characterized by the nominal operating voltage and the maximum number of input signals that it can sample. The usual standard number of inputs per module is 4, 6, 8, 10, 16, and 32.

The term "input devices" characterizes the devices that are able to carry commands or information from the industrial automation system or the operator of the control system, which in this case is the PLC. As an example, the pressing of a button from the operator of a machine is "command" to the control system (PLC), while the activation of a limit switch from a moving part of a machine informs the PLC that the component has received the desired end position. Typical input devices are:

- Limit switches
- Selector switches
- Buttons of all kinds
- Rotary switches
- Switching relay contacts
- Photoelectric switches
- Proximity switches
- Keyboard contacts
- Any kind of converters
- Arithmetic switches

Input devices have been described extensively in Chapter 2, where the major types of sensors, actuators, and power switching devices were analyzed. However, in this chapter, we will only mention "numerical switches", which are able to insert numbers in the PLC in a simplified approach without using a computer or other digital programming device. The numeric switches carry a numbered thumbwheel switch, through which the numerical data to be transferred to the PLC by using a limited number of digital inputs can be selected. A typical kind of this numerical switch

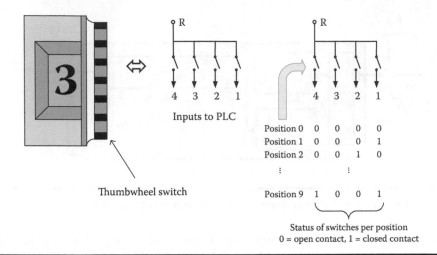

Figure 6.11   Thumbwheel switch for arithmetic data insertion in a PLC via digital inputs.

is presented in Figure 6.11, which contains four switching contacts that can be opened or closed depending on the position of the switch, i.e., according to the number selected. For example, in position 2, contact 2 is closed, while the rest of the others are open. The numerical switch in Figure 6.11 is able to provide ten different choices (numbers 0–9) by utilizing only four digital inputs. The combination of two such numerical switches allows for 100 options (numbers 0–99) by using only eight digital inputs.

For performing the functionalities indicated in Figure 6.10 inside a digital input module, corresponding electronic circuits that are required in their simplified form are presented in Figure 6.12a. If the input signal is DC voltage, only the insertion of a resistor circuit for converting the voltage level is required. If the input signal is AC voltage, then the input circuit contains an additional AC to DC converter or bridge rectifier. The smoothing capacitor C stabilizes the output voltage of the rectifier, while the resistors reduce the voltage to the desired low level, typically 5 V. The voltage of the Zener diode determines the minimum voltage value that could be detected. This low-level DC voltage is applied to the light emitting diode (LED) that is structurally integrated with a phototransistor. The last element is an electronic circuit powered by an independent internal power source (5 V DC) of the module. When light is emitted from the LED and falls on the phototransistor, the latter is conductive and allows the current path to the conductor to indicate the state of the logic signal. In conclusion, the existence of a voltage signal in the circuit entrance—e.g., the ON state of the input device—is transferred as DC voltage of a low or logic level to the microprocessor. Simultaneously with the creation of the logic signal, the indicative LED for the input device status is also supplied with power.

With the electrical isolation through the optical isolator, a complete decoupling is achieved between the electronics section for the input signal reading and the corresponding microprocessor. In this way, it is ensured that a possible short circuit, incorrect connection, transient spikes of the grid voltage, or even an additional noise that it is not removed from the modulation unit, will not damage the microprocessor. Less reliable methods use a transformer or a reed switch as an isolator. In this case, a transformer presents thermal losses that create thermal temperature problems, while

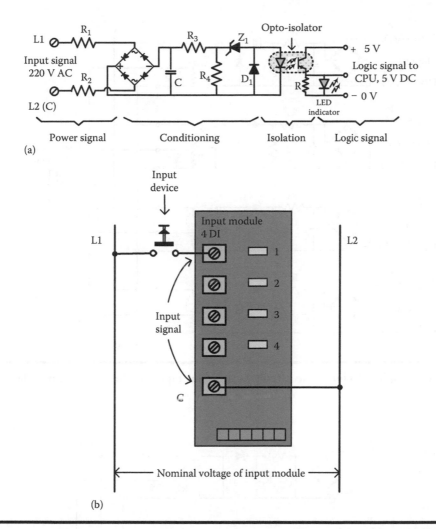

Figure 6.12 Simplified electronic circuit for an AC voltage signal of a digital input module (a) and an external electric connection of an input device (b).

reed switches have a limited mechanical life. Some PLCs in the market are available with input modules that do not have optical or other type of isolation and should be avoided. As mentioned above, in a digital input module, the number of inputs varies depending on the type of unit. When the number of inputs in a module is relatively large, then providing a proper power supply should be considered. The inputs of a module can either have a separate power supply, as presented in Figure 6.13, or require a common one as shown in Figure 6.14. The separate power supply of the inputs has the advantage that we can connect to the same module various input devices supplied by different power sources. When the inputs have a common negative or low voltage node, as presented in Figure 6.14, then all the input devices should be powered from the same power source, e.g., from the same phase of the three-phase supply system. For this reason, in many PLCs, the inputs of a module are grouped into groups of 2, 4, or 8 inputs with a separate power supply. In Figure 6.15, two grouping cases with a DC power supply are presented.

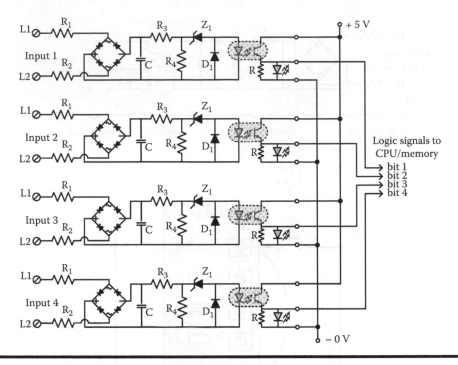

Figure 6.13 **Simplified electronic circuit of a digital input module with four isolated-supply inputs.**

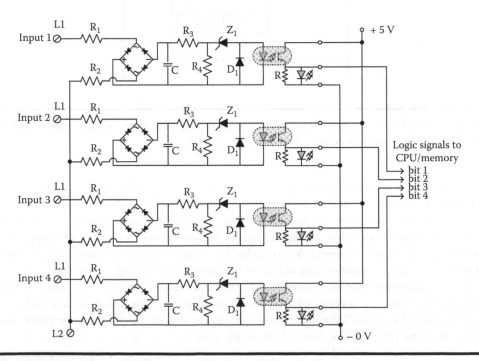

Figure 6.14 **Simplified electronic circuit of a digital input module with four non-isolated-supply inputs.**

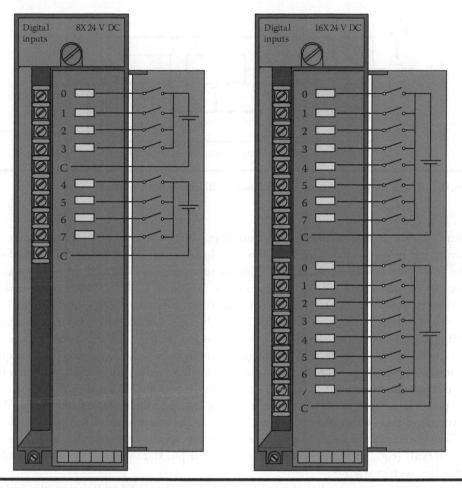

Figure 6.15 **Modules of digital inputs with DC power supply in groups of four and eight inputs. Each group contains non-isolated inputs.**

6.5 Digital Output Modules

In the introductory description of the PLC's operation, it has been indicated that the results of the logical processing of an automation program are stored in the output image table and are transferred subsequently, through the digital output module, to the output devices. The results of the program execution are logic signals that should be altered into power signals inside the digital output module, capable of energizing the corresponding output devices. In general, it can be stated that the digital output module processes the low voltage signals coming from the microprocessor in a completely opposite way than the digital input module. In Figure 6.16, all the operations that are carried inside the digital output module are indicated. The majority of these electrical circuits that perform these functionalities are similar to the ones in the digital input modules and are not going to be described again; however, the major differences will be highlighted.

When the logic signals from the output image table are transferred serially to the output module, then in the demultiplexing unit these signals are decoded and transmitted to the respective individual outputs of the output module. Since the logical signals are sent

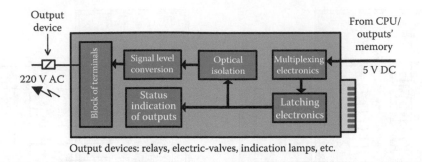

Figure 6.16 Block diagram of operations inside a digital output module.

periodically by the microprocessor to the output module (scanning process), the latter should have some kind of memory to retain or store the signals until the next transmission of signals from the microprocessor. The maintenance of these signals is performed by an integrated circuit that it is called "Latch" or "Hold" (e.g., the CD4508BE CMOS 4-Bit Latch Logic IC by Texas Instruments). As presented in Figure 6.16, after the demultiplexing of the logical signals it follows the latch unit that is not included in the digital input module. This is normal, since a potential change of the input signals' state and therefore a change in the input devices will be detected in the next scan cycle. If the change occurs in the dead time between two sequential scans, then the operation of the microprocessor is not affected, since it regulates when to take account of the change of status for the input devices. In contrast, in the digital output module, it is not possible to change the state of the output signals, and therefore the output devices, during the dead time between two sequential scans. The microprocessor's operating logic for the PLCs is based precisely on the principle that it changes the state of the output devices at regular intervals (e.g., per scan cycle) and that each output state will remain unchanged in the dead time between two scans.

The remaining units of the digital output module (e.g., the electrical isolation, the optical display, and conversion ones) are similar to the corresponding ones of a digital input module in terms of design and function. However, in the realization of the units, the manufacturers pay special attention, since the digital output modules control devices with relatively larger power levels and current. Thus, the digital output modules comprise switch power elements at current levels typically up to 1 A or 2 A, the function of which (ON-OFF) causes electrical and magnetic noise that is sufficient to destroy the other sensitive electronic components. In Figure 6.17a, the electronic circuitry of a digital output that implements the above basic functions is shown in simplified form. In the conversion circuit of the output signals, weak signals are coming from the processor (after the electrical isolation) and are converted into power signals with current and voltage levels capable of driving output devices that demand power consumption. Therefore, every digital output module requires an external power source, which supplies the output devices through the switching components of the converter circuit. As switching power elements, someone can find a power transistor, a triac, a printed circuit board relay, a reed switch, etc., depending on the type of digital output module and the quality of manufacturing. When the triac, presented in Figure 6.17a, changes its switching state, it is possible to create a large inductive voltage sufficient to cause damage to the electronic circuit. This risk is addressed by the utilization of the $R_S C_S$ surge suppressor. The varistor (Var) element also limits the transient phenomena and the noise. In series with the switch element, a fuse is introduced for protection from the overload of the output circuit, and

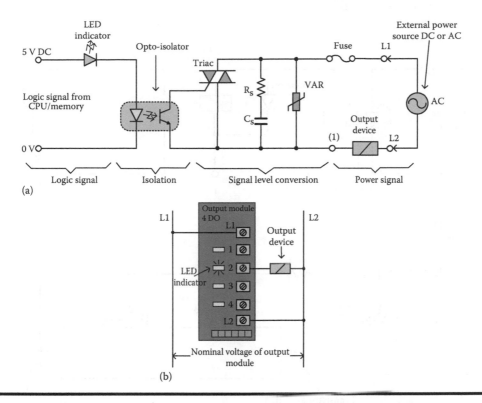

Figure 6.17 **Simplified electronic circuit of a digital output module with a triac as a switching element (a) and an external electric connection of an output device requiring AC voltage supply (b).**

therefore the output device. It is possible to have one fuse per individual digital output or a fuse every two or three or more outputs. Typical output devices are:

- Power or auxiliary relays
- Solenoids of electrovalves
- Indicating lights
- Devices for alarm indication
- Coils of electro-pneumatic valves
- Electronic devices that require an activation input or a change in their ON-OFF state

A typical external connection diagram of an output device to a digital output module is shown in Figure 6.17b. In Figure 6.18a and 6.18b, the alternative digital output circuits using relay and power transistors as switching elements are shown. The digital outputs with relay are generally suitable for AC and DC operating voltages of the output devices, which are able to provide additional electrical isolation and are resistant to impulse currents and overvoltages. The disadvantages are typical for all relays, e.g., the possibility of mechanical damage, the wear of the relay's contacts, and their limited mechanical life.

Digital outputs with a power transistor as switching element are suitable only for DC operating voltages of the output devices, have no mechanical parts, and therefore do not suffer from wear or noise, while they are characterized by a relatively high switching frequency. However, these are sensitive to impulse currents and overvoltage. Finally, the digital outputs with triac are suitable

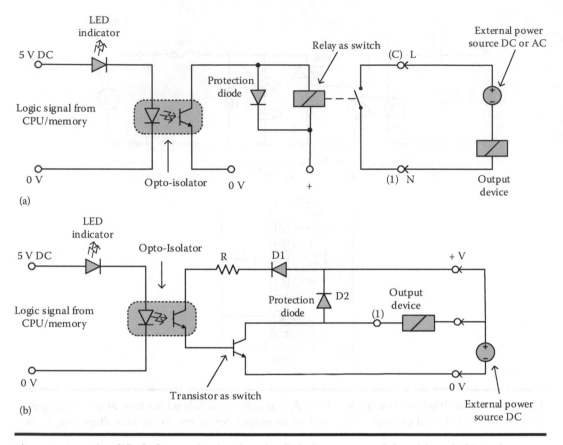

Figure 6.18 Simplified electronic circuits of a digital output module with switching elements: a printed circuit board relay (a) and a NPN power transistor (b).

for AC supply voltages, do not suffer from wear, but are more sensitive to impulse currents and overvoltage, and thus are always equipped with a special fuse.

As in the case of digital input modules, digital output modules may also have a separate or a common power supply. In Figure 6.19, the two power supply cases are presented, with the same characteristics and remarks as for the digital input modules, while Figure 6.20 presents two typical groupings of digital outputs with a DC power supply.

6.5.1 Technical Specifications for Digital Input/Output Modules

From the previous description of the digital input/output modules, it is obvious that there are some specific technical features that characterize these units, while the selection of the necessary I/O modules is made for each individual application. These technical characteristics can be summarized as follows:

1. The nominal operational voltage of the modules (value and kind)
2. The number of inputs and/or outputs per module
3. The kind of switching components that each digital output module contains (relay or electronic switching components)

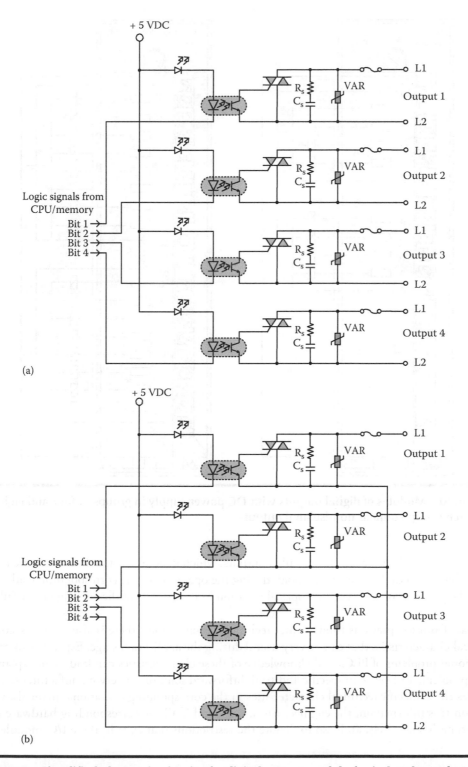

Figure 6.19 **Simplified electronic circuit of a digital output module for isolated-supply (a) and non-isolated-supply (b) outputs.**

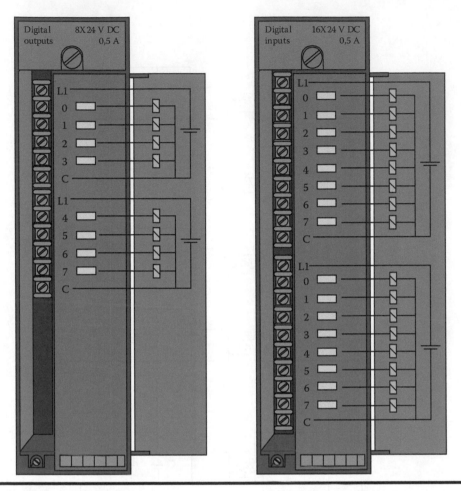

Figure 6.20 Modules of digital outputs with DC power supply in groups of four and eight outputs. Each group contains non-isolated outputs.

4. The nominal current of the module's outputs, which denotes the maximum electric current value that a digital output can hold, during the operation of the corresponding load
5. The presence or absence of an optical isolation in the circuits of the I/O digital modules

Apart from the previous fundamental technical specifications, there is also a set of secondary technical characteristics that can be very useful during the application stage. Especially in cases of operational problems of PLCs, only knowledge of these characteristics can lead to an explanation of the problem, and hence to specific and straightforward solutions. Every manufacturer of a PLC provides these specific technical characteristics in the corresponding operational manuals, where, based on this information, the engineer can understand if all the corresponding hardware is utilized properly and safely, and also to denote the restrictions that exist in these I/O modules that limit specific types of electrical connections. Some typical examples of these secondary technical characteristics are the following:

- *Range of voltage for the digital input, where the input signal is recognized as a logical "0".* As an example, in a digital input module, with a nominal operation voltage of 24 V DC, every input voltage from −30 V until +5 V is recognized as a logical "0".
- *Range of voltage for the digital input, where the input signal is recognized as a logical "1".* For example, in a digital input module, with a nominal operation voltage of 24 V DC, every input voltage from +15 V until +30 V is recognized as a logical "1".
- *Input current for a logical "1".* This indicates the minimum consumed current from a digital input that the input device should supply to the input circuit of the unit so that the latter recognizes the input signal as a logical "1". This current is usually at a level of about 10 mA.
- *Current consumption of a digital input or output.* This indicates the electrical current that is being consumed by the input or output circuit from the power supply of the PLC. In this case, there should be no confusion of this current with the current that is being supplied to the PLC from external sources of power. The values of the consumption current vary depending on the module type. As an example, a unit of 32 digital inputs can consume an electrical current of 30 mA, a unit of 16 digital outputs an electrical current of 160 mA, and a unit of 32 digital output an electrical current of 200 mA.
- *Leakage current of a digital output.* This indicates the maximum value of the electrical current that flows in a digital output circuit even when the circuit is in OFF state, which means a logical "0". This value has significant importance when the output device contains solid state components. In this case, it is possible for the flow current to be able to falsely energize the device, even in the case that the output is in the OFF state. A solution to this problem will be discussed in Section 6.10.6.
- *Switching frequency of an output.* This indicates the maximum frequency that the switching component on the digital output circuit can change state, or equivalently it expresses the frequency that the output can open and close. The switching frequency of an output can be different for the case of a resistive or conductive load.
- *Region of permitted rippling of the output voltage.* This indicates the minimum and maximum value of the operational voltage of a digital output module, where between these values the output circuit could operate without a fault. At this point it should also be noted that the operational voltage and the corresponding rippling refer to the external power source that is going to be connected to the output module. For example, in a digital output module, with a nominal operational voltage of 24 V DC, the allowed variation can be 20–30 V DC, while for 120 V AC, this could be 92–138 V AC.

Apart from the previous technical specifications, PLC manufacturers may indicate additional ones that concern very specific cases, which will not be further analyzed and presented at this point. In general, modules of digital inputs or digital outputs, with different technical characteristics, can be at the same PLC that comes in a modular form. This means that a PLC of this type is able to control an industrial system with different types of I/O modules, depending on the allowable cost and the requirements of the specific industrial application.

6.6 Analog Input/Output Modules

Since the beginning of their history, PLCs contained only units of discrete inputs/outputs and thus only devices of an ON-OFF type, which were able to be connected to PLCs and, correspondingly, to be controlled. Based on this restriction, at that time, PLCs were able to control only specific parts of the industrial process (the digital ones). Every analog control scheme requires the utilization of analog inputs and outputs, and thus with the corresponding huge improvements in the area of electronics, specific analog modules equipped the PLCs to provide the ability to interact with analog signals, a step that created a significant impact in the area of industrial control.

6.6.1 Analog Input Modules

An analog input module contains all the necessary circuits for the connection of analog signals and their corresponding alteration into digital ones. The basic component of an analog input module is the analog to digital (A/D) converter that converts the analog input signals into a digital value, indicated in the simplified diagram of Figure 6.21a. Subsequently, this digital value is transferred to the analog input's storage memory where the central processing unit can access it, based on the requirements stated in the execution program in the PLC. When the analog input module contains a rather large number of inputs and thus has a corresponding high cost, it is not financially profitable for the manufacturers to offer separate modules for every kind of analog signal that the industrial application might include. For this reason, the analog input module is equipped with a signal adapting block that is a simple electrical circuit, able to transform the input analog signal to a necessary signal compatible with the utilized A/D converter. Thus, if in such a module, the type of the analog input is altered (e.g., a change of the input sensor), the only thing required is

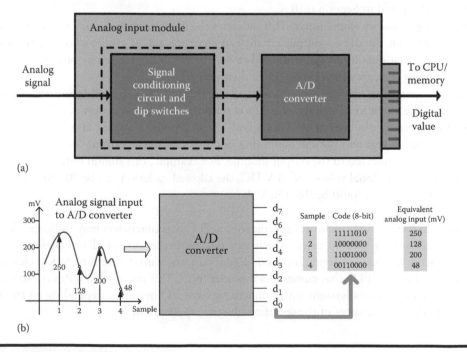

Figure 6.21 Simplified block diagram of an analog input module (a) and the basic principle of A/D conversion (b).

the alteration of the signal adapting block. Specific analog input modules (instead of the alteration block) carry micro-switches (DIP switches) for selecting the type of analog input signals. In modules with a small number of inputs, there is no signal adapting block or DIP switches, and thus each module is manufactured for a specific type of analog signal. The analog signals supplied in an analog input module are standard and compatible with the corresponding output signals of the analog input devices (sensors or transducers). In Table 6.1, the most common analog input devices and the standard analog signals that we find in the analog input modules are shown.

Another significant operational difference between the analog input modules (except for the A/D converter) and the digital input modules is that the former are equipped with a local memory for storing the digital values of the analog inputs. In Figure 6.22, a more detailed block diagram of an analog input module is depicted. In this case, a thermocouple (e.g., a couple of platinum-platinum of R type) detects the temperature of an object and, subsequently, applies

Table 6.1   Usual Analog Devices and Standard Ranges of Analog Signals

Analog Input Devices	*Analog Signals*
Temperature sensors	4–20 mA
Pressure sensors	0–5 V DC
Flow meters	0–10 V DC
Transducers of any kind	±10 V DC
Electronic devices with analog output	±50 mV DC
	±500 mV DC

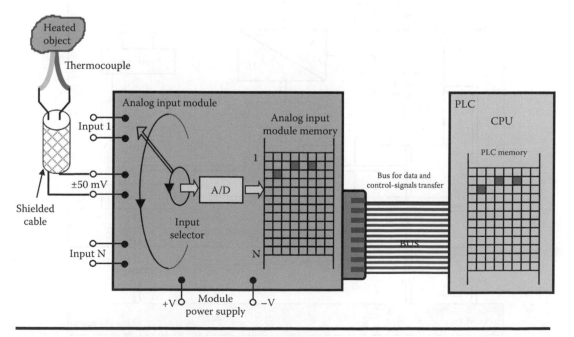

Figure 6.22   Analog input module which contains an A/D converter common for all input channels and memory for temporary storage of digital values.

this analog voltage through a shielded cable in an analog input of the module. The utilization of shielded cable is suggested in cases of transmitting weak electrical signals in order to block the corrupting electromagnetic noise generated from nearby conductors, and has the potential to cause unwanted and important operational errors (wrong corrupted measurements). Every analog input module contains only one A/D converter that is common for all the analog inputs. This approach is followed by the majority of manufacturers, mainly for purposes of reducing the cost of the hardware that is significantly increased if each of the analog inputs has its own A/D converter. Since all the input signals are applied at the same time in the input module, the input selector is responsible for routing a separate analog signal to the A/D unit in every clock cycle. The digital values of the analog signals provided from the A/D converter are stored in the memory of the input module, which is of a restricted size and, on many occasions, is referred to as a data register. With timing and control signals transmitted from the central processing unit, the content of the module's memory is transferred through the data transfer bus to the memory of the PLC where the values are processed according to the executed specific program instructions. Some manufacturers use the term "input channel" rather than the analog input one. In Figure 6.23, the

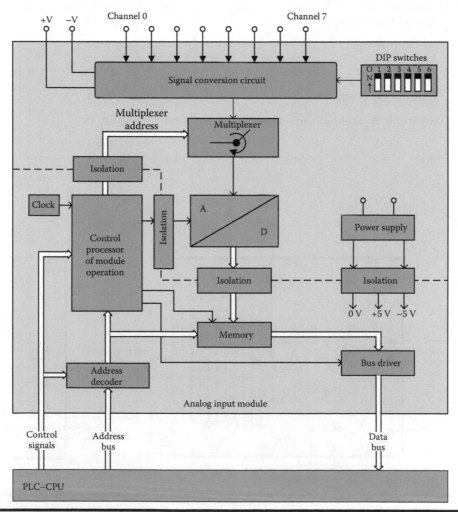

Figure 6.23 Applied block diagram of an analog input module operation (Siemens).

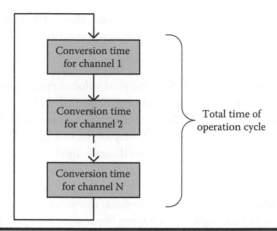

Figure 6.24   Determination of the cyclic operation time of an analog input module.

functional block diagram of a real analog input module with eight input channels is presented, while the complexity of the design with respect to the digital input module is straightforward. The overall functionality of the module is orchestrated by its microprocessor, which checks the operation of all the related components, like the multiplexer, the A/D converter, and the memory. The multiplexer and the A/D converter are electrically isolated from the rest of the electronic units. As depicted in Figure 6.23, the analog input module can be considered to be separated into two parts, totally electrically isolated, with analog signals on one side and digital ones on the other, so that the microprocessor of the module, as well as the central processing unit of the PLC, can be protected. The communication buses among the module and the PLC are utilized for the transfer of data, addresses, and control signals. The time duration of one operational cycle for the analog input module is independent of the scanning time of the PLC. The time needed for the conversion of an analog input, from the A/D converter, is dependent directly on the conversion method (e.g., a successive-approximation or integrating method). Since the conversion of the analog signals takes place in a serial manner in a one-by-one channel, the overall operational cycle time is the summation of the corresponding conversion times of all the active input channels, as shown in Figure 6.24. Due to the fact that every A/D conversion lasts around 10–50 μs, the operational cycle of a module with eight analog input channels will last less than 1 μs, a time that is much less when compared with the scanning time for most of the commonly found real-life applications and the usual scanning times of PLCs.

6.6.2 Accuracy in the Conversion of Analog Signals

In the modules of analog inputs, depending on the manufacturer and the model, various levels of resolution can be found that are a direct measurement of the accuracy in the conversion of the analog signals. The resolution of an A/D converter expresses the smallest difference that this converter is able to detect, with respect to the varying input signal, and is related to the number of bits of the binary word that represents the digital value of the analog signal. Since the number of bits has a direct connection to the maximum number and the size of steps into the range of variation of the input signal, the determination of the desired resolution for an A/D converter is a very important factor for the selection of the corresponding hardware. An A/D converter that utilizes data words of 8 bits will have a resolution of 1 step in 256 steps of the full-scale variation of the input signal.

If it is desired to have a greater accuracy of a control action, then another A/D converter should be selected that will have 10 bits, and thus a corresponding resolution of 1 step in 1024 steps of the full scale. More specifically:

- 8-bit converter: Resolution = 256 different digital values between the minimum and maximum of the range, thus 0.004 (2^{-8}=0.004) step of the full scale
- 10-bit converter: Resolution = 1024 different digital values between the minimum and maximum of the range, thus 0.001 (2^{-10}=0.001) step of the full scale
- 12-bit converter: Resolution = 4096 different digital values
- 14-bit converter: Resolution = 16384 different digital values, and so on

Thus, the bigger the resolution of an A/D converter, the more accurate the digital representation of an analog signal is. For example, if an analog input signal is varying from 0 to +10 V, the step size of the accuracy, in the case of an 8 bit and 10 bit, the A/D converter is defined as:

- 8-bit A/D converter: 10/256 = 39.1 mV per step or 25.6 steps per volt of the voltage input signal
- 10-bit A/D converter: 10/1024 = 9.76 mV per step or 102.4 steps per volt of the voltage input signal

6.6.3 Analog Output Modules

The analog output modules receive numerical data in digital form from the central processing unit of the PLC, transforming them into an analog voltage or current, in order to control the operation of an analog device. The digital values of the analog signals are usually words of 16 bits that are created from the processing of the program instructions, and are temporarily stored in the input/output memory of the PLC, before being transmitted to the analog output modules for their conversion into analog signals usually of 4–20 mA current or –10 V DC to +10 V DC. This process is highlighted in Figure 6.25. Even if the digital word can have a length of 16 bits, from these bits, only 11 are utilized for the digital representation of the analog signal and consist of the useful converted data, while the rest are utilized for the transfer of other general information or timing signals. A similar strategy is also followed in the digital representation of the analog input signals. For example, one bit can express the situation of a digitized measured value over a range or the

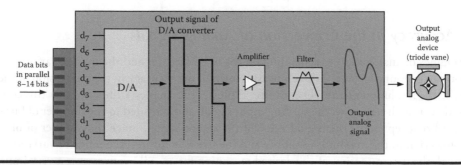

Figure 6.25 Block diagram of operations inside an analog output module for converting a digital value signal to analog.

existence of a fault (e.g., cut wire). The most significant bit of the word, expresses the sign (+, −) of the numerical value. A similar utilization of bits takes place in the case where the digital word has a shorter length than 16 bits, an operation that affects negatively the accuracy of the A/D converter.

When the digital word d_0–d_7 reaches the D/A converter, a corresponding constant electrical voltage is produced at the output of the converter. This means that the output of the converter continues to be discrete (1024 levels of voltage for a word of 10 bits) and when the digital input word is varying, the result is to have a corresponding piecewise constant voltage in the output of the converter which is also varied, as indicated in Figure 6.25. In order to have the varying output signal, but with the piecewise constant like a real analog signal, a specific circuit of an amplifier is driven and filtered, to smoothen the stepped voltage (or current). The accuracy of the analog output is dependent on the frequency update of the digital word input. This update is mainly dependent on the scanning cycle of the PLC, since the data produced from running the program related to the analog output module is transmitted only once in every scan cycle. In Figure 6.26, the functional block diagram of a real analog output module is depicted. As in the case of analog input modules, this part of the digital logical circuits is isolated from the part of the analog ones

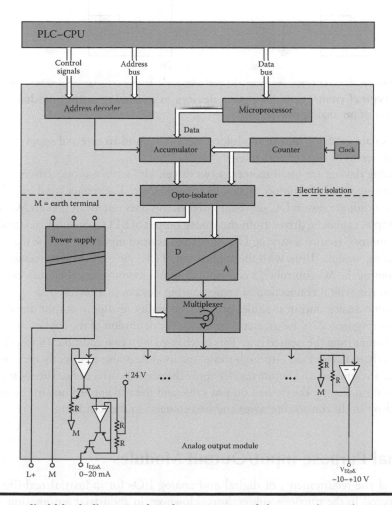

Figure 6.26   Applied block diagram of analog output module operations (Siemens).

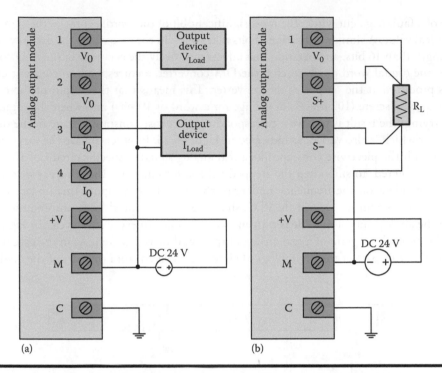

Figure 6.27 **Typical connections of analog devices in an analog output module (a), and four-wire connection of an analog device (b).**

via an optical isolator. Analog output modules generally require an external supply unit with some requirements in terms of current and voltage.

Typical analog devices are small motors, servo valves, DC servo motors, drivers for high power motors with varying speed of rotation, analog instruments, etc. The voltages and the currents for an analog output module are always DC signals. However, this does not mean that an AC controller of a motor, for example, cannot be driven from the analog output of a PLC. Such AC controllers (e.g., the soft starters of motors) require a varying DC signal as a control input that can be the analog output of a corresponding module. Thus, with the application of a varying DC output voltage (e.g., 0–10 V DC) in the input of the AC controller, a corresponding AC motor speed alteration can be achieved. In Figure 6.27a, the typical connection of analog output devices in a corresponding output module is presented. Some analog output modules provide the ability to supply output device with a four-wire connection (Figure 6.27b) for greater accuracy in the definition of the applied voltage that is, in many cases, different from the desired one. This is achieved with two additional wires (S+, S−) that are connected directly to the R_L load. Through those, the module is able to directly measure the voltage of the load (like a voltmeter) and automatically apply the needed corrections. The most common reason for the deviation, among the desired output value and the real applied one in the output device, is the voltage drop in the connecting wires and bad contacts in the terminals.

6.7 Special Purpose Input/Output Modules

The majority of the requirements of digital and analog I/Os for industrial real-life applications have been covered in the previous subsections. However, in industrial automation applications, there are times when specific signals are needed that a PLC is not able to process by utilizing the previous modules. Some classical examples from this category are the following:

■ Applications that require a fast response of the PLC input: In these cases, there are devices in industrial automation that are producing signals that are varying faster than the scanning capabilities of the PLC. This means that the PLC, based on the conventional I/O modules, is no longer able to detect rapid changes of the input signal and act on them.

■ Applications related to the control of rotation of special motors, e.g., the case of stepper motors: These motors, and especially their driving circuits, require the supply of special pulse trains, where their frequency affects the rotation speed and number of pulses affecting their angle of rotation. In this case, conventional I/O modules are not capable of providing such signals.

■ Applications with complicated control problems, such as analog PID control with specific response requirements.

■ Applications where the transmission of data related to human machine interfaces (HMIs), such as data for monitors, printers, personal computers, etc., is needed: In addition to this, special equipment is needed for the communication of a PLC with another PLC, or for their connection to an industrial communication network.

In all these cases, and in other similar ones, special input/output modules are required for special manufacturing that is also provided by the majority of PLC manufacturers. A typical characteristic of these modules is the fact that they are utilizing their own microprocessor that operate independently in order to perform the required task, but are simultaneously connected to the main processing unit of the PLC to exchange data and control signals.

6.7.1 Fast Input Response Modules

Fast input response modules are utilized when the PLC needs to detect pulses of a very short duration. If the scanning time of a PLC is for example of 1 ms, then the frequency of state change for a discrete input cannot be more than 1 KHz, and in the case that the scanning time is 100 ms, which is another typical value, then the frequency of the input cannot be more than 10 Hz. Thus, the PLC is not able to detect pulses of a frequency e.g., 50 KHz or 500 KHz. The fast input response module contains a very fast microprocessor that it is able to count pulses of a very fast speed, independently of the scan cycle of the PLC. This measurement is stored in the module's memory, and is transmitted to the main processing unit during the exact moment of sampling of the rest of the inputs, which means once in every general scanning loop. These modules are very useful in applications that use fast counting, fast rotation of drums, position decoders, etc.

6.7.2 Stepper Motor Control Modules

Stepper motors are utilized in applications of micropositioning of objects, mainly due to their accuracy during their rotation that, in most cases, is altered to a corresponding linear movement. The stepper motors can rotate in very small steps (e.g., 1° or even smaller) and with high accuracy in both directions of rotation. The power supply and driving unit of these stepper motors requires the application of pulses as control signals. For each input pulse, the motor produces an equivalent amount of rotation, the so-called "step", as described in Section 2.1.1. The frequency

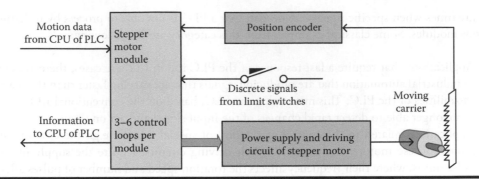

Figure 6.28   Closed-loop control of a stepper motor with a specific I/O module.

of this pulse train defines the speed of motor rotation. A second pulse input is utilized to select the direction of rotation, while a third one defines the full or half step rotation. Such control signals are not usually produced by most PLCs, and thus it is necessary to utilize specific stepper motor control modules. Based on the program's logic at the PLC, when the stepper motor should be moved, a small block of data related to the magnitude, the speed, and the direction of the rotation is transmitted to the control module of the stepper motor that handles the generation of proper signals for driving the motor. The overall control system is of an open-loop architecture, which means that the PLC has no information (feedback) of the real movement that the stepper motor has executed.

The complex structures of micropositioning with 3–6° axis of movement (one stepper motor for each motion axis) are usually equipped with sensors of position measurements (e.g., encoders) that transmit information concerning the real position of the motor, which allows for the further application of control signals for tuning the performed movement to be equal to the desired one (closed-loop control architecture). In Figure 6.28, a closed-loop control system, where the stepper motor control module is connected to the main processing unit of the PLC and to the controlled process, is indicated. The stepper motor control module receives commands from the main processing unit of the PLC that are related to the rotation of the stepper motor, and transmits positioning information back to the PLC when it is asked. The stepper control module receives the real position of the controlled moving object from a position encoder, compares it to the desired position (in the form of a pulse train) that has been transmitted to the motor controller, and creates a corresponding positioning error. Based on this error and the applied controller, a corresponding corrective action is produced to control the movement. Depending on the number of motors that one module can control, different multiple combinations of movements can be generated that can be stored in the stepper's control module, either in the form of a movement's library or as alternative motion programs. The central processing unit of the PLC can activate a program of movements by transmitting a specific control signal to the stepper motor control module.

6.7.3 Three Terms (PID) Control Modules

In every industrial production process, there is always a need for continuous control of a physical variable such as hydraulic pressure, temperature, the rotation speed of a motor, etc. Therefore, the PLC units should contain specific hardware to enable this. Most of the currently available PLCs in the market contain an embedded software program for applying analog control of three terms (PID control), which is stored in the memory of the PLC and is called from the user's program as a subroutine. The user defines the desired parameters and the analog inputs/outputs for which this routine will be executed, and the latest is inserted in the main PLC program. Since the control algorithm includes mathematical operations, the scanning time of the PLC can be significantly increased, and thus it might cause improper control of the overall process. For these cases, PLC manufacturers provide independent special PID control modules, which contain their dedicated microprocessor and thus do not affect the overall scanning time of the PLC. These modules contain their own analog inputs and outputs that are directly connected to the controlled process. From the central processing unit of the PLC, only the initial values, the desired set points, etc., are transmitted to the PID control module. Subsequently, the microprocessor of the control module operates in parallel to the PLC to execute the desired control task. When the PLC needs the variables of some values which exist in the memory of the control module in order to perform the proper control of the overall process, these are transmitted from the I/O memory of the control module once during the PLC scan cycle. More information on the continuous analog control of three terms (PID) is provided in Chapter 9.

6.7.4 Communication Modules

When there is a need for a PLC to communicate with another PLC, or with other devices such as monitors, printers, or remote I/O modules, or in general with another network of PLCs for the exchange of specific information, a communication module is needed for completing this task. The simple, compact-type PLCs are usually equipped with a standardized communication port for connecting to a PC, and some of them have an additional communication port for connecting to a specific Industrial network. The communication modules in general are equipped with their dedicated microprocessor, which is called a communication processor (CP). This communication processor is responsible for implementing the communication protocol for communicating with the external devices, either independent or networked and with the central processing unit of the PLC. The last, according to the executed program, sends/receives alphanumeric data to/from external, independent devices through the communication processor.

6.8 Central Processing Unit

The overall operation of a PLC is realized and orchestrated from its central processing unit (CPU). The heart of the CPU is the microprocessor, while in many cases the terms "central processing unit" and "microprocessor" are used interchangeably. However, the correct terminology indicates that the microprocessor is the specific integrated circuit that the PLC is utilizing, while the CPU, except for the microprocessor, contains additional necessary components and electronic circuits for the full operation of the PLC. Figure 6.29 indicates the overall architecture of the main units of the PLC and their interconnections within the CPU in a rather simplified way. This figure does not contain the power supply that simply supplies all the necessary voltages for the proper operation of all the units and components. The CPU also contains the digital circuits that store or recall data from memory, as well as the necessary circuits for the communication of the microprocessor with the PLC's programming unit. The operation of the microprocessor is controlled from a program called "operating system" or "executable program". The executable program is stored permanently in a memory unit that it is always of a ROM type (read only memory) since, from the moment that it is developed by the manufacturer and stored in the memory, modifications are unnecessary and not allowed. As a definition of the executable program, is a special program,

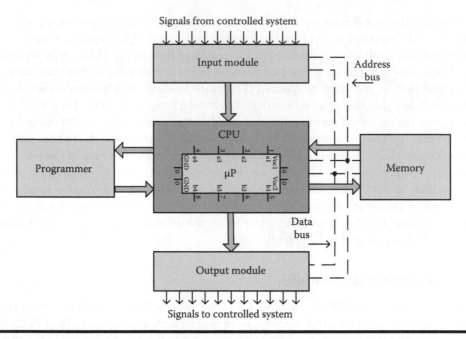

Figure 6.29 Operational interconnection between a CPU and the other units of a PLC.

written in assembly language, which is able to drive the microprocessor to perform the internal functionalities such as processing, control of other internal units, and communication.

More specifically, the CPU of a typical PLC is able to perform the following basic operations:

1. *I/O operations.* These operations allow the PLC to communicate with the external world and include mainly the scanning of the inputs and the updating of the outputs.
2. *Logical and arithmetic operations.* These operations contain all the logical functions of Boolean algebra (AND, OR, NAND, etc.) and basic arithmetic operations (such as addition, subtraction, multiplication, etc.). In the same category, all the specific operations such as timing, counting, and comparison can also be included.
3. *Reading, writing, and special handling operations.* These operations contain actions that are applied on the content of the memory locations and may concern data or instructions.
4. *Communication operations.* These operations contain specific functions that the PLC should execute when it is communicating with external peripheral I/O modules, specific modules, another PLC, etc.

In order for the CPU to execute the previous operations, it continuously communicates with the memory and the other components through the data bus and the address bus (Figure 6.29). The aim of the address bus is to activate, at the proper time instant, I/O points or memory locations that will subsequently utilized from the data bus for the transmission of data. The microprocessor selects an address which is decoded. In this way, the proper corresponding I/O point or memory location is selected through the address bus. The selected I/O point or memory location will then receive or transmit data. The data bus, which consists of bidirectional channels, is utilized for the transmission of data from the memory to the I/Os and conversely, from the CPU to the memory of registers, timers, counters, etc. In Figure 6.30, the basic components of a microprocessor are indicated, as well as their interconnections with external components. The memory with the operating system contains,

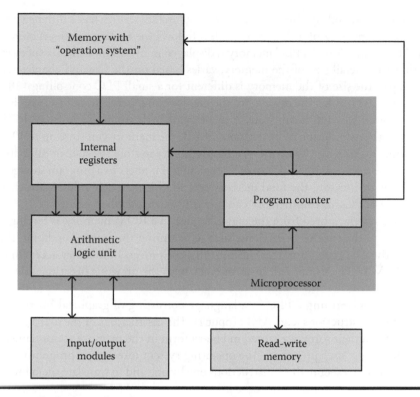

Figure 6.30   Typical parts of a microprocessor used in a PLC.

in the form of a program, the executable instructions. Based on these instructions, the operation of the microprocessor is performed. The arithmetic logical unit (ALU) executes all the arithmetic and logical operations, while the program counter supervises and controls the program instructions step by step. When the user's program contains jumping instructions (e.g., GOTO instructions), the program counter records the execution path of the program at every time instant. In the internal registers of the CPU, data are being stored based on instructions from the executable program, and are utilized from the ALU when required. In the writing and reading memory, the user's program is stored in addition to specific operational parameters. All the operational steps of the microprocessor are driven from a clock that guarantees the proper and timely execution of the instructions.

The operating system is also responsible for the execution of various diagnostic functions from the CPU. These diagnostics can be categorized into two forms: the ones that the system is executing itself in every initialization of the device, and those that are user-initiated by specific instructions. Typical diagnostic functions during the initial operation of PLCs are checking the proper operation of the random-access memory (RAM), the microprocessor, the battery, and the power supply. The user is able to initiate specific diagnostic tests that are mainly related to the memory or other units, e.g., the communication ports.

6.8.1 *Memory Organization*

The executable program, the user's program (the automation program), the states of the I/Os, and various data of non-permanent nature are stored in the memory of a PLC. The term "memory organization" is described as the fragmentation of the memory in various sectors where each one has a specific aim and corresponding size. Even if every PLC manufacturer has a different organization of the memory, the memory of a PLC more or less follows the same general structure. In Figure 6.31, a typical fragmentation of a PLC memory is displayed in five sectors. The size of every memory sector, and thus the overall size of the memory, varies based on the size and characteristics of the PLC. For example, the size of the memory is different for a small PLC controlling 128 I/Os from a PLC controlling 1024 I/Os. The capacity of the memory, independently of its type, is defined by the number of digital bits that it can store. A memory of 1 kB size is able to store 1024 B, which means that under the assumption of instructions of 1 B, the memory can store up to 1024 program instructions, while if this is an I/O memory, it has the ability to store the state of 8192 discrete I/Os. Therefore, the same memory size indicates different issues if it refers to a program storage or to other internal units. For this reason, the total memory of a PLC is not such a dominant characteristic, such as in the case of PCs, as this memory can be easily extended when needed.

The operating system, stored from the manufacturer in a ROM memory, is forcing the microprocessor to execute the various operations, such as scanning the inputs, updating the outputs, the execution of the user's program, etc. Subsequently, the process of memory access in the various sectors of the RAM will be examined, in accordance with the previous functionalities.

■ The user is programming a PLC in a language consisting of graphical instructions ⊣├ or alphanumeric instructions (e.g., AND Input 1). The instructions of the user's program, independently from their form, are stored in binary form in the memory sector named "automation program" or "user program". The operating system (executable program) is forcing the microprocessor to execute these instructions one by one and to translate them into equivalent

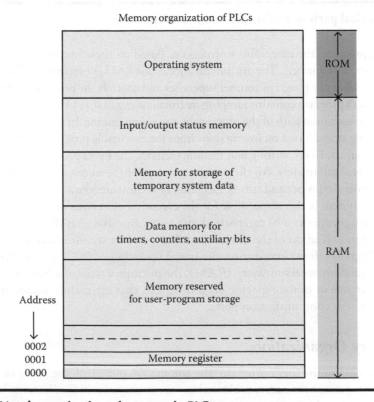

Figure 6.31 Usual organization of memory in PLCs.

instructions in assembly. The software often contains instructions that require data from other parts of the memory. In this case, the operating system is driving the microprocessor to collect these data for further processing. For example, if one instruction refers to a digital input state, the microprocessor is responsible for tuning the operation to get this value from the memory, specifically from the "input image table".

■ Every time that the operating system requests the microprocessor to scan the current states of the inputs, it stores this information in the input mapping memory, also called the input image table. Subsequently, as the microprocessor executes the automation program, various state updates for the output devices are produced. These states are stored from the microprocessor in the output mapping memory, also called the output image table. When the execution of the automation program's instructions is finished, the output modules are updated, which means that the stored states in the output mapping memory are transferred to the outputs.

■ As the operating system executes the various operations through the microprocessor, it is normally required to temporarily store some information or current results. For this purpose, a specific memory sector from the RAM is dedicated, where the user has no access.

■ When the automation program contains instructions that are related with counters, timers, auxiliary bits, and data functions, then another specific sector of the memory is needed for storing the corresponding parameters. For example, the CPU has to store the number and type of time units (e.g., ms, sec, min, etc.) in the case of timers, or the limits and the counting step for the case of counters. When an automation program's instruction is executed that concerns a counter or a timer, the operating system drives the microprocessor to seek the corresponding data in this memory sector.

At this point, the role of the auxiliary memory bits should also be explained. From the design of classical automation circuits, it has been indicated that the implementation of the logic according to which the controlled system operates, requires the utilization of auxiliary relays multiple times. The main role of these auxiliary relays is to represent an operational state of the automated system, and to provide necessary NC and NO contacts, which are inserted properly in the automation circuit. In programmable automation, the role of an auxiliary relay takes a simple memory location of a single bit. The activation and deactivation of an auxiliary relay is equivalent to the storage of a logical "1" or "0" respectively, in the memory sector dedicated for this purpose. The utilization of the relay's auxiliary contacts is equivalent to calling functions of an auxiliary memory bit through the automation program instructions. Since this call can take place as many times as necessary in an automation program, this is referred to in programmable automation as an "infinite number of contacts" situation. The memory locations with the auxiliary bits are sometimes mentioned also as "internal coils" or "logical coils" in correspondence to the coils of the auxiliary relays. Every auxiliary memory bit in the corresponding memory location has its own address, so that it can be defined uniquely through the related instructions.

After the description of the I/O modules, CPU structure, memory, and communication and data exchange buses, the overall functionality and construction of one PLC can be addressed. This component-based approach is indicated in the block diagram of Figure 6.32. In short, the overall operation of a PLC involves the following operations:

1. The CPU sequentially defines the addresses of the inputs through the address bus, transfers the states of the inputs through the data bus, and stores them in the input mapping memory.
2. The CPU sequentially executes the automation program instructions, stored in the corresponding memory.

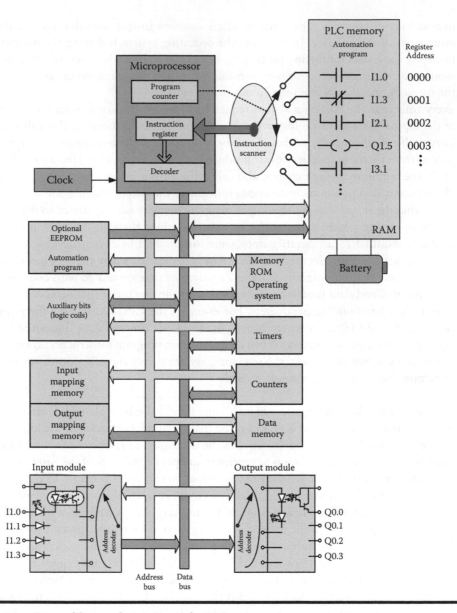

Figure 6.32 General internal structure of a PLC.

3. The results from the execution of the output activation or deactivation instructions are stored in the output mapping memory.
4. When the automation program instructions refer to auxiliary bits, timers, counters, and other internal units, then the microprocessor is referring to corresponding memory locations through the data and address buses.
5. The CPU, by sequentially defining the addresses of the outputs through the address bus, transfers the output states from the output mapping memory to the corresponding outputs through the data bus. This output update creates a corresponding change in the operational states of the output devices.
6. All the previous actions are driven from the operating system stored in the ROM.

The battery allows the retention of the automation program in the RAM memory, even when the power supply has been interrupted, and it is possible to equip the PLC with an extra memory of EEPROM type (see Section 6.8.2) for permanent storage of the automation program.

6.8.2 Memory Types

As has been shown up to now, there are various types of memories that a PLC utilizes (e.g., ROM, RAM, etc.) In this section, the most common memory types found in a PLC, will be described, including their typical characteristics and their specific utilization in the PLCs.

6.8.2.1 Read Only Memory (ROM)

The non-volatile memories (ROM) are used for the permanent storage of functional data and programs, so that only the action of reading of the memory contents is allowed. In general, the information stored in a ROM memory of a PLC is put there by the manufacturer, and is mainly related to the way that the PLC operates. Specifically, the operating system of the PLC is stored in the ROM memory and contains the set of instructions for operating the PLC. The ROM instructions cannot be erased, which means that they have the ability to retain their stored data, even in the case of a power supply cut, thus they do not require an emergency power supply.

6.8.2.2 Random Access Memory (RAM)

RAM memories are sometimes referred to as read or write memories, and are designed in a way that data and programs can be directly written or read from, without restrictions. These memories consist of the most flexible type of memory, since the user is able to access all of the memory locations and add new data directly by erasing the previously written information as many times as needed. In the RAM memory of a PLC, the data are stored electrically through the programming device and, in some cases, it is not needed to stop the operation of the PLC. Usually in the RAM memory, the automation program of the user is stored as well as any other data that are produced or acquired during the execution of this program.

In a potential power supply loss, even an instant one, the RAM memory loses all of its contents. This means that in a case of power failure, the PLC will lose the automation program and thus it is necessary to load the program again. To deal with this case, RAM memory is combined with a battery that can provide necessary power to the memory for retaining the current data for a time window of about one year.

6.8.2.3 Erasable Programmable Read Only Memory (EPROM)

Today, EPROM memory is not used in computers and PLCs and has been replaced by EEPROM chips. We refer to this kind of memory only because an engineer may encounter it in old PLCs still operating in some industries. The EPROM memory, in contrast to its name, is a ROM type memory that can be reprogrammed after a complete erase by the application of ultraviolet radiation. Unlike RAM, EPROM does not lose its stored information in a potential power loss, and it does not allow user access to change the information already stored there. The integrated circuit of the EPROM memory carries a small window over the memory position, and when ultraviolet radiation is transmitted there for a few minutes, the contents of the memory are erased. This process demands to remove the EPROM memory from the PLC and to transfer it to a specific erasing unit. This process is an overall disadvantage, since the operation of the PLC needs to be terminated. After erasing the EPROM, a full automation program can be stored to the memory through the utilization of the programming device.

6.8.2.4 Electrically Erasable Programmable Read Only Memory (EEPROM)

The EEPROM memory is a type of memory that can be erased by the application of an electrical voltage through proper electrical connections to the memory's IC through the programming device. After this erase, the EEPROM memory can be rewritten and it does not lose the stored information in a potential power loss. Although it shows the same flexibility as RAM memory, it is generally slower (slower data access to the stored information) than RAM memory.

When an automation program is loaded for the first time in a PLC, the testing phase of the overall control system follows. During this time, it is quite common to require changes and alterations in the automation program, changes that can be directly applied to RAM memory by the sequential download of the modified program. After the finalization of the automation program and its proper operation, this is transferred to the EEPROM (if such a memory is available in the PLC) which constitutes the final and definitive memory for the normal operation of the PLC.

Many larger PLCs have the capability to accept, except from basic RAM, external memories in specific slots that are available for this purpose. These memories are usually of EEPROM or FLASH type, can vary in capacity, and are utilized for program or data storing. A FLASH memory is similar to an EEPROM memory, with the main differences between them concerning the user, the access, and erase manner of the stored data. In the first one, blocks of thousands of bytes can be erased at a time, while in the second one, a byte at a time is possible to be accessed and erased. Finally, many PLCs can be connected to external storage devices, such as hard drives for the storing of big automation programs or sets of data.

6.8.3 Addressing I/Os and Other Internal Elements

Every digital input or output of a PLC is characterized by a unique name that will be utilized during the automation logic programming. This name is nothing else but the address of the digital I/O that expresses the corresponding position of the I/O between others. Thus, every digital I/O has address that defines position in the specific I/O module, as well as its position in the I/O mapping memory. The type of the addresses that are provided to the I/Os of a PLC varies with each manufacturer, since until now there has been no standardized way in addressing them. The same unstandardized addressing situation is found in the other internal components of a PLC, such as the auxiliary bits, timers, and counters.

Every PLC manufacturer uses its own addressing system, which may be based on the decimal or octal system, use numerical or alphanumeric data, etc., while subsequently, reference will be provided to the most common addressing methods. The decimal I/O addressing system uses numbers 1–8 for the first group, 9–16 for the second, 17–24 for the third, and so on. The octal system uses only eight digits from 0–7, that is, numbers 0–7 for the first group, 10–17 for the second, 20–27 for third, and so on. The distinction between inputs and outputs is done by adding a letter, such as X, I, or E for inputs, with I as the predominant one; and Y, O, or Q for outputs, with Q as the predominant one. Other addressing systems do not use letters but only five- or four-digit numbers. For example, each interconnection point with a code between 0000 and 0999 may be an output, while with a code between 1000 and 1999 may be an input.

The digital I/Os of a PLC, with a byte memory structure, are also grouped by octets. In this case, a digital I/O is uniquely characterized by the octal bit corresponding to it, as shown in Figure 6.33. For an octal system of addressing, the first digit represents the number of the byte and the second digit is the number of the bit into the byte. Thus, digital input I 6.3 corresponds to the bit 3 of the byte 6, with respect to the input mapping memory and, at the same time, corresponds to

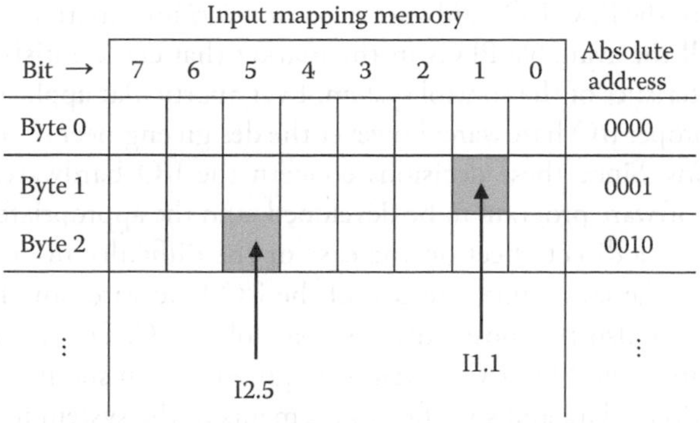

Input mapping memory

Bit →	7	6	5	4	3	2	1	0	Absolute address
Byte 0									0000
Byte 1							▲		0001
Byte 2			▲						0010
⋮			I2.5				I1.1		⋮

Figure 6.33   Grouping and addressing of digital inputs according to an octal system.

the fourth input of the sixth input group, with respect to the input module. In addressing systems with a five-digit number (e.g., 110 00), the first digit determines whether the connection point is an input 1 or an output 0, the second digit expresses the number of the base with various I/O modules, and the third digit is the number of the I/O module. Very small PLCs, with 10–20 I/Os, follow the decimal addressing system. In some large PLCs, the addressing system may be "flexible", i.e., the I/O addresses are set by the user via DIP switches or by programing the addressing data into an EEPROM memory. In the case of non-flexible addressing systems, each I/O module and every corresponding I/O has a fixed address. The procedure of determining the specific I/Os for the PLC, where the corresponding I/O devices will be connected, known as I/O address assignment, is an important task that may or may not simplify the automation programming task, the diagnosis of errors, and dealing with general logical programming problems.

The other internal elements of a PLC are addressed in an analogous or even simpler way. The auxiliary bits or logic coils and the corresponding memory locations are encoded with the letter M or F (from the English terms "memory bit" or "flag") followed by a two- or three-digit number with decimal or octal numbering. Also, with "Txxx" the memory locations for the timers are defined, with "Cxxx" the memory locations for the counters, and with "Sxxx" usually the general content memory locations, where "xxx" is an integer number, the maximum of which varies according to the size of the PLC.

6.9 PLC Expansion and I/O Configuration

Small PLCs with a compact structure have a predetermined set of inputs and outputs, which is not subject to any kind of change. The opposite is true in the case of large PLCs, whose CPU has the ability to control a few hundred (or thousands) of inputs or outputs. In the latter case, the I/O system (which is the PLC hardware including any kind of I/O modules) is not fully predetermined by the manufacturer; instead, the engineer can configure it according to the application's needs. The need for a large number of I/Os served from a CPU requires the extension of the PLC in additional racks of I/O modules. Multiple possibilities of expandability for PLCs of this category include the possibility of local or remote I/O modules, the variable number of I/O modules that can be supported in both cases, as well as the selection of I/O modules with a different number of I/O channels in each, which define the problem of proper design and configuration of the I/O system of a central PLC.

The overall design of the PLC I/O hardware needs *a priori* information on the relevant technical characteristics of all the available PLCs in the market that could satisfy the required control and functional characteristics of the control system. Every particular application dictates partially the way to design the proper I/O hardware; however the design engineer should make a significant number of the decisions. Since these decisions concern the I/O hardware, they also indirectly affect the automation software program to be developed, and the appropriate or inappropriate I/O system design, which has a direct effect on the ease or the difficulty in developing the required automation software. At the same time, the cost of the I/O hardware can often be more than the cost of the CPU, which is also the most expensive part of a PLC. For the above reasons, a careful and thorough design of the PLC I/O system is required, which should be based on a detailed recording of the operational data and specific requirements of the system to be automated.

6.9.1 Local and Peripheral I/O System

In applications where a large number of I/Os are controlled by a central PLC, it is expected that a part of the I/O devices will be at a long distance from the PLC. In this case, a cost-effective and technically convenient solution for connecting the devices is to remove the I/O modules from the PLC, place them close to the I/O devices, and to remotely connect them with the PLC through a "communicative type" link, as shown in Figure 6.34b. Otherwise, the remote I/O device should be connected directly to the respective I/O modules via multiple power lines, as indicated in Figure 6.34a, a solution that not only greatly increases the installation costs, but also makes the connection more susceptible to electromagnetic noise interference and functional failure. When the PLC has to be extended to more than one rack for I/O hardware support, it is possible to have a communicative type link between the CPU and the additional racks.

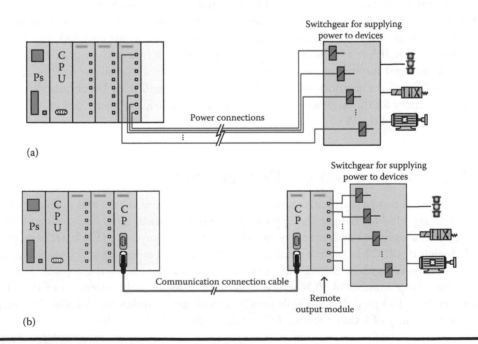

Figure 6.34 Two possible ways for connecting remote I/O devices in a PLC: through power lines (a) and through communication processors (b).

The I/O hardware of a PLC is designated as "local" when the corresponding I/O modules are located in the immediate neighborhood of the CPU. The distance that defines the designation "local I/O hardware" is determined by the distance of the interconnecting cable between the CPU and the I/O hardware, generally defined in the order of a few meters (20–50 m). When it is desirable to place the I/O modules at a relatively greater distance for the reasons explained above, then the I/O hardware is designated as "peripheral" and special communication equipment is required for the operation of the remote modules. The spacing in the placing of the peripheral I/O hardware is in the order of a few kilometers (1–10 Km). Figure 6.35 presents the characteristics and differences between the local and peripheral I/O hardware, while it is obvious that the local I/O modules are no different

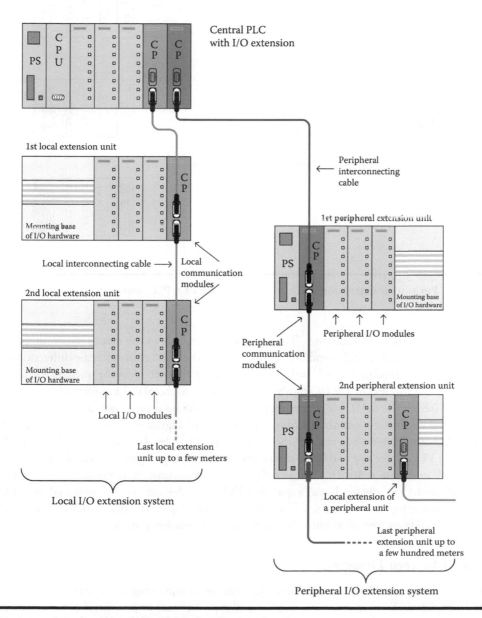

Figure 6.35   Local and peripheral I/O hardware of a PLC.

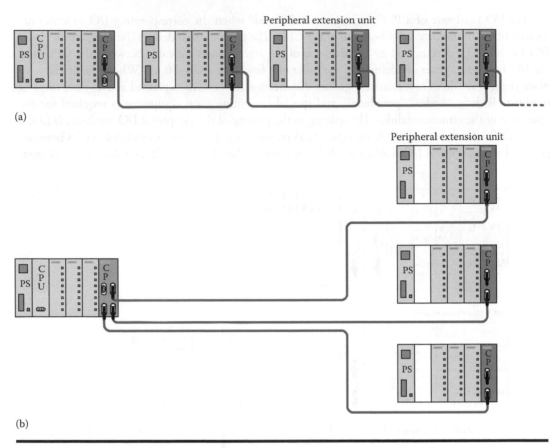

Figure 6.36 Peripheral expansion of the I/O hardware of a PLC, with series topology (a) and star topology (b).

than the corresponding peripheral ones. The difference exists only in the means of communication, in the communication units (CPs) and connecting cables. The main reason for the differentiation of the hardware at a local and peripheral level is the voltage drop that occurs in the interconnecting cable when it is long. For this reason, the I/O hardware expansion peripheral units require additional power supplies (PSs) (see Figure 6.35), which are not necessary in the local expansion units.

PLC manufacturers do not always follow the layout of Figure 6.35, while, on the contrary, there are differences that do not undermine the general principles. In addition to the simple local extension of a PLC, there are two basic peripheral extension topologies, the serial and star topology presented in Figure 6.36. In each peripheral expansion unit, it is possible to have a local extension of the I/O hardware. Local expansion I/Os modules are usually mounted in the same electrical enclosure together with the CPU and the power supply. The peripheral I/O hardware is installed in separate enclosures at corresponding remote extension points.

6.9.2 I/O System Design

With the design of the I/O system of a large PLC, the application engineer needs to determine the local and peripheral I/O modules, their number and type per local or peripheral category, the communication hardware and any other auxiliary equipment, and generally the overall I/O hardware needed so that the PLC is connected to the controlled process. The design of the I/O system includes counting I/O devices and the recording of their operational characteristics; grouping I/O devices based on various criteria such as operating voltage, position, operational function, etc.; incorporation of the I/O devices into the local or peripheral I/O system; and determination of the number and type of required I/O modules and the required complementary communication equipment.

The design of the I/O system begins with counting the I/O devices and examining their operation. For example, we measure the power relays, various simple coils, light indicators, signaling devices, control buttons, selector switches, sensors and, in general, any device that can be an input or output to the PLC. By taking the positions and the characteristics of the I/O devices into consideration, the ability to incorporate them into similar groups is investigated, either because they exist on the same machine or in the same production process; because they are topographically close and have a similar functionality; or because they have the same technical characteristics (e.g., nominal operating voltage). After determining the groups with the I/O devices, the selection of the groups that will constitute the local I/O system and those that will constitute the peripheral I/O system will follow. A prerequisite for defining the local and peripheral I/O system is the determination of the location in which the PLC will be installed, a decision taken on the basis of general technical and economic criteria.

An example would be the above design steps that are necessary for the configuration of the I/O system of a central PLC, which will be better understood by means of a hypothetical automation system. Figure 6.37 presents the topographic diagram of the departments of a hypothetical industrial process. The PLC will be located in the area of the first department and will be connected to its devices with local I/O equipment, while for connecting to the corresponding second and third departments, peripheral I/O equipment will be utilized. Based on the I/O devices that have been

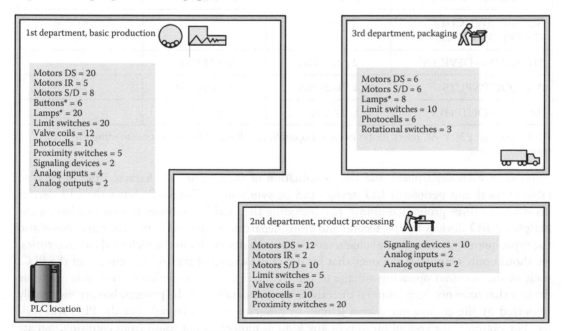

1st department, basic production

Motors DS = 20
Motors IR = 5
Motors S/D = 8
Buttons* = 6
Lamps* = 20
Limit switches = 20
Valve coils = 12
Photocells = 10
Proximity switches = 5
Signaling devices = 2
Analog inputs = 4
Analog outputs = 2

PLC location

3rd department, packaging

Motors DS = 6
Motors S/D = 6
Lamps* = 8
Limit switches = 10
Photocells = 6
Rotational switches = 3

2nd department, product processing

Motors DS = 12   Signaling devices = 10
Motors IR = 2   Analog inputs = 2
Motors S/D = 10   Analog outputs = 2
Limit switches = 5
Valve coils = 20
Photocells = 10
Proximity switches = 20

(*) These devices are additional over those required for handling, indicating operation of motors
DS = direct starting, IR = inversible rotation, S/D = star-delta starting

Figure 6.37   A small production procedure consisting of three departments controlled by a central PLC.

Table 6.2   Local and Peripheral I/O Devices of the Industrial Production Application Shown in Figure 6.37

I/O DEVICE (KIND)	LOCAL DEVICES FIRST DEPARTMENT QUANTITY/ VOLTAGE	PERIPHERAL DEVICES SECOND DEPARTMENT QUANTITY/ VOLTAGE	PERIPHERAL DEVICES THIRD DEPARTMENT QUANTITY/ VOLTAGE
MOTOR DS	20 / 230 V AC*	12 / 230 V AC	6 / 230 V AC
MOTOR IR	5 / 230 V AC	2 / 230 V AC	–
MOTOR S/D	8 / 230 V AC	10 / 230 V AC	6 / 230 V AC
BUTTON	71 / 230 V AC	50 / 230 V AC	24 / 230 V AC
	6 / 24 V DC	–	–
INDICATION LAMPS	71 / 230 V AC	50 / 230 V AC	24 / 230 V AC
	20 / 24 V DC	–	8 / 24 V DC
VALVE COILS	12 / 24 V DC	20 / 24 V DC	–
LIMIT SWITCHES	20 / 230 V AC	5 / 230 V AC	10 / 230 VAC
PHOTOCELLS	10 / 230 V AC	–	6 / 230 V AC
PROXIMITY SWITCHES	5 / 230 V AC	20 / 230 V AC	–
ROTARY SWITCHES (TWO POSITIONS)	–	–	3 / 230 V AC
SIGNALING DEVICES	2 / 24 V DC	10 / 24 V DC	–
ANALOG PNPUTS	4 / 4–20 mA	2 / 4–20 mA	–
ANALOG OUTPUTS	2 / ±1V	2 / ±1V	–

(*) The voltage 230 V AC refers to the nominal operation voltage of the relay coil feeding the corresponding motor.

counted in each department and the examination of their technical characteristics, a detailed table of local and peripheral I/O devices can be synthesized. Table 6.2 shows the I/O devices divided into three groups (columns), one including the local I/O devices and two including the peripheral I/O devices for the second and third departments, respectively. This table shows also the type, quantity, and technical characteristics of each device. From the technical characteristics, we should only emphasize the ones that influence the choice of the I/O equipment of the PLC, such as the nominal operating voltage of the I/O devices. Also included in the table are those devices that have not been counted directly from consideration of the process, but are necessarily imported by the automation system (circuit or program of automation) that the PLC will realize. For example, the control buttons of any kind of motors do not result from counting, but are imported necessarily by the automation circuit, since the latter provides the possibility of manual control of the motors in addition to any automatic operation. The same is valid for the indication lights of a motor's operation. To meet the I/O equipment requirements identified in Table 6.2, PLCs with similar capabilities are assumed. These capabilities of PLCs concern the RAM (> 4 K), the local or peripheral expansions, the maximum number of inputs or outputs that they can cover, their specific control and communication modules that are available, etc. It is also assumed that the PLCs have I/O modules with different numbers of inputs or outputs per module (e.g., 4, 8, 16, and 32, with I or O per module), both digital and analog ones.

The next design step for I/O system configuration is the classification of each I/O point (I/O device) per group (column) in Table 6.2. The various I/O points of Table 6.2 are classified according to whether they are digital or analog I/Os, the level of the nominal operating voltage (230 V AC, 24 V DC, etc.), and the type of point (input or output). The result of such a classification is presented in Table 6.3, from which it is possible to precisely determine the required I/O equipment, namely the I/O modules needed, the mounting racks of the modules in the local and peripheral system, the communication modules, the power supplies, etc. The existence of I/O modules with different I/O densities, allows for the adoption of many alternative solutions, from which the most economical and functional should be selected. An additional characteristic from the various PLC manufacturers is the number of I/O modules that could be placed on an extension hardware rack. Based on the module capacity per rack and considering the inevitable occupation of positions by power supplies and communication modules, the required expansion units in both the local and the peripheral I/O system can be determined. Since it is beyond the scope of this book to provide a detailed description of all the alternatives, it will be assumed only digital I/O modules with a density of 16 inputs or outputs per module, analog I/O modules with a density of 4 analog inputs or outputs per module, and expansion bases or racks with a capacity of 8 modules per base, will be assumed. Based on these data, as well as the data in Table 6.3, we can easily estimate that the I/O modules presented in Table 6.4 are required. Since the 24 V DC digital inputs are only 6 in the local I/O system, a special input module with 8 digital inputs is selected in this case. The final equipment specified in Table 6.4 is not the optimal one. Several combinations of modules with different I/O densities can be made. For example, a total of fourteen 24 V DC digital outputs to the local I/O system will remain unused for the future, a number that can be considered excessive. Thus, it would be possible to select 2 modules with 16 outputs per module, and 1 module with 4 outputs, so that with 36 total outputs it could more economically cover the need for 34 outputs (2 digital outputs for future use). Also, in the peripheral I/O system of the third department, the one expansion base will be almost empty, since it will contain only two I/O modules. A different choice of the I/O density of the modules may provide a more functional design. Figure 6.38 shows the electrical diagram of the interconnection of all the I/O equipment, including the central PLC. It should be highlighted that the specific configuration of the I/O equipment corresponds to the hypothetical choice made for the I/O density per unit. Different density options will give us different configurations. The examination of all possible configurations and the search for the most economical of them is now taking the form of a techno-economical study that the engineer should be able to carry out. At this point the design configuration of the I/O system for this example is finished.

The above I/O system design, but also any similar one for any other industrial application, should be made in relation to the existing PLC systems in the market. This means that the design engineer should always be updated on the existing PLCs and their corresponding size and capabilities, so that during the selection study, the final design will converge towards the most economical PLC selection. However, it is obvious that, independently of the final selected PLC, the final design cannot have exactly the required size of the controlled industrial process. In general, there will always be some redundant equipment, either I/O modules or expansion slot positions, that can be accepted as further availability for future use.

Apart from the cost comparison between the various PLCs for choosing the most cost-effective one, one must account for the cost of the equipment required to install the PLC. Both the PLC's

Table 6.3 Calculation of the Required I/O Digital and Analog Points for the Industrial Production Application Shown in Figure 6.37

MODULE TYPE I/O	LOCAL CONNECTION FIRST DEPARTMENT NUMBER OF I/O POINTS		PERIPHERAL CONNECTION SECOND DEPARTMENT NUMBER OF I/O POINTS		PERIPHERAL CONNECTION THIRD DEPARTMENT NUMBER OF I/O POINTS	
DIGITAL OUTPUTS 230 V AC	Relays for motors DS	20	Relays for motors DS	12	Relays for motors DS	6
	Relays for motors IR	10	Relays for motors IR	4	Relays for motors S/D	18
	Relays for motors S/D	24	Relays for motors S/D	30	Indication lamps	24
	Indication lamps	71	Indication lamps	50	TOTAL	48
	TOTAL	125	TOTAL	96		
DIGITAL INPUTS 230 V AC	Buttons	71	Buttons	50	Buttons	24
	Limit switches	20	Limit switches	5	Limit switches	10
	Photocells	10	Proximity switches	20	Photocells	6
	Proximity switches	5	TOTAL	75	Proximity switches	3
	TOTAL	106			TOTAL	43
DIGITAL OUTPUTS 24 V DC	Valve coils	12	Valve coils	20	Indication lamps	8
	Signaling devices	2	Signaling devices	10		
	Indication lamps	20	TOTAL	30		
	TOTAL	34				
DIGITAL INPUTS 24 V DC	Buttons	6	Buttons	–		–
ANALOG OUTPUTS ±1 V		2		2		–
ANALOG INPUTS 4–20 mA		4		2		–

Table 6.4 Calculation of the required I/O modules for a preselected density of I/O points per module

Module Type I/O	Local System FIRST Department Number of Modules		Peripheral System SECOND Department Number of Modules		Peripheral System THIRD Department Number of Modules	
MODULES OF 16 DIGITAL OUTPUTS 230 V AC	8	3	6	0	3	0
MODULES OF 16 DIGITAL INPUTS 230 V AC	7	6	5	5	3	5
MODULES OF 16 DIGITAL OUTPUTS 24 V DC	3	14	2	2	1	8
MODULES OF 8 DIGITAL INPUTS 24 V DC	1	2	–		–	
MODULES OF 4 ANALOG OUTPUTS ±1 V	1	2	1	2	–	
MODULES OF 4 ANALOG INPUTS 4-20 mA	1	0	1	2	–	
TOTAL NUMBER OF MODULES	21		15		7	
NUMBER OF EXTENSION RUCKS	3+1	4	3	5	2	6

Note: The numbers in narrow columns express the redundant inputs or outputs which are available for future use, except for the last row where empty places in the ruck express the future installation of modules.

central unit and the peripheral units are placed in industrial-type electrical enclosures. These industrial enclosures should be of an appropriate size for the spacious installation of the PLCs' units, and their dimensions should be calculated to meet the maximum temperature criterion in their interior. Additionally, it should still be possible to place all the components in such a way to provide easy access. Thus, in the event of a fault or maintenance, the replacement of some components should not be time consuming. For large PLCs, the cost of installing the equipment in a way that meets the manufacturer's specifications and international regulations is quite important.

With the extension of a central PLC to a peripheral I/O system, it is possible for one portion of the controlled production process to shut down for maintenance, conversion, etc., while the remainder of the process is working properly. With the peripheral I/O system, the placement of multiple PLCs in different parts of an industrial process can be avoided, without implying that this is always possible or desirable. Instead, in the case of very large industrial processes or multiple small processes under operational coordination, independent PLCs are installed and interconnected through a communication network. The relatively smaller PLCs, interconnected over an industrial network, are able to replace a large central PLC and have the advantage that the likelihood of a simultaneous failure of all the PLCs is very small. On the contrary, a failure in the central PLC is quite possible, and this has the potential that the whole process will be out of control.

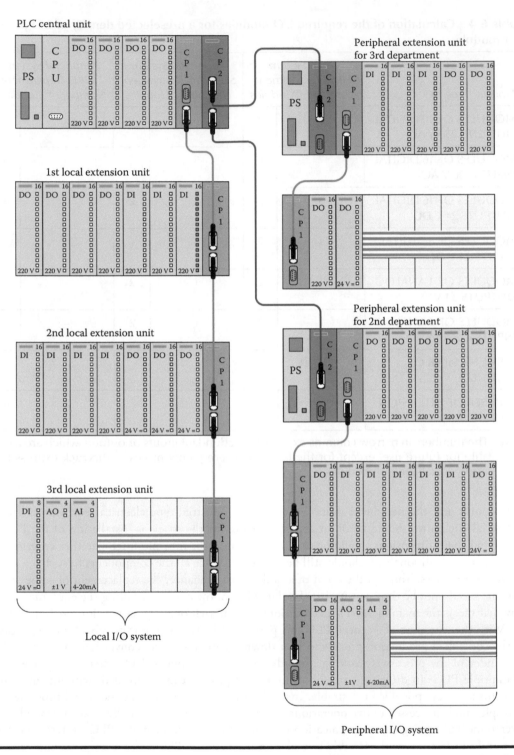

Figure 6.38 I/O hardware system configuration for the industrial production application shown in Figure 6.37.

6.10 On the Installation of PLCs

The installation and operation of PLCs should follow certain rules and standards in order to avoid operational problems as much as possible. In this case, particular attention should be paid to the installation of the PLCs in harsh industrial environments, since more causes for malfunctions exist. The primary source of installation instructions for a PLC is always the manufacturer of the PLC the directions of which should be followed closely. In addition to the installation instructions from a particular PLC manufacturer, there are some more general issues regarding the proper functionality of the PLCs that are addressed in the following subsections.

6.10.1 Electrical Enclosure for the PLC Installation

International standards and standardization associations* have adopted a number of specifications that industrial electrical enclosures should meet. These regulations are related first to the environment where the PLCs will be installed and second to the degree of protection required by the contained electrical equipment. For example, the National Electrical Manufacturers Association (NEMA) defines the enclosure type 12 as suitable for electronic control devices. More analytically, enclosures of this degree of protection are constructed without knockouts for indoor use to provide protection to personnel against access to hazardous parts, against ingress of solid foreign objects (falling dirt and circulating dust, lint, fibers, and flying), and to provide protection with respect to harmful effects on the equipment due to the ingress of water (dripping and light splashing). The International Protection Rating, often referred to as an Ingress Protection (IP) rating, is a set of codes used to define specific levels of enclosure protection. These codes consist of the prefix IP, followed by two numbers that express classifications used to measure levels of protection. The overall IP ratings according to IEC 60529 standard are displayed in Table 6.5.

A key issue in the installation of a PLC that should not be overlooked is the heat dissipation within the electrical enclosure in which the PLC will be installed. In this case, the temperature inside the electrical enclosure must not, for any reason, exceed the maximum operating temperature set by the manufacturer of the PLC, which in most cases is about 50°C. For this reason, the type of electrical enclosure and its dimensions should be calculated according to the equivalent thermal load of installed power devices, where the possible ventilation will be examined if it is required.

To avoid electromagnetic interference within the PLC's enclosure, the high-power electrical equipment should be completely separated from the low-power electronic or control equipment. The power equipment includes:

- Power relays
- Transformers
- Frequency converters
- DC power supplies
- Any other power device

---

* International Organizations: IEC: International Electrotechnical Commission; CENELEC: European Committee for Electrotechnical Standardization; National Organizations: VDE Germany; BSI England; UTE France; NEMA USA.

Table 6.5   Ingress Protection (IP) Ratings of Electrical Enclosures according to IEC 60529

Ingress Protection (IP) Two-Number Ratings			
1st Number Protection against SOLIDS		*2nd Number Protection against LIQUIDS*	
0	No protection	0	No protection
1	Protection against objects over 50 mm (e.g. hands, large tools)	1	Protection against vertically falling drops of water or condensation
2	Protection against objects over 12 mm (e.g. fingers)	2	Protection against falling drops of water up to 15° from vertical
3	Protection against objects over 2.5 mm (e.g. wires, small tools)	3	Protection against water spray up to 60° from vertical
4	Protection against objects over 1 mm (e.g. wires, specific fine tools)	4	Protection against water spray from all directions
5	Limited protection against dust	5	Protection against low pressure water jets from all directions
6	Complete protection against dust	6	Protection against high pressure water jets from all directions
		7	Protection against temporary immersion in water
		8	Complete protection against long periods of immersion in water under pressure

Example: IP 67 Enclosure means totally protected against dust and immersion

The control equipment (low power) mainly includes the PLC, but also any other special purpose electronic device, e.g., an electronic stepper motor drive card. A galvanic link is created when two or more electrical circuits share a common part of an electrical conductor, as shown in Figure 6.39, which is usually the case for earth and chassis connections. In the circuit of Figure 6.39a, a voltage drop will be created in the common conductor with Z impedance when the relay C is energized. This voltage drop interferes with the signal in the second circuit containing the PLC, which has undesirable side effects. The simplest solution to this problem is to aim for a short a length as possible of the common part of the conductor, a property that is able to reduce the interference due to the galvanic coupling. This situation is explained in Figure 6.39b as well as in the following example.

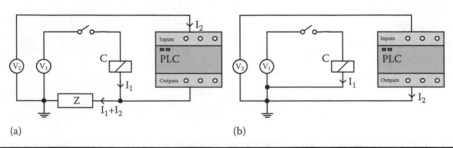

Figure 6.39   Galvanic coupling in electric circuits with a common return conductor. (a) voltage drop among Z impedence, (b) minimum length of common conductor.

Example 6.1   Voltage Drop Interference

The value of the Z impedance determines the magnitude of the voltage drop. For a cable with an impedance of 250 nH/m lead and a switching current of 1.2 A (after the closing of the switch), with a rise time of 0.1 μS, the interference voltage will be $V_{interference} = L(di/dt)=3$ V.

Except for the separation of the equipment in low and high power, special care should be taken to ensure that the power lines (cables) are lead into the electrical enclosure in separate plastic channels from the corresponding conductors that are carrying the control or the measurement signals. Many industrial enclosure manufacturers separate the equipment and the conductors even at their carrying power level. For example, they can separate the 230 V AC cables from the 24 V DC conductors that usually feed the control units to reduce the probability of electrical interference. In Figure 6.40, there are two examples of a right and a wrong separation of equipment within an industrial electrical enclosure. Generally, if some simple rules are applied, it can be ensured that the equipment is operating properly within the industrial electrical enclosure, and that the causes of electromagnetic interference are avoided as follows:

1. In the case of lighting inside an industrial enclosure, always use a light with filament tube (Linestra).
2. The cables inside the industrial electrical enclosure are separated into the following three categories:
   a. Power cables (e.g., power supply cables, motor cables, etc.)
   b. Signal and control cables (network cables, digital cables, closed loop cables, etc.)
   c. Measurements cables (e.g., analog signal cables from sensors)
3. The power, control, and data cables should be placed at the largest possible distance between them to avoid capacitive and inductive coupling.
4. Always separate the AC cables from the DC cables.
5. The distance between the power cables and the cables carrying digital signals should be at least 10 cm. Moreover, the distance between the power cable and the cables carrying analog signals should be at least 30 cm.
6. During the installation of individual conductors, the supply and return conductors should be placed together in the same plastic channel in order to reduce the interference of the electrical fields, based on their different current direction.
7. In each industrial enclosure that contains a PLC, an electrical socket should also be placed so that, in case a programming device is used, the PLC will be protected from currents generated from uneven electrical fields, and in case the ground cables for the programmer and the PLC do not have the same earth potential voltage (are not short circuited).
8. External conductors, collected by different devices and inserted in the industrial enclosure in order to be connected to the PLC, are never connected directly to the terminals of the I/O modules, but to additional auxiliary terminals that are typically located at the bottom of the industrial enclosure.

Figure 6.41 presents a small industrial enclosure, where there is a modular form PLC in its interior, with all the cables well arranged in their respective driving plastic channels.

6.10.2 Electromagnetic Interference

Recently, electromagnetic compatibility (EMC) has become more and more important, due to the variety of electrical and electronic devices in both industrial environments and everyday environments. Electromagnetic compatibility is denoted as the ability of electrical equipment to operate

Bad separation of equipment and wires

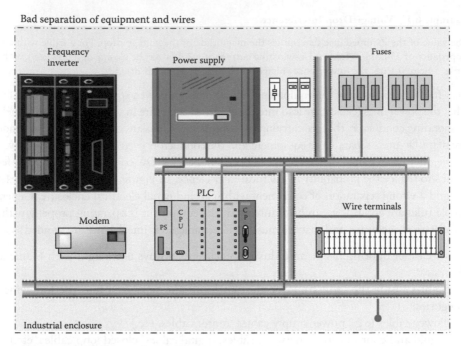

Correct separation of equipment and wires

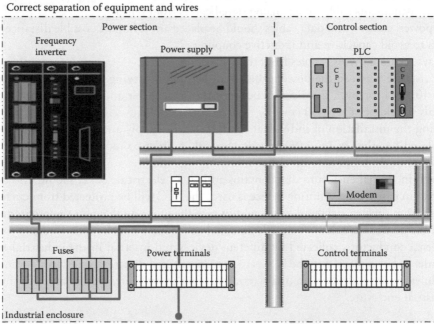

Figure 6.40 Power equipment and wire separation in an industrial enclosure including a PLC.

Figure 6.41   A small-size industrial enclosure with a modular PLC.

without to create electrical interferences in the operation of other devices, and also without being influenced from the electrical interference produced by other devices.

Modern automation and control systems are able to combine a multitude of such devices that contain power electronics, classical electrical devices and components, digital circuits, and digital data processing units. In all these devices, there is a wide variety of currents, voltages, and field densities that cause various malfunctions that are responsible for the downtime of machines or even of entire systems. Due to the high cost of a potential loss of production, nowadays, electromagnetic compatibility issues are taken indispensably into account in the design of new industrial plants and are a quality criterion for both the individual devices as well as the whole automation system.* Two more definitions complement a basic approach to this issue:

- Electromagnetic interference (EMI) is called the operational influence of an electric circuit, an electrical device, or a living organism from the electromagnetic environment.
- Electromagnetic emission (EME) is called the phenomenon of electromagnetic energy production-emission from a source.

Electromagnetic interference may occur in the form of inductive or electrostatic interference, and generally interference due to the existence of an electromagnetic field. Figure 6.42 illustrates the production of EMI in the form of galvanic, inductive, capacitive, and radiative coupling inside an enclosure containing a PLC and various kinds of conductors. PLCs, as well as the PCs for industrial use, should operate reliably in an environment that is generally considered to be full of generated and transmitted electromagnetic interference signals. They should also be designed so that they do not affect other devices. The latter issue, however, is a matter for the manufacturers of PLCs, and cannot be influenced by the design engineer. The design engineer can only reject a PLC that does not meet the EMI levels or provide technical information about itself. In a programmable automation system, there are several paths between the PLC and its electromagnetic environment, through which interfering signals can be propagated, as depicted in Figure 6.43. The power supplies, the I/O modules, the communication interfaces, and even the metallic environment of the industrial enclosures, all can act as coupling paths for all of the above interference

---

* (Norms and guidelines on electromagnetic compatibility, DIN VDE 0870 and European Directive 89/336/EEC).

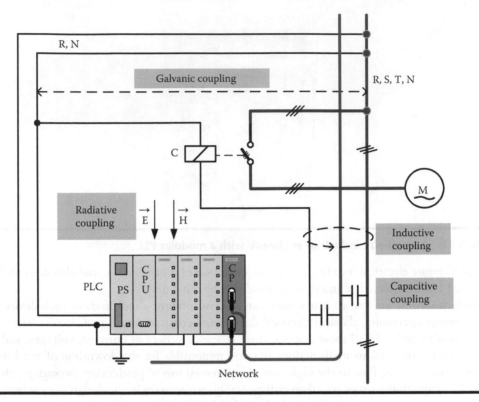

Figure 6.42   Various forms of electric or magnetic coupling and interference creation.

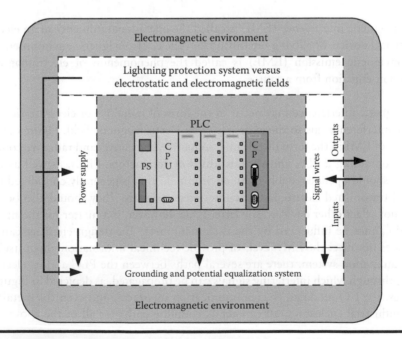

Figure 6.43   Passageways of electromagnetic interference between a PLC and the existent electromagnetic environment.

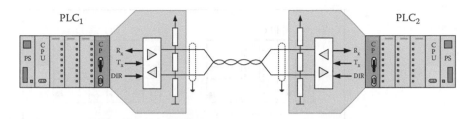

Figure 6.44   RS 485 connection between two PLCs.

types. For this reason, in cases of debugging, it is not easy to identify when something is the source of electromagnetic interference, which therefore causes general malfunction. The frequency of an interference signal may range from a few Hz to 100 MHz, while the currents and voltages generated are not easily measured. It is therefore preferable, in order to avoid the debugging task for identifying the problematic hardware, to design the entire automation system from the beginning according to the rules for avoiding electromagnetic interference. For example, the connection cables of analog inputs and outputs must always be fitted with a protective copper or aluminum braided shielding. Digital I/O modules already have a basic protection against interference, due to the optical isolators that are inserted in the corresponding circuits, as we have seen in Sections 6.3 through 6.7. The communication cables should also withstand electromagnetic interference and be connected according to the manufacturer's instructions. In Figure 6.44, the RS 485 connection interface is shown between two PLCs through a twisted pair cable with a protective braided shielding that has to be grounded at both ends.

6.10.3 *Grounding*

The elementary prerequisite for a seamless and safe operation of an industrial installation with one or more PLCs is the proper grounding of all the involved plant units, specifically PLCs. Poor grounding of a PLC is often the cause of operating problems that:

1. Are not easily identifiable and cannot even be categorized if they are due to software (incorrect programming) or hardware failures
2. Can only be explained by very specific measurements; many times, the programming engineer wastes a lot of time searching for the cause of the malfunction in the program, while in reality it is a hardware problem
3. Are not only relevant to the functionality of the PLC, but also to the safety of the personnel
4. Are directly connected to the sources of electric noise and interference that are obscure, multiple, and multiform

Thus, in order to be sure from the beginning that similar problems will not be encountered, special attention should be paid to the implementation of the ground circuit. A primary source of technical information on how to ground a PLC comes directly from its manufacturer. With the help of electrical diagrams, the method of power supplying and grounding should be clearly defined, not only for the central unit and the power supply of the PLC but also for the different I/O modules. A second basic source of technical information is the industrial electrical installation regulations. Due to the fact that the electrical grounding is quite a huge field in electrical

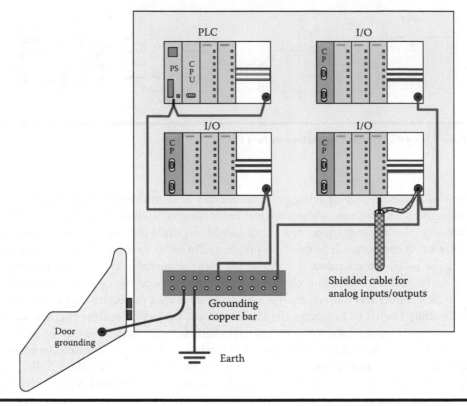

Figure 6.45 Grounding configuration in an industrial enclosure with a PLC and additional groups of I/O modules.

engineering, further analysis on this issue will not be provided here; however, the most important general rules and design guidelines are as follows:*

1. All PLCs should be firmly grounded in accordance with the manufacturer's specifications. Many PLCs have individual grounding points that should all be grounded; for example, a ground point on the power supply and a second ground point on the PLC's mounting rack. All connections of the ground conductors should be screwed and never soldered.
2. The industrial electrical enclosure within which the PLC will be installed should be evaluated for good grounding.
3. Every ground conductor should have a resistance less than 0.1 Ω, and resistance between the earth bus-bar of the industrial enclosure and the ground must be less than 0.1 Ω. The grounding conductors should indicate a current flow only during the event of a failure.
4. The entire grounding "chain" must be continuous and there should not be a discontinuity point, which means that a part of the equipment is substantially ungrounded, although will appear to have grounding connections. In Figure 6.45, an indicative grounding chain is

* DIN VDE 0160.

presented from a ground electrode to bus-bar, from bus-bar to mounting metallic base, from mounting base to PS module of the PLC, from mounting base to braided shielding, etc.

5. The grounding and its quality of implementation is a particularly important issue when the system is equipped with lightning protection.

6.10.4 Electromagnetic Shielding of Cables

Shielding generally has different forms and may involve different things, such as shielding of three-dimensional housing, a single location, individual devices, and simple wires. In many cases the shielding is created by itself, for example, a small or large metallic enclosure in which a piece of electrical equipment could be placed; this is in itself a kind of shielding from the electromagnetic environment.

In the section on electromagnetic interference, the need to protect the cables from electromagnetic interference has already been mentioned. Figure 6.44 in particular has been presented as the screening of a RS485 two-pin interface. In this section, all the basic rules of correct shielding will be listed only for cases of industrial automation signals (low voltage signals like analog measurements and data).

The purpose of the shielding is to stop the electromagnetic lines of the interfering field and, as a result, block the electromagnetic interaction between adjacent cables. The protective metallic mesh absorbs the interference by playing the role of the escape line. In other words, the shield surrounding the inner signal carrying conductor acts on EMI in two ways, which either can reflect the energy or pick up the radiated EMI and conduct it to the ground. In either case, the EMI does not reach the signal conductors. There are two basic types of shielding typically used for cables, the foil type and the braided type. For each type, there are also different kinds of shielding construction, the most important ones being:

1. Foil shield (using a thin layer of aluminum attached to a polyester carrier)
2. Braid shield (a woven mesh of tinned copper wires which provides coverage between 70% and 95%, depending on the tightness)
3. Double foil and braid shield (better coverage and protection)
4. Triple shield (foil-braid-foil shielding combination for very noisy environments)

In each of these items, the shielding mesh has a different impedance that is frequency dependent. In general, the following rules can be established:

1. The less impedance the mesh presents, the better the shielding becomes. In this case, the mesh may receive large discharge currents.
2. For low-frequency magnetic fields, the twisting of the conductors is a fairly effective method.
3. Data cables (network) and cables for analog measurements must always be shielded.
4. In the analog signal cables, the protective mesh must be grounded at only one end (obviously on the side of the PLC) as shown in Figure 6.46. This is required in order to avoid creating ground loops with corresponding currents that may pose a problem to the PLC's processor.
5. In all other types of cables for data transmission, networks, etc., the shielding mesh must be earthed at both ends, as shown in Figure 6.44. Only the double ground at both ends of the shielding mesh ensures that the conductors are shielded from inductive and high frequency interference.
6. The shielding mesh should never be used as equipotential bonding conductor.

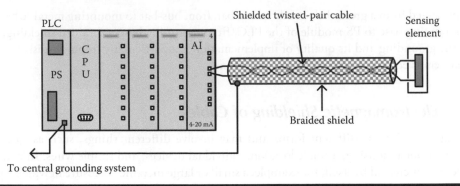

Figure 6.46 shows the connection-contacting of the shielding mesh of four cables for analog signals in detail, with the grounded mounting rack in a Siemens PLC. — *labels:* Shielded twisted-pair cable / Sensing element / PLC / Braided shield / To central grounding system / 4-20 mA

Figure 6.46   Grounding of a shielded cable connecting an analog sensor to a PLC.

Figure 6.47 shows the connection-contacting of the shielding mesh of four cables for analog signals in detail, with the grounded mounting rack in a Siemens PLC.

6.10.5 Lightning Protection

In the past few years, lightning protection was mainly confined to the protection of buildings. Today, this situation has changed since industry now uses automation equipment that contains a large number of microprocessors and generally integrated circuits for control and monitor purposes. These units are much more sensitive to overvoltages than traditional relays and other conventional automation devices. In these cases, overvoltages of a few volts above the nominal ones are enough to destroy the corresponding electronic components, and for this reason it is necessary that the automation installation is protected from lightning discharges.

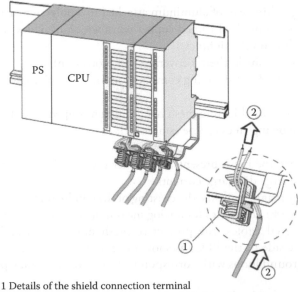

1 Details of the shield connection terminal
2 Wiring of the shielded cable to analog I/O through terminal

Figure 6.47   Mechanism for the quick connection of a braided shield of a cable to the grounded rack of a PLC (Siemens).

The lightning protection and the corresponding measurements for the evaluation of the technical parameters are divided into two parts, external and internal lightning protection in accordance with the regulations VDE 0185 and IEC 62305. This protection is further classified into four zones of lightning protection with the lightning protection of types "Room" and "Device" that correspond to zones two and three. Lightning protection is generally an important issue in electrical installations, and thus it is not possible to present it in this book in full detail.

Internal lightning protection is aimed at reducing the effects of lightning current and the resulting electric and magnetic fields on metal installations on all kinds of electric appliances within a closed space. Internal lightning protection must exclude the generation of dangerous sparks between electrical or telecommunication installations and metal structures, and include:

■ The equipotential bonding through the grounding system
■ The shielding and overall isolation between the installation sections
■ The use of devices for protection against overvoltages caused by lightning (surge arresters, surge protectors, and surge suppressors are terms used for such devices)

The equipotential procedure reduces potential differences across devices and conductors induced by the lightning current, and ideally means that an identical state of electrical potential for all equipment items is valid. This is achieved by bridging the metallic structures with the electrical and telecommunication installations through connecting conductors, as well as with the external conductive parts. All the incoming and outgoing cables in the space that should be protected should be equipped with an equipotential system. This rule includes power cables and the low-power signal cables, as well as the metallic channel cable trays shown in Figure 6.48. Also, all the cables between two buildings must be shielded, and subsequently, to bridge the corresponding shielding. The closed metallic cable trays facilitate the shielding and the equipotential connection. In Figure 6.49, it is presented as a way of edging two or more PLCs of lightning protection devices in a communication network, either directly to the bus cable or at the T-branch.

At this point, it should be noted that there are a plethora of lightning protection devices for communication network cables in the related market. Manufacturers of these devices generally provide the information required for the correct connection method and for their appropriate location along the communication interface cable. In Figure 6.50, a detailed approach of how to interface two PLCs with lightning protection is shown, located in adjacent metallic industrial enclosures in the same interior space. Lightning protection in the power supply section of the whole installation is also depicted.

6.10.6 Input Devices with Leakage Current— Impedance Adjustment of I/Os

During the installation, interconnection, and operation of PLCs, special care should be provided to the issue of resistance adaptation of I/O circuits. Today, many of the input sensors and, in general, the input devices that cooperate with the PLCs, are made up of solid-state electronic circuits. Every input device containing a solid-state switching element (e.g., a triac, a transistor, etc.) that is connected in series with the high impedance digital input of a PLC, may cause an erroneous (i.e., no real signal) activation of the input. The same problem also occurs when

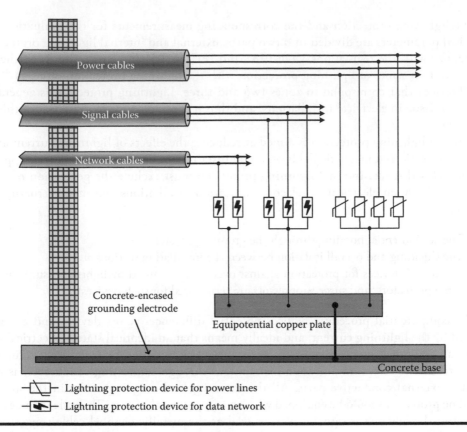

Figure 6.48 Lightning protection through the potential equalization and use of suitable protective devices for all kinds of cables.

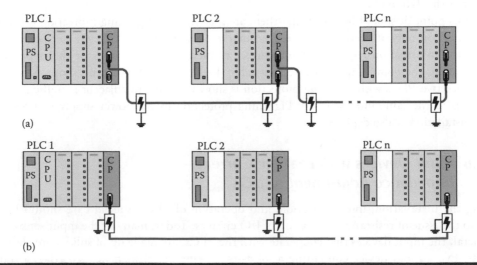

Figure 6.49 Protective devices against impulse voltage in a communication bus at the ends of the line (a) and before the T connections of the bus (b).

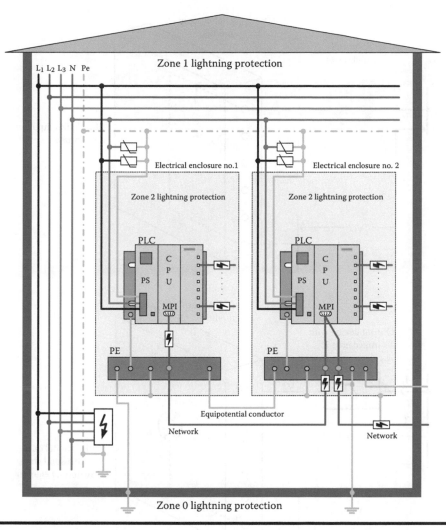

Figure 6.50 **Electric connection pattern of two PLCs inside corresponding enclosures for protection from impulse voltages.**

the solid-state switch, contained in a digital output module, is connected to a high impedance output device. In this case, it is possible to actuate the output device incorrectly without a real trigger signal from the CPU.

In both cases, the cause of the error is the fact that the switching solid state elements, presented either in the output circuits of the PLC or in the input devices that are connected to the PLC, produce a leakage current when they are in the OFF state. The leakage current, which may range from 1 to 20 mA, is capable of causing an inappropriate activation of either the input of the PLC or the load that is connected to the output of the PLC. These two cases are presented in Figure 6.51. In order to address this problem, a resistive load (or, in general, an impedance) should be inserted in parallel to the input circuit impedance (Z) or to the high impedance output device (Z), a process that it is called "impedance adjustment of the input or the output". The calculation of the required resistive load is rather simple and, in both cases, will be described in what follows in the form of numerical examples.

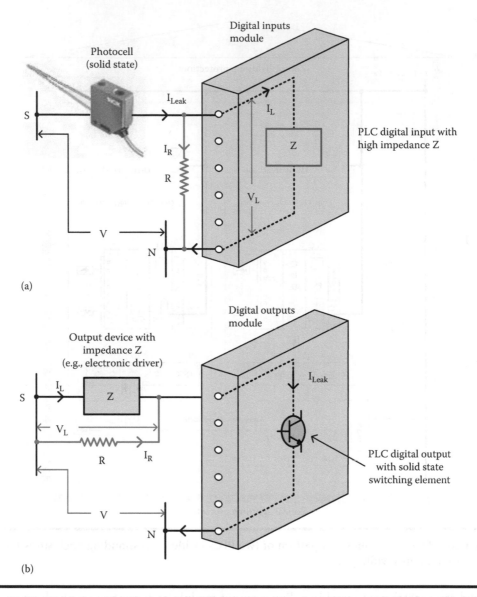

Figure 6.51   Resistive adaption of a digital input (a) or output (b) is applied when the leakage current causes their fault activation.

6.10.7 Input Impedance Adjustment

Figure 6.51a presents the connection of a photocell, which is a solid-state device, to a digital input of the PLC. The fact that the leakage current I_{leak} of the photocell is constant falsely actuates the input of the PLC, which means that the voltage $V_Z = I_{leak} Z$ is greater than the one that the manufacturer of the PLC defines as the maximum voltage for the input to be considered logic "0". The resistor R is connected in parallel to the input circuit of the PLC, i.e., to the impedance Z of the input to substantially absorb the leakage current. The notations shown in the figure are the following:

V = Nominal operating voltage of the input module (PLC)
V_L = Maximum voltage under which the input remains OFF (PLC)
I_R = Current flow through the resistance R
I_L = Current flow through the input when it is under voltage V_L
Z = Input impedance (PLC)
R = Input resistance adjustment (PLC)
I_{leak} = Leakage current of the photocell (Photocell)

The notations concerning the (PLC) or (Photocell) refer to the technical characteristics of the corresponding device. The rest of the presented variables will be calculated in what follows. It is assumed that from the technical characteristics of the digital input module, the following values can be extracted:

■ Maximum allowable voltage in the OFF mode, V_L = 60 V
■ Input impedance Z = 40 KΩ
■ Nominal operating voltage of the input stage V = 230 V

Then, the maximum allowable leakage current of the input can be calculated as I_L= V_L/Z = 1.5 mA.

In the case that the leakage current of an input device is greater than 1.5 mA, this will cause a faulty activation to the input of the PLC and thus it is necessary to insert the resistor R as shown in Figure 6.51a (everything related to the impedance adjustment has been marked in gray). From the technical characteristics of the photocell, it is assumed that I_{leak} = 12 mA (quite larger than the permitted 1.5 mA). Then the resistor R, which as mentioned above interferes to absorb the leakage current of the input, while preventing the current to flow to the input, can be calculated as:

$$R = V_L / I_R = V_L / (I_{leak} - I_L) = 5.71 \text{ K}\Omega$$

which is the maximum permissible value. However, since the resistor R will be under a voltage 230 V when the photocell is energized (ON state of the input device), and should have the equivalent power (for 230 V and not for 60 V) or:

$$P_R = V I_R = V^2 / R = 9.26 \text{ W}$$

that is the minimum value. Therefore, the impedance adjustment will be achieved with a resistance of approximately 5 KΩ and 10 W.

6.10.8 Output Impedance Adjustment

In Figure 6.51b, the connection of an output device, with a very high impedance (Z) in a digital output of the PLC, is presented. The symbols presented in this figure have the same meaning as the case of Figure 6.51a, except for the voltage V_L, which now expresses the maximum allowable voltage for which the output device remains in the OFF state. The technical characteristics of the digital output module is assumed to have the following settings:

■ Leakage current in the OFF state of output, $I_{leak} = 8$ mA
■ Minimum current of the load (Z) at the ON state of the output, $I_{ON} = 60$ mA.

Therefore, the output of the PLC shows a leakage current of 8 mA when it is in the OFF state. Every output device that can operate with a current smaller or equal to 8 mA will require the insertion of an adjusting resistor to ensure that it will remain OFF when the output of the PLC is deactivated.

On the other hand, since the output device should absorb a current of at least 60 mA, to ensure that the output will remain ON, every output device that does not require an operating current of more than 60 mA will need to be adjusted with a resistor in a parallel connection in order to absorb the extra current from the output. Thus, there are two conditions that need to be met, with the second one to ensure the satisfaction of the first one. In this case, the calculation of the resistance R will be based on the second condition.

For example, it is assumed that the output of the PLC is connected to the input of another PLC, with the same above characteristics for "communication" purposes (1 bit information transmission from the first PLC to the second). Then it is derived that:

$$I_L = V_L / Z = 1.5 \text{ mA} < 8 \text{ mA}$$

If no adjustment is made, the output device—that is, the input of the second PLC—will be activated, while the output of the first PLC is in the OFF state due to the leakage current. In order not to have the operating current I_{ON} exactly on the limit of the 60m, the resistor R is chosen to absorb all the I_{ON} current. Thus, in this case it is derived that:

$$I_R = I_{ON}, R = V/I_{ON} = 3.83 \text{ K}\Omega, \text{ and } P_R = V I_{ON} = 13.8 \text{ W}$$

Thus, with the selection of a 3.8 KΩ and 15 W resistance, it is ensured that the current is $I_{ON} > 60$ mA. At this point it should also be checked whether, in the OFF state, the voltage at the ends of the load—that is, at the input of the 2nd PLC, denoted by V'_L is $V'_L < 60$ V. Thus:

$$R_{total} = R\|Z = 3.47 \text{ K}\Omega \text{ and } V'_L = I_{leak} R_{total} = 27.76 \text{ V} < 60 \text{ V}$$

Instead of simple resistance, which is usually greater than the 5 W and therefore produces heat in the ON state (230 V), a combination of R, C series elements can be used that have the advantage of producing much less heat can be used. The RC combination should of course have the corresponding impedance that has been calculated for the simple resistance, and its calculation will be based on the relationships,

$$Z = \sqrt{R^2 + X_C^2}, \quad X_C = \frac{1}{2\pi f C},$$

where f is the network's frequency and C the capacitance of the capacitor.

6.10.9 Parallelizing Digital Outputs of PLCs— Transitional Protection of I/Os

Another problem is created with the possible parallelism of two digital outputs of a PLC in order to increase the power supply to the output device. Let us assume that the digital outputs of a PLC have a nominal operating current in the ON state of 0.5 A, while an output device that we want to connect to a digital output of the PLC requires a current of 0.8 A. In this case, it is obvious that the output of the PLC cannot directly supply the output device. Thus, it is possible to connect two digital outputs of the PLC to the output device (0.5 A + 0.5 A > 0.8 A) and program at the PLC the simultaneous activation of the two outputs. However, such a solution should be avoided, because the switching elements of the PLC digital outputs that have been presented are triac, power transistors, relays, etc., and cannot ensure a simultaneous operation due to the various electrical differences they present. Moreover, different initial conditions or differences in the logical update of the outputs may cause additional delays in activating the two outputs at the same time. Such a delay is able to cause a single digital output to accept the total load of the output device (0.8 A, for the examined example) that will resulting in a voltage drop of that digital output.

A last issue that the automation engineer should not overlook, is that of the "transitional protection" of the digital inputs and outputs when their operation is combined with external control or detection devices. In particular, Figure 6.52 shows an output device (inductive load) that is connected to the digital output of the PLC via a push button, in order to control the operation of the output device when the PLC has activated this output. Such a connection is possible upon the closing or opening of the push button contact, generating an electrical noise and, therefore, requiring transitional protection. Thus, when it is needed to insert a switch or any kind of a switching contact of a sensor between an output device and the PLC, a surge suppressor should be connected in parallel to the load (e.g., to the output device) in order to achieve transitional protection.

In Figure 6.53, the output device (inductive load) is actuated by both the PLC's digital output 1, and the external push button that bypasses the PLC and allows the output device to be activated

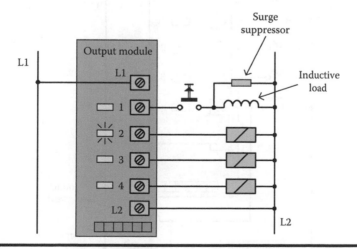

Figure 6.52   The insertion of a switching element in the circuit of a digital output in a series requiring the placement of a surge suppressor.

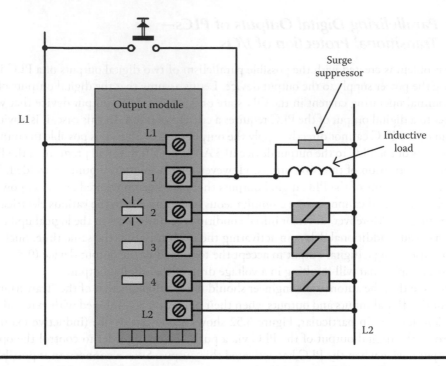

Figure 6.53 The insertion of an external switch parallel to the digital output of the PLC requiring the placement of a surge suppressor.

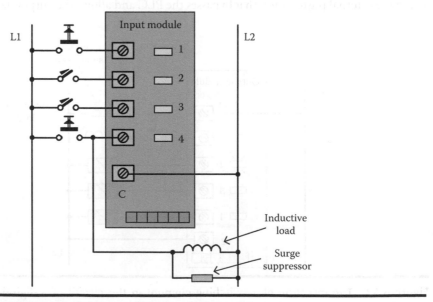

Figure 6.54 When a button or switch is common to a PLC (such as an input device) and a conventional automation circuit, then protection via a surge suppressor is required.

whenever it is needed, regardless of the PLC's logic. In this case, transient protection is also needed and the insertion of a surge suppressor is required.

Finally, Figure 6.54 presents an input device (push button) that, except for the activation of the input 4 of the PLC, also causes the activation of an external inductive load. Although the operation of the latter is independent of the PLC's logic, it may cause electrical noise or interference, and consequently a transient protection is required by the utilization of a surge suppressor. In all the previous cases, the typical form of the equivalent circuit of a surge suppressor is an RC series circuit with elements of R = 1 KΩ and C = 0.5 μF.

6.10.10 Starting the Operation of a PLC and Fault Detection

The procedure of the first-time operation of a PLC should contain careful steps in order to exclude destructive damage of the I/O modules, CPU, power supply, or any other equipment involved. Before putting a PLC into operation for the first time, the entire installation should be examined and verified in accordance with the manufacturer's specifications and the national or international regulations. Additionally, the good grounding of the system should be specifically checked. Before powering up the PLC, the following actions are needed:

1. Ensure that the supply voltage for the PLC corresponds to the one needed from the PLC. Many PLCs are capable of receiving more than one electrical supply voltage, such as 230 V AC and 110 V AC. The selection is made either with a special switch or with a jumper in the appropriate position. In many cases, it is quite common that the PLC is set for an AC voltage of 110 V AC and, in the end, supplied with 230 V AC with catastrophic results.
2. Check that all the power and communication cables are correctly positioned in their respective slots. The communication connectors fitted with locking screws should be carefully installed in their correct positions.
3. Examine whether all the I/O modules are correctly positioned on the mounting base for the case of a modular PLC, and that there is a communication "bridge" between each one and the next one.
4. Put the operational mode switch of the PLC in the STOP position.
5. If the electric installation includes a general emergency switch (e.g., an emergency STOP button), make sure it is OFF or in the open contact position.
6. Ensure that all the output devices are not powered.

Once the previous steps are completed, supply power to the PLC and check that the status of all the LEDs on the central processing unit are as the manufacturer of the PLC specifies them. Also check that the status of all the input devices, ON or OFF, are in conformity with the logic of the automation program. This simply means that some input devices are normally closed contacts, so that the corresponding LEDs of the digital inputs are lighted (ON state). Then, the input devices that are in the normally open contact state should be checked by causing a manual activation of each of them. For example, in the case of a button, it should be simply pressed, or in case of a photocell on a conveyor belt, an object should be interfered to check the corresponding activation. Whenever an input device of this category (e.g., a NO contact) is activated, the corresponding LED of the digital input of the PLC should also be activated. In this way, it is ensured that the input devices are functioning properly, if the wiring of the input devices was implemented correctly, and whether the input device and input address match the one contained in the automation program. Subsequently, the PLC is set to the run mode (RUN) and the outputs are checked so they behave according to the logic of the automation program. This test can be based on the indicating LEDs of the digital outputs, the state of the power relays provided, or the power supply of the loads that were interrupted according to action 6, above. In order to control the correctness of the automation program, it does not need to operate large loads (such as large motors) at the stage of the first test of the PLC.

If the above check reveals an operating problem in either the general operation of the PLC, or in one or more inputs or outputs, then the debugging process of either the program or the hardware is followed. This process, in order to have a quick and effective outcome, should be systematic and include the following three basic steps:

■ The isolation of the operational problem from other possible ones
■ The identification of the operational symptom or the equipment fault
■ The careful fault correction and restoration

If the problem concerns the actual operation of the PLC and the CPU, then the indicating LEDs should be checked in order to diagnose the operating status of the PLC and follow the manufacturer's debugging instructions. The CPU and the AC power adapter may have some or all of the following indication LEDs.

■ DC or AC POWER ON
■ TEST MODE (A PLC can be in STOP, RUN, TEST, programming mode, etc.)
■ PROCESSOR FAULT
■ MEMORY FAULT
■ I/O FAULT
■ LOW BATTERY FAULT

If the problem involves a digital input or output, then some basic actions for the debugging process are as follows:

1. Suppose a digital input device (e.g., a limit switch) is switched on (the contact of the switch is closed) and causes a digital output to be activated. If this is not the case then if, with an activated input device, the indicative LED of the digital input is activated, then the problem is due to the digital input module or possibly to the logic of the executed automation program. If, with an activated input device, the indicative LED does not light up, then the problem is due either to the input device or to its wiring. In this case, measuring the voltage at the corresponding digital input of the PLC, as shown in Figure 6.55, will help derive a final conclusion.
2. If the device assumed that a digital output of a PLC is enabled, but the corresponding output device connected to it is not activated, and the digital output LED does not light up, then the problem exists in the digital output module. If the digital output indicator LED lights up but the output device is not activated, then the problem exists either in the wiring of the output device or on the device itself.

In both cases, measuring the voltage at the digital output of the PLC, as shown in Figure 6.56, will help to accurately locate the fault.

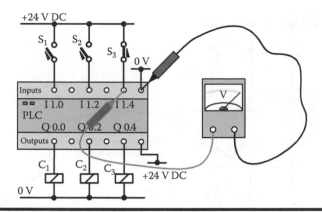

Figure 6.55   **Measurement of the voltage that an input device must supply in the corresponding digital input of the PLC.**

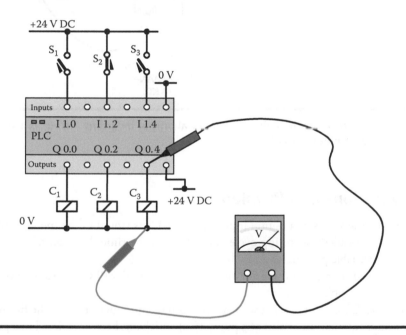

Figure 6.56   **Measurement of the voltage of a digital output with a connected load.**

Special attention should be provided when the output of the PLC is utilizing a triac as a switching element. In this case, the triac has a high internal resistance when not conducting (about 1 MΩ), so the insertion of a voltmeter, which also has a high internal resistance (about 6–10 MΩ), results in a voltage divider generated when the digital output has no load. In this case, the indication of the voltmeter may be close to the nominal voltage of the digital output even when the triac is OFF. For the voltmeter indication to be correct, a 10 KΩ resistor should be connected in parallel to the instrument's terminals, as presented in Figure 6.57. Then the total resistance of the instrument will be much smaller than that of the triac, and the indication will be almost zero for a non-conducting triac.

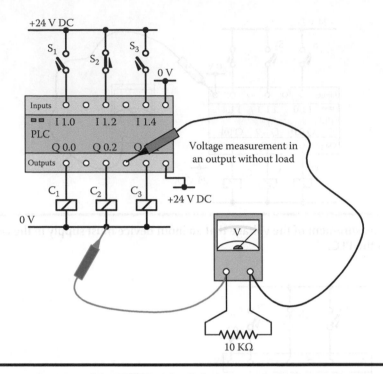

Figure 6.57   Measurement of the voltage of a digital output including a triac as a switching element and without a connected load.

Review Questions and Problems

6.1. Explain how the status of a digital input signal 230 V is stored in the memory of a PLC.

6.2. What is the so-called "scanning cycle" of a PLC? Is the time duration of scanning cycle a constant or variable parameter and why?

6.3. Describe the advantages, if there are any, of using a modular PLC instead of a compact PLC in an industrial application.

6.4. One of the PLC's advantages is the easy detection of equipment faults. The figure shows the conventional automation circuit of a pump (C2) which has been translated in a program and runs in a PLC. Suppose that the float level switch has been damaged and its contact remains open, although the liquid level in the tank is above the float switch. Explain why the detection of this fault will be easier in the case of programmable automation.

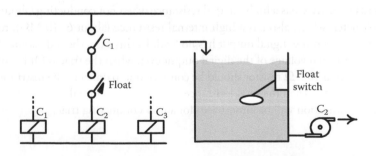

6.5. What type of digital output (DO) module do you have to select for the activation of a solid state output device in order to avoid operating problems due to leakage current?

6.6. The figure shows a typical simplified electric circuit of a digital input module. Describe the operations that take place inside the module between the input signal and the logic signal which is transferred to a CPU.

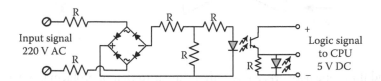

6.7. The classical scanning cycle of three steps (input scanning, program execution, and output updating) is interrupted when the user program contains instructions for a peripheral read–write. For example, an instruction for a direct-in-the-module reading of an input will give the result shown in the figure. In your opinion, when is the scanning cycle faster? When the user program contains instructions for peripheral read–write, or included no instructions?

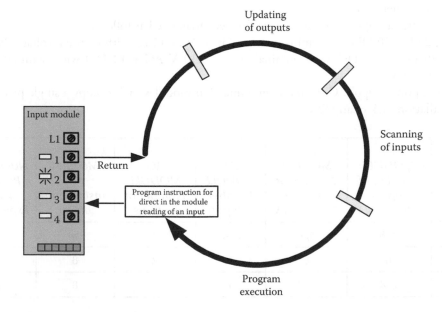

6.8. An analog valve for flow control accepts an analog signal of –10 V DC up to +10 V DC from an analog output of a PLC. The PLC controls the liquid flow based on a proper algorithm. The analog output uses 12 bits for a digital value representation and 1 bit for sign. Complete a three-column table which will give (for each opening position of the valve from 0% up to 100% in steps of 10%) the corresponding voltage value and its digital representation that is created by the analog output of the PLC.

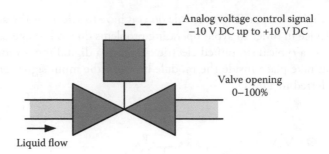

6.9. In a liquid flow pipe network, a flow valve, similar to that of problem 6.8, has been inserted and permits the flow of 100 L/min when it is at its maximum opening. What will be the stored digital value of the analog output adjusted for 60 L/min flow?

6.10. A small industrial process is controlled via a modular PLC. The automation task controls a system with 32 digital inputs, 28 digital outputs, 4 analog inputs, and 4 analog outputs. Given that peripheral equipment is not required due to the small dimensions of the industrial building, determine the required local equipment by selecting the PLC from the three available in the market, which will have the major number of empty places for future installation of I/O modules. The technical characteristics of the three available PLCs are shown in the table. Please create a diagram showing the structure and interconnection of the equipment units.

The digital inputs and outputs are further categorized as follows:

32 DI = 20 DI with nominal voltage 230 V AC + 12 DI with nominal voltage 24 V DC

28 DO = 16 DO with nominal voltage 230 V AC + 12 DO with nominal voltage 24 V DC

The power supply, CPU and communication modules each occupy a single place in the installation rack of any PLC.

	DI MODULE Number of Inputs 230 V or 24 V	DO MODULE Number of Outputs 230 V or 24 V	AI MODULE Number of Inputs	AO MODULE Number of outputs	Modules per Installation Rack	Number of Possible Local Expansions
PLC-A	4, 8	4, 8	2, 4	4	6	2
PLC-B	6	6	2	2	8	2
PLC-C	2, 4	2, 4	4	4	8	2

6.11. In the technical characteristics catalog of a digital input module with a nominal operating voltage of 24 V DC, it is written,

"Input current for signal '1' = 8.5 mA at 24 V DC"

Explain what exactly expresses this magnitude (8.5 mA) and if it has any possible importance.

6.12. In a PLC with many digital I/Os and a large executed program, the scanning cycle duration is 50 msec. The digital input device connected to the PLC is an SPST contact output of a sensor which opens and closes with frequency 100 Hz. Explain if it is possible to count the number of ON/OFF changes of the sensor output during a time interval of 1 hour.

6.13. Four machines, C_1–C_4, operate according to the automation circuit shown in the figure, either manually through buttons $Start_i$/$Stop_i$ and the rotational switch RS_i in place M or automatically through command from the digital $Output_i$ of the PLC and the rotational switch RS_i in place A. In order to accomplish this operation, a digital output module is available from the manufacturer that defines the connection in the diagram shown in the figure (module view). Design how the outputs of the PLC will be connected to the conventional automation circuit in order for the desired operation to be achieved. Note that the conventional circuit may be modified suitably to facilitate the connection to PLC, but it will not affect the desired operation.

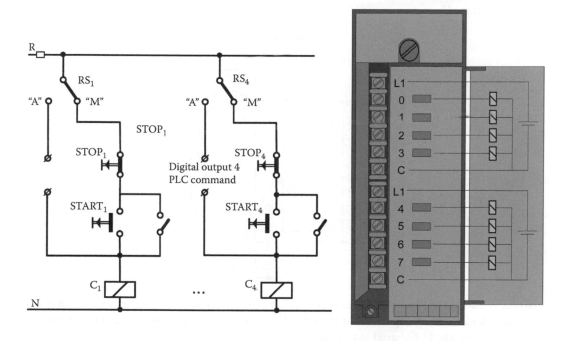

6.14. Write the differences between a PLC and a general purpose computer from a hardware and software point of view.

6.15. Write the basic differences between a digital input module and an analog input module of a PLC.

6.16. The manufacturer of a digital output module with nominal operating voltage of 24 V DC gives a connection pattern of the output devices as shown in diagram (e) of the figure. Which one of the four connections (a, b, c, or d) of the output devices realized by a technician (shown in the figure) has a faulty connection? Explain your answer.

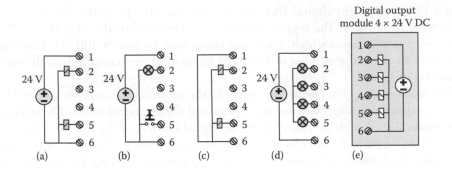

(a) (b) (c) (d) (e)

6.17. The activation of the power relay C from a PLC is performed via a power transistor or a triac which are in the digital output module. Suppose that the power transistor or triac has been burned and is at short-circuit status. What problem arises concerning the function of the button STOP which must stop the operation of power relay C, and how it can be addressed?

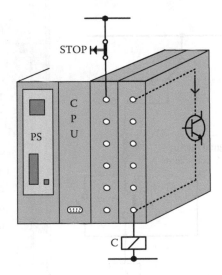

6.18. A modular type PLC is going to control a small system with 8 digital inputs, 8 digital outputs, and 1 analog input. What structural units (modules) do you have to buy in order to implement the PLC-based automation system?

6.19. Which is the maximum time delay that may happen between the activation of a digital input and the activation of a digital output which depends logically on the first?

6.20. The operation of the electromechanical equipment of a road tunnel is going to be controlled via a central PLC. The electromechanical equipment includes general purpose lighting, acoustic and luminous signaling, a ventilation system, a fire safety system, and a large number of sensors and meters. Although the machinery is relatively distributed along the whole length of the tunnel, they may be grouped in the three points as shown in the figure. The kind and number of I/O devices are also indicated in the figure. Configure and design the I/O hardware system in detail, selecting the PLC placement first. In particular, determine the local and peripheral hardware, the kind and number of required I/O modules, the communication pattern between I/O modules and CPU, and the method of grouping I/O points and modules, giving the corresponding diagrams.

The technical characteristics of the PLC are:

a. Each digital module contains 8 I/O points.
b. Each analog module contains 4 I/O points.
c. The maximum cable length for peripheral equipment is 900 m.
d. The maximum cable length for local equipment is 30 m.
e. The maximum cable length between a digital I/O module and digital I/O devices is 100 m.
f. Each installation rack may contain up to 6 I/O modules.

The automation and control system of the road tunnel will have a central selector switch for selecting the manual or automatic operation of the system. All handling and indicating devices must be installed at one of the two ends of the tunnel (on a control panel) independently of the PLC location. All digital inputs and outputs have the same nominal operating voltage of 24 V DC.

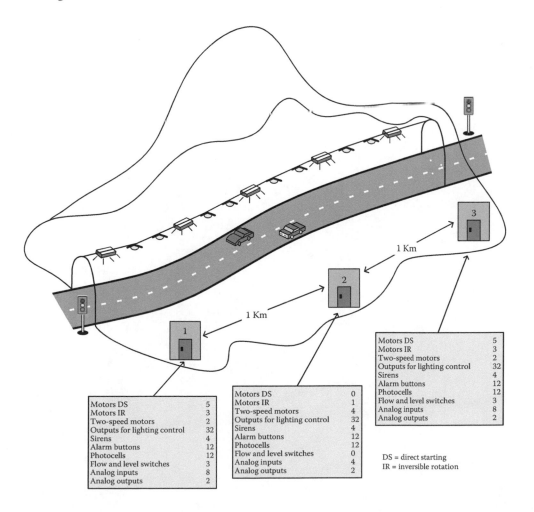

Panel 1	
Motors DS	5
Motors IR	3
Two-speed motors	2
Outputs for lighting control	32
Sirens	4
Alarm buttons	12
Photocells	12
Flow and level switches	3
Analog inputs	8
Analog outputs	2

Panel 2	
Motors DS	0
Motors IR	1
Two-speed motors	4
Outputs for lighting control	32
Sirens	4
Alarm buttons	12
Photocells	12
Flow and level switches	0
Analog inputs	4
Analog outputs	2

Panel 3	
Motors DS	5
Motors IR	3
Two-speed motors	2
Outputs for lighting control	32
Sirens	4
Alarm buttons	12
Photocells	12
Flow and level switches	3
Analog inputs	8
Analog outputs	2

DS = direct starting
IR = inversible rotation

Chapter 7

Basic Programming Principles of PLCs

7.1 Introduction to Programming of PLCs

Since a PLC is basically a digital device that monitors, processes, and generates data from an industrial process, it requires a software program with instructions to perform the desired functions. This program has two separate parts with two different objectives. The first part is the operating system that has been described in Chapter 6, which is responsible for all internal functions of the PLC, and is developed by the manufacturer of the PLC and cannot be modified in any way by the user. The second part of the PLC's program is related to the logical and other functions that the user desires to execute and needs to be written in a language that the PLC understands. This second type of programming will be the focus of Chapter 7 and will include the existing programming languages, basic issues on the standardization of languages and programming norms, individual language instructions, and corresponding applications.

The programming of PLCs has evolved through various stages in parallel to the technological (hardware) evolution of these devices. Initially, PLC programming included only digital input and output instructions that performed only logical functions corresponding to the case of conventional electromechanical relays. The corresponding language, known as relay ladder logic (RLL), was a mixed form language that combined graphic symbols and short-syntax alphanumeric strings. The name "ladder" was derived from the fact that a vertical line expressing the "beginning of the logic" (or, metaphorically, the high voltage potential) is placed on the left, and the line with the "end of the logic" (or, metaphorically, the low voltage potential) is placed on the right. Between these two vertical lines, simple logical elements (or combinations of them) form a branch, and are called "logical contacts" or "relay logic" as a whole, as shown in Figure 7.1. Between the two vertical lines, there is a number of such branches in the form of a ladder. Subsequently, the evolution in the field of PLCs was massive. Within a few years, PLCs included functions and operations of timers and counters by introducing corresponding instructions in the ladder language. They also introduced the ability to carry out four basic mathematical operations, while nowadays PLCs have reached a point that their programs support all mathematical functions, from absolute variable value up to the exponential, logarithmic, and trigonometric functions, for example.

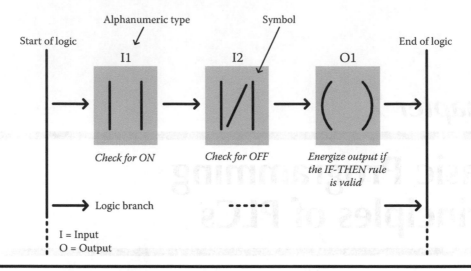

Figure 7.1   The generic form of Ladder logic.

However, this evolution of timers, counters, and basic arithmetic operations was not enough. The world of mechanical automation systems demanded more and more computing power, memory, and specific functionalities. It was also looking for the possibility of introducing and using subroutines, so that the program had a structure in harmony with the distributed nature of large automation systems. With these conditions, PLC manufacturers introduced other programming languages in a graphical or text form and at a higher level of programming, while retaining the ladder language up to now, mainly because of some key advantages, namely:

■ It is a symbolic or graphical language with a very simple way of representation that closely resembles the structure of classic automation circuits, and this is why it is not only familiar to all engineers dealing with industrial automation, but to technical electricians as well.
■ It has very short execution times for instructions, as well as short execution time of a program branch, so that all branches are considered running almost in parallel mode, just like in conventional automation circuits.
■ It provides the ability for online programming (thus allowing the user to make changes to the program while the system is running) and the program can be compiled in real time.

At this point it should also be highlighted that ladder language is not suitable for all kinds and sizes of industrial applications. If an engineer had to design a large automation system and write the required automation program starting from an empty page, then the ladder language would not be selected. For this case, the utilization of a flow chart and high-level programming language will be needed, such as the sequential function chart (SFC) language and the function block diagram (FBD) that will be described subsequently.

Until about 1990, PLC manufacturers were developing their proprietary programming languages with complete incompatibility between them. As an example, ladder languages from two different manufacturers may have had a similar form, but under no circumstances were they compatible. Furthermore, there were cases where two different PLCs of the same company were manufactured to use ladder language but with a different notation that did not allow the exchange of programs or the direct transfer of a program from one PLC to the other. Under these circumstances, the term "ladder diagram" expressed a set of similar but incompatible languages from all PLC manufacturers, rather than a common programming language for PLCs. On the other hand, this situation forced the PLC user to use a specific manufacturer since he needed to use specific tools for software development, learning, maintenance, and upgrading. In this unfortunate situation, the International Electrotechnical Commission (IEC) tried to impose standardization by defining the IEC 61131-3 standard, which is described in the next section.

7.2 The IEC 61131 Standard

The standards that are generally created by international associations and committees play an important role in the development of all industrial technologies. The standards generally help in achieving compatibility, transparency, and interoperability across different products; increase users' confidence in their products; and help in developing tools and methodologies under a common umbrella of technological specifications. Thus, the IEC came to address the chaotic situation in the incompatibility of programming languages for PLCs with the establishment of the open standard IEC 61131, the first part of which was issued in 1992 and referred to PLCs in general. The third part of this standard, the so-called IEC 61131-3, was issued in 2003 (second edition) and referred to the programming languages of PLCs. Nowadays, all major PLC manufacturers have accepted this standard, and their products are developed accordingly. It must be noted that the IEC 61131-3 standard, by definition, is not an additional programming language; instead, it supports modern software engineering methods in order to lead PLC operation in improved languages and programs, creation of usable and interoperable code, and in easy debugging.

The IEC 61131-3 standard incorporates five different approaches to programming control systems, and especially for the case of PLCs. Each approach is based on a simple language that targets a particular type of automation problem or application. These languages can be easily combined to generate a mixed code, or used in other types of control systems than PLCs, such as smart sensors, programmable drivers, process controllers, SCADA software, and so on. The main objective of the standard is to normalize the existing programming languages, rather than to prevent the possible development of new languages. Every PLC manufacturer is free to develop either extensions of a language that complies with the standard, or a new language from scratch if it is deemed necessary. On the other hand, the IEC 61131-3 standard has introduced function blocks that can be programmed in a non-standard language, such as C^{++}. The five embedded languages for the PLCs are the following ones:

1. Ladder Diagram, LD, or LAD—Graphic Language.

    A program written in a Ladder Diagram language is similar to the classic wired automation circuits. It uses logical symbols such as:

    $\dashv\vdash$ X = Open logical contact if the variable X is in a logic "0" (e.g., digital input, output, logic coil, etc.)

    $\dashv/\vdash$ X = Closed logical contact if the variable X is in a logic "0".

    $\dashv(\ )\vdash$ = Internal logic coil or digital output at a logic of "0" or "1" if the logic just before in the branch (see Figure 7.1) is "0" or "1", respectively.

    The horizontal branches between the two vertical lines at the beginning of the logic and the end of the logic, presented in Figure 7.1, include logical elements such as the above or combinations of them which are called "rungs" or "networks" which make up the "ladder".

Each branch ends on the right side in a single logic coil or output. Some manufacturers allow the parallelism of two or more logical coils or outputs in the same branch. In addition to the above simple elements, the LAD language has block symbols ⊒f⊢ for all the basic functions that are needed in an automation system, such as timers, counters, arithmetic operations, etc.

2. Function Block Diagram (FBD)—Graphic Language.

Because there is not a generally established translation of the FBD language, it is possible to refer to this with the terms "logic components language", "logic diagram language", "logic gates language", and other similar terms. Literally, the function block diagram language contains symbol blocks of functional operations, which implement various functions from the simple AND function from Boolean algebra up to PID control. Every block has the name of the function that it implements and accepts on the left side the inputs and on the right side the output, which carries the result of the function as presented in Figure 7.2. In fact, both the inputs and the outputs contain an additional set of assistive functions, for example the output ENO of the block ADD_I for adding integers, that inform us if the result is out of limits.

The input and output variables of the PLC are connected to the inputs and outputs of the functional blocks respectively by interconnection lines or are simply declared before them. The output of a block can be connected to the input of another functional block. Any interconnection line is directed in the sense that data are transferred from left to right, while both ends of the interconnection line should be of the same type. This means that the output of the logical block AND in Figure 7.2, which is a discrete variable (0 or 1), cannot be connected to the input 1 of the block ADD_I, which should be an integer variable.

3. Instruction List (IL)—Text-Based Language.

The instruction list is a low-level language that looks like the assembly language, but it is in a higher level than the LAD and FBD, in the case that the classification criterion is the translation of a program from one language to another. Due to the fact that the programs from other languages can always be translated into the corresponding IL language (the opposite is not true), it is considered as the main PLC language. IL is also known as the statement list (STL) or Boolean language. Because the most basic IL commands are just the basic functions of Boolean algebra (such as AND, OR, AND NOT, OR NOT, etc.) abbreviated, the last name will be adopted subsequently as the more descriptive one.

An instruction in the IL language consists of two parts: the operation of the instruction and the name of the operand for which the function will be implemented, as shown in Figure 7.3. The name of the utilized variable is directly related to the addressing system adopted by the PLC manufacturer referred to in Section 6.8.3. At the application level, both parts of the instruction are abbreviated as far as possible, as presented in Figure 7.3. The programming

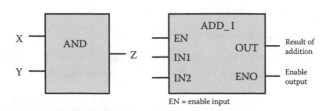

Figure 7.2 Programming elements of FBD language.

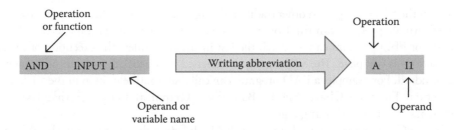

Figure 7.3 Indicative form of an instruction in IL language.

instructions are written one after the other as a list, hence the language received the IL name from IEC 61131-3. An example of a small Boolean program is as follows:

```
LD    I0.0
A     I0.1
=     Q0.0
LD    I0.4
O     I0.5
=     Q0.1
```

4. Structured Text (ST)—Language Based on Text.

ST is a high-level language whose syntax resembles that of Pascal language. For programmers that are already familiar with programming in a high-level language, such as Basic or C, it will be very easy to program in the ST language. In general ST supports complex and embedded instructions such as:
 – Repeat-until, while-do
 – Execution under conditions (IF-THEN-ELSE, case)
 – Functions (SQRT (•), SIN (•))

Any of these programs starts with the declaration of the variables and then follows the main text of instructions, separated by a semicolon. Variables can be clearly defined as numeric values (e.g., integer, real, etc.), internally stored variables, or inputs and outputs. Empty spaces can divide the instructions from the variables, and it is allowed to parse and insert comments to increase the readability of the program. An example of the ST language is:

```
FRD(bcd_input, delay_time);
t.PRE:=delay_time;
IF (test_input) THEN
        t.EnableTimer:=1;
ELSE
        t.EnableTimer:=0;
END_IF;
TONR(t);
Set:=t.DN;
```

This program reads the "bcd_input" in a form of a BCD code format and is used to change the delay time of the TON timer. When the "test_input" input is activated, the timer starts counting and after a specific amount of time the "set" is activated. An important difference between the ST language and other traditional languages is the way the flow of a program is controlled. An ST program can run from start to the end many times within a second. Conversely, a LAD or FBD program cannot reach the last instruction unless the execution of the previous instructions is complete. The ST language is designed to consociate with the other languages of the standard. For example, a LAD program can call a subroutine written in the ST language.

5. Sequential Function Chart (SFC)—Basically a Graphical Language that Uses Program Segments Written in IL Language.

   The SFC language is based on the GRAFCET (graphical function chart), which is a software tool of French origin and is very similar to Petri networks. The GRAFCET language (IEC 848 standard) was established in 1977 by a French working group to provide a graphical tool for describing and defining sequential industrial processes that are controlled by logical controllers. The basic SFC programming components (symbols) are:

   1. Steps or system operating states associated with actuator actions `STEP`—`ACTION`

   2. Transitions related to logical conditions of system variables such as the condition "IF t = 30° C" or the wait condition "WHEN TIMER IS ON" ———

   3. Directional interconnections between steps and transitions.

   Figure 7.4 presents a general example of an SFC language program in order to understand its utilization.

   – In each branch only one step can be active.
   – When a transition is satisfied, which means that all the conditions for this transition are met, then the previous step is deactivated and the next one is forced to be executed.

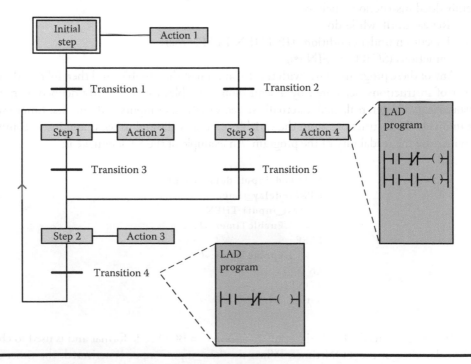

Figure 7.4 The form of a program in SFC language.

– When a step is active, it means that the automation program is in this state and that the corresponding action to the step is performed.
– Each transition and step represents or contains a whole (small or large, simple or complex) automation program written in one of the other languages (e.g., LAD or Boolean).

Since the SFC language, in addition to the previous basic elements, contains elements such as the "OR branch" and "simultaneous steps branch", it is suitable for applications with repetitive industrial processing cycles. With these elements of SFC, someone can organize and synthesize large automation programs that include sequential and/or parallel (i.e., simultaneous) control sub-processes. Figure 7.5 presents an SFC program for a two-door security system where one door requires a three-digit security code and the other one requires a two-digit security code to open.

The IEC 61131-3 standard, with the five programming languages described above, helped to clear the existing differences in the PLC manufacturers' software and thus helped to achieve an overall standardization in programming. However, there are still necessary steps to be realized for reaching full compatibility, especially for transferring programs from one PLC to another from a different manufacturer. From the five previously described languages, the main focus of Chapter 7 will be LAD, FBD, and IL, since mastering these three will make it very easy to cope with the scheduling needs in SFC language, where its graphical programming relies on the three

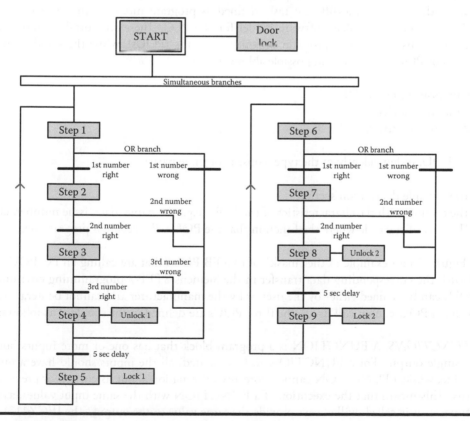

Figure 7.5 The SFC program for a two-door security system or a two-mechanism security system in the same door.

previous ones. In the following, some general remarks on these languages will be presented from a user's experience point of view:

- The LAD language is easy to be understand by all and is preferred by engineers for some simple applications.
- Engineers who have some preference in digital logic design, digital electronics, or digital circuit design often choose FBD language.
- For those engineers that do not have the previous experience, it is difficult to choose FBD language.
- A large majority of engineers choose IL language.
- When the application is complex or contains enumeration, comparisons, and time counting functions, then the vast majority of engineers do not choose LAD language.

7.3 Structural Programming

The IEC 61131-3 standard, in addition to the programming languages that have been presented, defines the concepts and components of structural programming further, which replaces the unusable and simple listing of instructions within a single unified program. The term structural programming means that the automation program is characterized by components that are different from each other and with a different task, defined as program modules. In this way, the IEC 61131-3 introduces a complete software model for the PLC. These structural components are referred to in this standard as program organization units (POUs), while the standard defines three types of POUs as (name and possible abbreviation):

- Functions F, FC, FUN
- Function Blocks FB
- Programs P, PB, PROG, OB

Each POU, independently of the type, consists of three parts:

- The variable declaration section
- The section with the characteristics of the POU (e.g., the name, date, issue number, etc.)
- The part of the code where the function that the POU will execute is programmed

In Figure 7.6 an example of the "mask" of two FB POUs that are calling an FC POU is presented with the corresponding data transfer in the Siemens STEP-7 programming environment. The POUs can be defined either by the user or by the manufacturer and cannot be iterative. This means that a POU cannot call itself. The three POUs are defined in the standard as follows:

- *FUNCTIONS.* A FUNCTION is a program block that has one or more inputs but only a single output. For a FUNCTION to be executed, all the inputs should have a numeric value, while a FUNCTION cannot store any information about its current or previous status. This means that the execution of a FUNCTION with the same input values, as many times as it is called, will always provide the same value to the output. The IEC 61131-3 has set the standard FUNCTIONS that are supplied by the manufacturer of the PLC and the

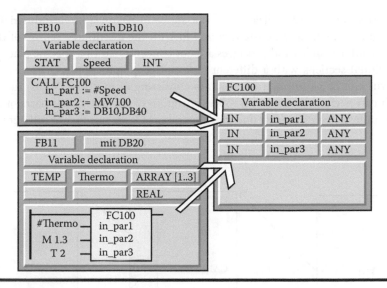

Figure 7.6   Two different POUs (FB10 and FB11) call the same FC100.

user-defined FUNCTIONS. For example, typical FUNCTIONS are AND, OR, ADD, ABS, SQRT, SIN, etc., some of which are extensible, which means that the user can decide on the number of the FUNCTION's inputs. The user-defined FUNCTIONS, once defined, can be called as many times in the program, while they can be programmed by the user in any of the standard languages.

■ *FUNCTION BLOCKS (FB).* A FUNCTION BLOCK is a program block that has one or more inputs and one or more outputs. In addition, FB can store information about its status in contrast to the FUNCTION. This means that the output values depend not only on the input values but also on the stored information. To achieve this, each FB is directly linked to a data block (DB) where the values are stored, while to execute an FB, all inputs should have a numeric value.

FBs have an operation that is related more to the integrated circuits (ICs) that implement a specific function. FBs contain data, as well an algorithm, and therefore can have a memory. Also, they have a well-defined interface and an invisible content, such as an IC has, which allows for the introduction of programming levels of different priorities. A temperature control law or a PID controller is a representative example of a user-defined FB. Once defined, it can be used as many times as needed in either the same program, different programs, or even in different projects, which is a feature that makes FBs absolutely reusable.

FBs, as in the case of FUNCTIONS, may be standard FBs developed by the manufacturer of the PLC or by the user. Counters, timers, and flip-flops are key examples of standard FBs. FBs can be programmed in any language from the IEC standard and, in many cases, in C language also.

■ *PROGRAMS.* PROGRAMS are program blocks that are constructed by POUs, such as FBs and FCs, as well as by simple instructions. They do not have inputs and outputs, all the utilized variables should be universal, and are programmed in all the standard languages. PROGRAMS can only exist within the PLC and not within an FB or an FC.

With the introduction and utilization of the above POUs in a large industrial system, the automation program can get a structure that will fit with the distributed nature of the industrial processes. In general, the industrial processes are not concentrated in a specific place; rather, they consist of individual sections with a different task and geographical placement each. Figure 7.7 presents the structure of an automation program consisting of multiple POUs, such as OB, FB, and FC, which can perfectly correspond to the operation of the individual parts of an industrial process. The complexity of a program does not depend on a large number of different FBs, but the

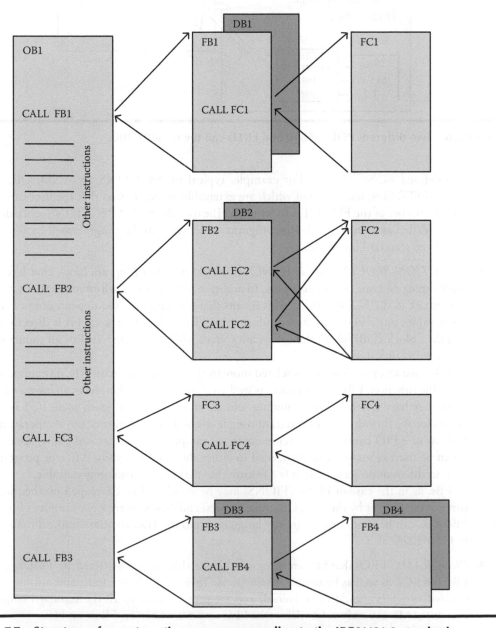

Figure 7.7 Structure of an automation program according to the IEC61131-3 standard.

reuse of the same FBs. The reuse of ready-made FUNCTIONS and FBs, developed by users, significantly reduces the programming and debugging time required and therefore the corresponding development costs.

Regardless of the IEC 61131-3 standard and its influence on the PLC manufacturers, the software development for PLCs is following, with only a little delay, the software being developed for PCs. Therefore, those who are already familiar with programming languages such as C and C++ will not have a particular problem with the concept, definitions, and implementation of structural programming.

7.4 Basic Programming Instructions

It should be made clear from the beginning that what it is called "software for PLC programming" is a complex software environment with many general operations offered to the user, such as the downloading/uploading of an automation program to/from the PLC, online monitoring of the program execution, run-stop controlling of the PLC via the software environment, hardware selection and configuration, and many others, as well as writing an automation program. Under the term "programming of a PLC" two main tasks are involved:

- The configuration and the general setup of the PLC, which includes the establishment of communication between programmer device and PLC
- The programming of the PLC functional logic operation

In many cases, engineers are only focused on the second part of pure programming, while ignoring the overall configuration of the PLC, a factor that is a key issue to the success of the overall automation. Most PLC manufacturers are offering a joint software package for programming PLCs of different types and technical features (at least for a family of PLCs). Furthermore, as has been presented, many PLCs offer a series of hardware configurations (I/O modules, CPU, memory, etc.) that require careful selection. The automation program that is going to be loaded to the PLC is hardware agnostic, which means that the software is not aware about the kind of available hardware which cannot be detected automatically. Thus, for the proper execution of an automation program, it should be stated as precisely as it can be for the exact type of available hardware. In particular, it should state the type of CPU, the kind and number of I/O modules, the type of communication processor if it exists. Also, it should be able to recognize or select the I/O addresses. This work is performed by the user within the user-friendly software environment of the PLC and is called "configuration of the PLC hardware". The remainder of this chapter will focus on the second task of basic programming of the automation logic included in the automation program.

The software environment of a PLC, regardless of the manufacturer, includes a large number of programming instructions; however, since the main objective is the comprehension of the way that PLC programming works, which does not extend to all possible PLC instructions, what follows will be a proper set of instructions from the three basic languages, namely Boolean, FBD, and LAD. With these instructions, listed in Table 7.1, the automation engineer will be able to handle most of the needs of industrial applications.

In the first three columns of Table 7.1, the syntax of each instruction is provided for the three programming languages: Boolean, FBD, and LAD, respectively. In the places where there is an empty cell, this indicates that the specific instruction does not exist in the corresponding

Table 7.1 **Programming Instructions in Three Basic Languages**

	Language				
Boole	*FBD*	*LAD*	*Operand*	*Instruction Operation*	
A	⊣A⊢	⊣ ⊢	I, Q, M, T, C	Look for 1 <u>and</u> perform AND with previous RLO	Boolean Logic Instructions
AN	⊣/A⊢	⊣/⊢	I, Q, M, T, C	Look for 0 <u>and</u> perform AND with previous RLO	
O	⊣o⊢	⊣ ⊢	I, Q, M, T, C	Look for 1 <u>and</u> perform OR with previous RLO	
ON	⊣/o⊢	⊣/⊢	I, Q, M, T, C	Look for 0 <u>and</u> perform OR with previous RLO	
A(				AND logic of a complex (bracketed) expression	
O(				OR logic of a complex (bracketed) expression	
)				Termination of bracketed expression	
=	⊣ ⊢	—( )—	Q, M	Assign value 1 to the operand if RLO = 1 Assign value 0 to the operand if RLO = 0	Activation Instructions
S	⊣S⊢	—(S)—	Q, M, C	Set, i.e., assign value 1 to the operand, if RLO = 1 Do nothing if RLO = 0 (Latch/Unlatch)[1]	
SX	⊣T⊢	—⊤T—	T	Activate timer T of kind X at positive going edge of RLO.	
R	⊣R⊢	—(R)—	Q, M, T, C	Reset, i.e., assign value 0 to the operand, if RLO = 1 Do nothing if RLO = 0	
L			IB, IW, QB, QW, MB, MW, Constant	Load the content of the operand or the value of constant in the accumulator, independently of RLO (Get/Put)[1]	Complemental Instructions

(Continued)

Table 7.1 (Continued) Programming instructions in three basic languages

Language			Operand	Instruction Operation	Complemental instructions
Boole	FBD	LAD			
T			IB, IW, QB, QW, MB, MW	Transfer the content of the accumulator to the operand independently of RLO	
CU	C_{xx} CU	C_{xx} CU	C	Increase the content of the counter C_{xx} by 1 on positive going edge of RLO	
CD	C_{xx} CD	C_{xx} CD	C	Decrease the content of the counter C_{xx} by 1 on positive going edge of RLO	
≥, ≤, ≠	CMP ≥1 IN1 IN2	CMP ≥1 IN1 IN2	Accumulator	Perform comparison between the contents of two accumulators	
JU	Label JMP	Label (JMP)	Label	Jump to another program instruction (instead of the next instruction) unconditionally	
JC	Label JMP	Label (JMP)	Label	Jump to another program instruction (instead of the next instruction) if RLO = 1	
CALL	DBxx FBxx	DBxx FBxx	FB, FC	Call a Function Block or a Function	

Notes:
(1) Alternative terms instead of Set/Reset and Load/Transfer respectively

IB = Input Byte	CU = Counter Up	FB = Function Block	
IW = Input Word	CD = Counter Down	FC = Function	
QB = Output Byte	JMP = Jump	DB = Data Block (Data File)	
QW = Output Word	JU = Jump Unconditionally	RLO = Result of Logic Operation	
MB = Memory Bit Byte	JC = Jump Conditionally		
MW = Memory Bit Word	CMP = Compare		

language. In the column of variables, the following notations are depicted with the corresponding explanations:

I	=	Input
Q	=	Output*
M	=	Auxiliary memory bit (logic coil)
T	=	Timer
C	=	Counter
FB	=	POU function block
FC	=	POU function
IB	=	Input byte
IW	=	Input word[†]
QB	=	Output byte
QW	=	Output word[†]
MB	=	Auxiliary memory byte
MW	=	Auxiliary memory word[†]
Label	=	Alphanumeric label[‡]

Each PLC contains at least two accumulators with widths of two digital words each, starting from the least significant byte on the right to the most significant byte on the left, as shown in Figure 7.8. The two words and the four bytes are marked with the initials L or H, representing "low" or "high" significance. The contents of the register 1 are modified by using the load instruction (abbreviation L). In the register, a byte, a word, or a double word can be loaded after the initial shift of the old content of register 1 to register 2, and the reset of register 1. Obviously, loading a numeric constant to the register creates an equivalent digital content of the constant in binary form, unless otherwise specified by the loading instruction (e.g., BCD code). The transfer instruction (abbreviation T) always transfers the contents of register 1 to where it is specified with the instruction.

In the column named "instruction operation", the action performed by the CPU is summarized when executing each instruction respectively. But before proceeding with the actions of the various programming instructions, it is necessary first to introduce and explain the term "result of instruction operation".

7.4.1 The Result of an Instruction Execution

The term "result of a logic operation" (RLO) is characterized by the logical result created in the CPU after an instruction execution. Every time that an instruction is executed, a new RLO is created that depends on the type of instruction and on the previous RLO, while the last one executed is erased afterwards. The concept of the RLO will be further explained with the help of the classic automation circuit presented in Figure 7.9. The activation of the relay C is dependent on the status of the three switching contacts S_1, S_2, and S_3. Initially, it is assumed that the relay is not energized, and the status of the circuit branch needs to be examined. In this case, it is assumed further that the status of the three switching contacts (their status is not known in advance) when testing the

---

* The normal letter for the output notation is "O", but since OR instruction in Boolean language uses the same letter, "Q" has been adopted in order to avoid any confusion.

[†] There are additional variables for double word (DW) with similar notation and handling that are omitted.

[‡] An alphanumeric word called "label" is written in front of an instruction to define the point of program transition after a jump instruction execution.

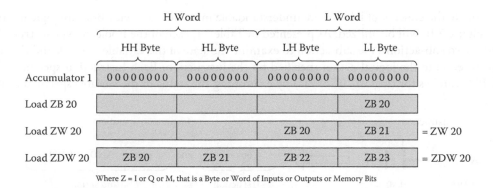

Figure 7.8   Content of accumulator 1 in bytes after some instruction execution.

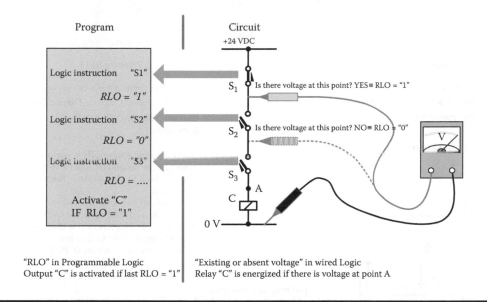

Figure 7.9   The RLO plays the role of existing or absent voltage metaphorically in programmable logic.

circuit is exactly as indicated in Figure 7.9. It is initially examined with a simple voltmeter if there is any electrical voltage at the power supply. If voltage exists, the next step is to examine if there is voltage just after the contact S_1. Since S_1 is a closed contact, the voltmeter will show us the trend. Continuing the examination of this circuit branch, the next point to be checked is the one immediately below the S_2 contact, as shown in Figure 7.9. Since the S_2 contact is open, the voltmeter will not show a voltage, and therefore it explains why the relay C is not energized. However, in the wired branch, the presence or absence of an electrical voltage, as well as the flow of the electric current, occurs naturally by itself and automatically as a consequence of the existence of a potential difference. In programmable logic, where the switching contacts are replaced by logical instructions, the role of the electrical voltage is carried out by the RLO. In the classic circuit, the relay C is activated if there is voltage at point A, i.e., in its ends as depicted in Figure 7.9. In programmable logic, the RLO is what determines—as a voltage—the result of the activating instructions. In this case, the output C (and hence the C relay connected to it) is activated only when RLO equals 1.

To make the concept of RLO more understandable in connection with Boolean logic instructions, Figure 7.10 will be utilized. As presented in Table 7.1, each of the Boolean logic instructions includes two sub-actions: the sub-action to examine the state of the variable (0 or 1), and the sub-action to execute the logical action specified by the instruction. Each instruction specifies what the CPU searches for (logical "0" or "1") by examining the status of the variable specified in this

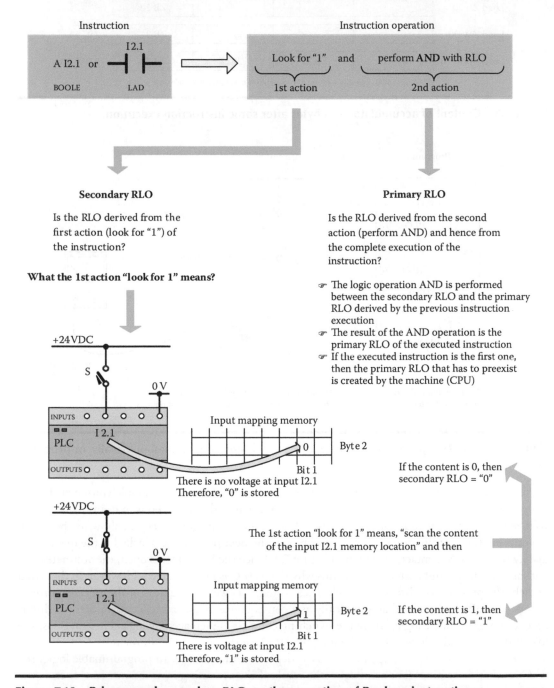

Figure 7.10 **Primary and secondary RLO on the execution of Boolean instructions.**

instruction. When the CPU finds what it is looking for at the corresponding memory location of the variable, then the returning search result is a logical "1". If the CPU finds the opposite of what it is looking for, then the conclusion is a logical "0". This result of the first sub-action is called "secondary RLO" because it is an intermediate logical result that in itself has no value unless it is combined with the primary RLO. For example, the statement "the contact S_1 is closed" has no valuable information, unless it is combined with an additional statement, e.g., "there is voltage before contact S_1 (= primary RLO)", which means that "the contact S_1 is closed and a voltage exists", and thus the relay will be energized if no other switching contact exists. The second sub-action of a Boolean logic instruction includes the logical action that should be performed between the secondary RLO (the contact S_2 is open) and the primary RLO that has been generated by the execution of the previous instruction (there is voltage after the S_1 contact). The result of the logical operation is that the new primary RLO (no voltage after contact S_2), replaces the previous primary RLO, which is generally lost. This result is kept temporarily only when there are parallel Boolean logical branches (OR). Therefore, the primary RLO is the RLO created after the complete execution of an instruction.

When the executed instruction is the first one in a series of instructions, (in the case of LAD language) or in an instructions' list (in the case of Boolean language), then because there is no RLO extracted by the above logical process, the CPU itself produces an RLO. This case is like stating that there is voltage at the +24-0 V DC power supply of the classic automation circuit. The first RLO that the CPU produces is such that it does not alter the Boolean logic of the first instruction, which means that it is a logical "1" if the first instruction is a logical AND, or logical "0" if the first instruction is a logical OR.

After executing an activation instruction, the primary RLO is erased permanently and begins the creation of a new RLO in the next branch of the program (for the case of LAD) or in the next instructions group (for the case of Boolean language). In the classic automation circuit of Figure 7.9, if the relay C is activated, it is not necessary to know that there is voltage at point A. This information is valueless because, in reality, it is inherent from the statement "the relay C is activated". The same applies if the relay is not activated because this is equivalent to the statement "there is no voltage at point A". In a similar approach, in programmable logic, the last primary RLO before the activation instruction does not need to be stored after executing the instruction, since the output C has been activated. The same is true also in the case where the last RLO is a logical "0", since then the output is also not activated as a final statement, and therefore it is not necessary to know that the RLO was "0" and thus it can be erased.

During the execution of the first sub-action of a Boolean logic instruction, the CPU searches for the state of the variable contained in the instruction at the corresponding memory location of the variable. If the variable is an input (I) or output (Q), then the CPU examines what is the content of the corresponding memory location in the I/O image table, in order to create the secondary RLO. Similarly, if the variable is an auxiliary bit (M), the CPU examines the storage memory of the auxiliary bits (logic coils), while for the case of timers (T) and counters (C), the CPU examines the corresponding memory dedicated for timers and counters.

7.4.2 Boolean Logic Instructions

The first instruction of the Boolean logic instructions group, depicted in Table 7.1, is the AND instruction. When the CPU executes this instruction, independently of the programming language,

$$\text{A I3.3} \quad , \quad \text{I3.3} \dashv\text{A}\rceil \quad , \quad \overset{\text{I3.3}}{\dashv\vdash}$$

it will examine whether a logical "1" is stored in the memory location of the PLC's input I3.3, which means, if it is true, then at the corresponding digital input I3.3 there is electrical voltage. Thus, in the case that the CPU finds a logical "1" then the secondary RLO (S.RLO) is "1". Subsequently, the AND function between S.RLO and the primary RLO (P.RLO) where P.RLO = "1" (this is generated by the CPU, since there is no other previous instruction), will yield RLO = "1". From the latter, it will also depend on the execution result of the instructions that follow.

The second instruction of the group is the AND NOT statement. When the CPU executes one of the instructions,

$$\text{AN I3.3} \quad , \quad \text{I3.3} \dashv\text{A} \quad , \quad \overset{\text{I3.3}}{\dashv/\vdash}$$

it examines whether there is a logical "0" in the memory location of the digital input I3.3, which means that there is no electrical voltage at the corresponding input I3.3 of the PLC. Therefore, if the CPU finds a logical "0", then the S.RLO is "1". Subsequently, the AND function between the S.RLO = "1" and the P.RLO = "1" (this is created by the CPU because it precedes no other instruction) will give RLO = "1". At this point, a basic differentiation of the programmable automation from the classical one should be highlighted. With the AND NOT instruction, it is possible to activate an output via an open input contact (when the contact closes, the output is deactivated), which cannot be implemented in conventional automation without the use of an auxiliary relay.

The third instruction in the same instruction group is the OR statement. When the CPU executes one of the following instructions,

$$\text{O I3.3} \quad , \quad \text{I3.3} \dashv\text{O} \quad , \quad \overset{\text{I3.3}}{\dashv\vdash}$$

it examines (as in the case of AND) whether there is logical "1" in the memory location of digital input I3.3. If this is the case then the S.RLO is "1" and then performs the OR function between the S.RLO = "1" and the P.RLO = "0" (which is created by the CPU, since no other instruction is preceded). The final result is RLO = "1" that waits for the execution of the next OR instruction (for Boolean language) or the "parallel" instruction (for LAD).

The next OR NOT instruction has a similar function to the AND NOT, with the difference that instead of the logical AND action, the OR operation is performed. The following instructions for creating complex expressions are only found in the Boolean language, and have a very simple function that will be presented through examples in Section 7.5.

7.4.3 Activation Instructions

The activation instructions mainly trigger a logical "1" to a specific variable included in the executed instruction in an analogous way. In the conventional automation circuit of Figure 7.9, the relay C is activated when there is voltage at its ends. If the final RLO before the activation instruction is logical "1", then the variable specified in the instruction is activated. All the activation instructions depend on the RLO and when this is logical "1", they cause activation of the variables based on the way that each instruction specifies.

The simplest activation command is the instruction "=" that it is based on the logic "activate while RLO is valid". That is, for as long as RLO = "1", the corresponding variable (output or auxiliary bit) contained in the instruction is activated. Once the RLO becomes a logical "0", the variable is deactivated. This logic is repeated continuously, as long as the operation of the PLC remains. The switching of RLO from "0" to "1" and from "1" to "0", and hence the activation/deactivation alternation of the variable cannot exceed the cycle scanning frequency of the PLC valid for the specific executed program.

The next activation command is the set (S) instruction that is based on the logic "activate the variable permanently". In particular, if the RLO = "1", the corresponding variable (output or auxiliary bit) contained in the instruction is activated, and it remains like this even if the RLO becomes a logic "0". For the variable counter C, the activation with the set instruction is achieved with the same conditions and aims to put the counter to the value previously loaded in the accumulator. This means that the execution of the instruction "S C7" will result in C7 = "value of the register content".

Given the above functionality of the set instruction, it is obvious that once a variable is activated, for example an output via the S instruction, there is no way to deactivate it as long as the PLC operates. For this reason, the reset (R) instruction, which has the inverse logic from the set one, has been introduced. In particular, if RLO = "1", the corresponding variable (output or auxiliary bit) is deactivated (logical "0") and remains off even if the RLO becomes a logical "0". As in the previous case, the reset instruction for a counter cancels the current content of the counter, giving to it a zero value.

Overall, it could be concluded that the pair of instructions S and R is the digital implementation of the electromechanical latch relay. It is also important to remember that the latch relay in this case has two coils, one for relay activation and one for deactivation. If the latch relay is energized and, subsequently the voltage from the normal coil is released, the relay does not switch off as a common relay would do, but remains continuously on. To deactivate it, an electrical voltage should be applied to the second coil.

Generally, with the set instruction, a timer is triggered and starts counting the time we have previously loaded into the accumulator, while it provides an output signal according to the implemented time function. The set instruction for the timers is treated separately for the following two reasons. First, each timer can implement various time functions (on-delay, off-delay, pulse, etc.). Therefore, together with the set trigger instruction of the timer, the type of time function that the PLC will implement should also be defined. This functionality is performed with a second code letter "X", which is marked in Table 7.1. In the position of X, a letter that uniquely determines the type of the specific time function is assigned. For this specific issue, there is no standardization among the PLC manufacturers. For example, in some small PLCs, the manufacturer states that the T1-T20 timers are ON-delay type, while the T21-T40 are the OFF-delay type and so on. Subsequently, the Siemens timing system and notations will be adopted in order to be able to program some simple applications.

Secondly, the set instruction for a timer cannot work as a simple set instruction for outputs and auxiliary bits. The reason is that if the RLO remains at a logical "1", it cannot continuously set the timer—something that is not problematic in the case of an output or an auxiliary bit—because it means that the time will be continuously updated and the timer will never count for the desired specific time. In order to overcome this problem, manufacturers have introduced the functionality of "RLO pulse rising" activation. This means that a timer is activated only when the RLO changes from a logical "0" to a logical "1" (pulse rise). If the RLO remains constant at a logic "1", the timer operation is not affected. To recount the time in a timer, the RLO should be changed from a logical "1" to a logic "0", and then again from a logical "0" to a logic "1", to create the required pulse rise.

The reset instruction applied on a timer causes its time zeroing if RLO = "1", without the need to have a pulse rising RLO. The reactivation of the timer, after the reset, requires a new pulse rise of the RLO that will result in a new execution of the "SX" instruction.

7.4.4 Complementary Instructions

Load (L) and transfer (T) instructions. With the L and T instructions, we can move various memory data of the PLC massively from one location to another. The data may be related to inputs, outputs, auxiliary bits, or other accessible memory locations. The density of the data movement can be a byte, a word, or a double word. The L and T instructions do not depend on the RLO, and thus they are executed continuously in each scan cycle. The instruction L can also refer to a constant numeric value while it transfers its defined content always to accumulator 1. From the other side, the T instruction always transfers the contents of accumulator 1 to the specified variable-memory location. The usefulness and utilization of these instructions will be illustrated in corresponding examples.

Counting instructions. The CU and CD counting instructions only concern the operation of the counters. When executing the CU and CD instructions, the content of the counter is increased or decreased by one, only if the RLO pulse rises. Increasing or decreasing the content of a counter is associated with the RLO pulse rise for a very simple reason, which will be considered in the following example. Let us assume that a conveyor belt transports bottles in a production process, while a photocell is placed on the conveyor belt for counting the bottles. The photocell is a digital input to the PLC, while there is a running program whose CU instruction refers to the input of the photocell. Whenever a bottle is passed in front of the photocell, the content of the counter is increased by one. Suppose now that for some reason the conveyor belt stops temporarily, and there is a bottle in front of the photocell that keeps it activated. For as long as the bottle remains in front of the photocell, the program will be executed many times. If there was no pulse raise condition in the CU instruction, then the PLC would list a lot of virtual bottles from the encountered stop situation, while in reality, there is only one (stopped) bottle in front of the photocell. Instead, with the RLO pulse raise condition, the stopped bottle should continue its movement for some time, turn off the photocell, and then be switched on again by the next bottle, to count another passing bottle.

In Table 7.1, the counting instruction symbols CU or CD, in the case of FBD and LAD programming languages, are simplified. Figure 7.11 presents the actual graphical form of the CU symbol or block in the LAD language (Step7—Siemens), where it accepts six logical connections or definitions. The input to the CU point of the block is the counting input. This input can be either connected or denoted to the counter C_{xx} where xx is the number of the counter. In the inputs of S and R, proper input signals will be declared or connected for performing the counter's set and reset instructions. However, as mentioned above with the set instruction, the counter is placed at a specific numerical value, and the latter should be declared somewhere in the program or in a memory position. This operation can be achieved with the proper connection to the PV input. To the CV output, the current value of the counter is provided by the

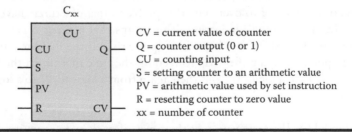

CV = current value of counter
Q = counter output (0 or 1)
CU = counting input
S = setting counter to an arithmetic value
PV = arithmetic value used by set instruction
R = resetting counter to zero value
xx = number of counter

Figure 7.11 Graphical symbol of instruction CU in LAD language.

CPU. For example, if we have a light electronic display for listing the current counting of the passed bottles, the required display information will be obtained from the CV output. Finally, the counter has a digital "0" or "1" output. This output is at logic "1" if the content of $C_{xx} \neq 0$. When the content is reset, the Q output of C_{xx} becomes a logic "0". It is possible to encounter the opposite logic where for $C_{xx} \neq 0$, Q = "0". This output can be used as a contact-based logic instruction anywhere in the LAD program with the form $\dashv \overset{C_{xx}}{\vdash}$. This instruction will provide RLO = "1" when $C_{xx} \neq 0$ according to the first definition. The same is valid for the corresponding instruction A Cxx in Boolean language, while a similar graphic symbol and features can also be found in FBD language.

Comparison instructions. It is possible, via comparison instructions, to compare two numbers in all their basic formats, such as integers, real, etc. The comparison of any kind ($\leq$, $\geq$, $>$, $<$, =, etc.) is always performed between the two contents existing in the first two accumulators always in the same order, e.g., content of register 1 > content of register 2. The comparison instructions are always used in conjunction with the load instruction. With two consecutive loading instructions, L N1 and L N2 (Boolean language), the two numbers are entered into the two registers and can be compared if a comparison instruction is followed. When the comparison is satisfied, the CPU generates an RLO = "1". In Figure 7.12, the graphical symbol for the comparison of two integer numbers in the LAD language (Step7—Siemens) is presented.

The input of the comparator is a discrete variable ("0" or "1") and enables the comparison when it is at logical "1". The output of the comparator generates a logical "1" when the comparison is satisfied, provided that the input is enabled. At the inputs IN1 and IN2 of the comparator, the two numbers to be compared are declared, which can be either variables (e.g., T, C, or Q) or constant values. In FBD language, a similar graphic symbol is utilized.

Program flow control instructions. These instructions, as their name suggests, help to change the execution flow of the instructions within the same program with or without a condition. With the JU label instruction (in Boolean language), the program flow switches to the instruction characterized by this label tag, bypassing the intermediate instructions, unconditionally and independently of the RLO. Conversely, with the instruction JC label, the transition occurs only when a condition is satisfied and thus RLO = "1". In LAD language, the jump instruction $\dashv\!\!-\!\!\overset{\text{Label}}{(\text{JMP})}$ is always executed, while the instruction $\dashv I \vdash\!\!-\!\!\overset{\text{Label}}{(\text{JMP})}$ is executed only when the digital input I3.3 is energized. Similar graphical symbols for the two above instructions are also found in FBD language.

When the CALL instruction (in Boolean language) is called at the point of program where the instruction is executed, another program block or POU according to the IEC 61131-3 standard terminology is in order to be executed. The POU can be an FB or an FC, in which it is possible to program a separate automation function that can either be repeated several times (subroutine) or

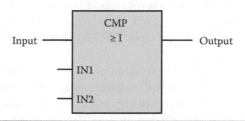

Figure 7.12   Graphical symbol of the comparison instruction (≥) of integers in LAD language.

to take place occasionally when a condition occurs. Some examples of call instructions for a POU are the following ones:

- CALL FC 10
- CALL FB25, DB1
- CALL FB25, DB2

After the FB or FC is called via the CALL instruction, the CPU leads the execution of the program to the next instruction after the CALL instruction. The calling of a POU is accompanied by the introduction of certain information and the declaration of variables, as described in Section 7.3. Since the analytical way to call a POU and the corresponding introduction of the required data depends on the manufacturer of the PLC and the specific software environment for programming, there will be no further analysis of this topic. In a similar way, the call of a POU in the FBD or LAD language is implemented by inserting the elements into a graphic block.

7.5 Programming According to the IEC 61131-3 Standard

In this section, detailed focus will be provided to all the instructions of Table 7.1, by presenting a specific way of how to utilize these instructions to program simple automation functions according to the IEC 61131-3 standard. These simple automation functions are small examples from industrial practice and, in any case, are not complete applications. However, large integrated applications consist of a large number of such simple and small automations and in many combinations, and thus it is important to know how to program all these simple automation functions rather than to know two or three large industrial applications. Some more integrated industrial applications of automation are presented in Chapter 10. In addition to the instructions of Table 7.1, some further instructions will also be presented that are relatively useful when the utilized PLC supports them.

7.5.1 General Highlights and Restrictions in PLC Programming

Parallel or serial execution of LAD instructions. For a program written in the LAD language, there are two ways to execute its instructions. The CPU can execute the instructions either serially in a row-by-row (rung) approach or parallel in a column-by-column approach, as presented in Figure 7.13. In serial execution, all the instructions are first executed in a row, and then the program proceeds to the next row. In parallel execution, the program executes the first instruction initially from all the rows (= column) of the program, and then it proceeds to execute the second instruction from all the rows and so on, while the activation instructions of the last column are executed at the end. In general, there has not been a standardized way to execute the instructions of a LAD program, while on the contrary, each manufacturer has followed the approach that is considered as most suitable for its digital system. For example, AEG Modicon follows parallel execution, while Allen-Bradley follows serial execution.

What matters for the parallel or serial execution of LAD instructions is whether the end result is the same. The answer to this question will be given with an example. Consider the LAD program with two rungs of instructions, shown in Figure 7.14. Suppose also that the digital input I0.0 is energized, then the execution of the instruction $\dashv\vdash^{I0.0}$ gives RLO = "1". Subsequently, we can follow the evolution and change of the RLO per executed instruction and sequentially per scan

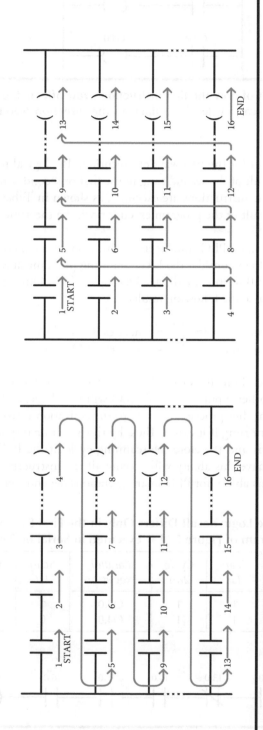

Figure 7.13 Serial (left) and parallel (right) execution of instructions in LAD language.

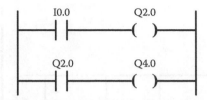

Figure 7.14   The logic result of how the first instruction's row affects the result of the second row in the first or second scan cycle, depending on the direction (serial or parallel) of the instruction's execution.

cycle (first, second, and so on) for the two ways of execution. The logical result of the first rung of instructions affects the result of the second rung in the first or second scan cycle depending on the way (serial or parallel) the instructions are executed, as shown in Table 7.2. Obviously, after the second scan cycle, the result of the program execution will be the same independently of the method of execution.

PLC's "intelligence" restrictions. PLC "intelligence" obviously refers to how "smart" the CPU is by means of its capabilities. Let's consider the LAD one-rung program shown in Figure 7.15 and suppose that all the inputs I1.0 to I6.0 are disabled (at logical "0"). During the execution of this program, the CPU will produce the following results:

The execution of the instruction I1.0 —|/|— generates RLO = "1".
The execution of the instruction I2.0 —| |— generates RLO = "0".

Subsequently, since the CPU realizes that the AND logic is no longer valid and thus independent of the state of the other inputs, the final RLO will be a logical "0". The CPU does not execute the other instructions, but proceeds directly to the activation instruction Q0.0 —( )—| and then in the next program rung if it exists. Thus, by this functionality it saves time from the non-execution of instructions and therefore the scanning cycle of the PLC becomes shorter. A PLC whose CPU does not have this ability will execute all the instructions of the rung even if there is no need to do so. This ability for PLC smart operations depends on the type of CPU and

Table 7.2   The Logic Result Differs Only in the First Scan Cycle if the LAD Program of Figure 7.14 Is Executed in Serial or Parallel

Serial Execution	Scan 1st	Cycle 2nd	Parallel Execution	Scan 1st	Cycle 2nd
Q2.0	1	1	Q2.0	1	1
Q4.0	1	1	Q4.0	0	1

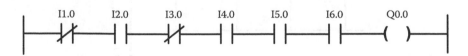

Figure 7.15   A LAD program rung, whose instructions are not required to all be executed if the CPU has the related capability.

the way that it is designed to operate within the PLC in combination with the other digital units, which is generally a matter for the manufacturer.

The next limitation in the intelligence of a PLC is the fact that the CPU can never activate a digital output that requires a "flow of the power in the reverse direction", which is the realization of logic from the right to the left that is in contrast to the philosophy of the ladder logic expressed in Figure 7.1. Let's consider the conventional automation circuit of Figure 7.16 that works perfectly without any problem. The relay F is activated if a simultaneous activation of one of the following combinations of switching contacts occurs:

ABC, ADK, EK, and EDBC

From these, the latter combination (EDBC) requires a "backward current flow", as it is called, which means the current flow is from the right to the left, which is something that does not create any problem in conventional automation and wired logic, and the circuit works correctly. However, if it is needed to implement the circuit of Figure 7.16 as a LAD program—although it is not feasible—then this implementation will most likely result in the hypothetical LAD program of Figure 7.17a. In this presented program and, in particular, in the switching element D, the CPU needs to apply a logic from the right to the left, which is not possible, based on the PLC's current capabilities. The term "not feasible" means that it is not possible to insert such a program, because once there is an attempt to insert a branch to the left, the software environment will indicate a

Figure 7.16 **A conventional automation circuit which can't be realized in a PLC with an identical logic.**

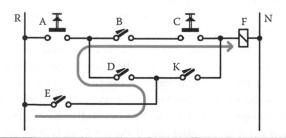

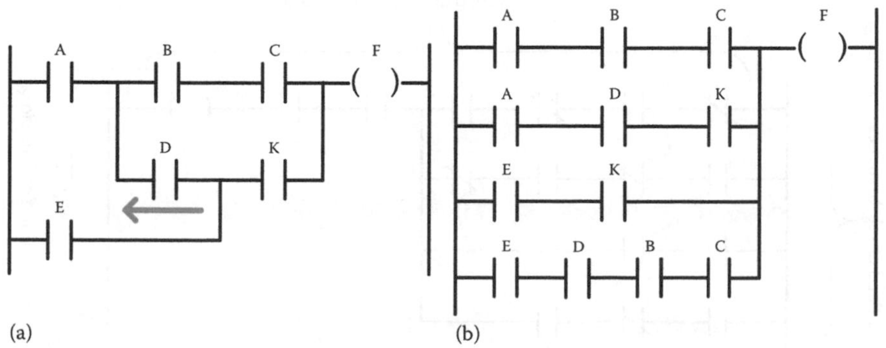

(a)   (b)

Figure 7.17 **Non-feasible logic in a LAD program (a) and the modification of the logic in order to have the same result (b).**

programming error. The solution to this problem is to look at what specifically the circuit of Figure 7.16 does, and to implement this functionality exactly without copying the structure and shape of the conventional circuit. This modification of the logic on the functionality is presented in Figure 7.17b and is an equivalent solution.

Number of LAD instructions per rung. Every manufacturer of a PLC introduces a limit to the number of LAD instructions that may exist in a program rung. Additionally, limits are being introduced on the total number of parallel sub-branches that can be present in the same program rung, as shown in Figure 7.18. Some typical limits are the 10 instructions and 7 sub-branches per program rung, as adopted by Westinghouse and Square D of the Schneider Electric group, 9 instructions and 8 sub-branches of General Electric, and 11 instructions and 7 sub-branches of Allen-Bradley. In each rung of a LAD program, some manufacturers allow only one digital output, such as the above companies, while others allow more than one output, such as Modicon, which allows up to 7 digital outputs per program rung. Restrictions are also present in the number of embedded loops in a program rung, a problem that can be solved by doubling the number of contacts' instructions. In the Siemens Step7© software initial version, there were 7 instructions per rung for the LAD and FBD languages, 2000 for the Boolean language, as well as 999 as a maximum number of branches and an almost unlimited number of sub-branches. In today's TIA portal software environment, there is no such limitation for real applications. In reality, the limitation on the number of branches is set by the size of the memory of the PLC, something that is valid for all PLC manufacturers.

The order of rungs. At this point, the issue of rung order in a LAD program will be examined, e.g., whether the order in which the rungs are inserted influences the final result of the program execution. It is obvious that in a conventional automation circuit, there is no such issue of the branches' order, since all branches are under the same electrical voltage and operate at the same time. Subsequently the LAD program of Figure 7.19a is considered, where it is assumed that the input I2.0 and the auxiliary bit M0.0 are initially deactivated and remain in this state without a change. The execution of the program will result in the permanent activation of the Q1.0 output and the simple activation of the auxiliary bit M0.0. The program of Figure 7.19b has the same rungs as in Figure 7.19a, but in the opposite order. The first (second) rung of the program of Figure 7.19a becomes the second (first) rung in the program of Figure 7.19a (7.19b). By examining the results from the execution of the program of Figure 7.19b, it can be found that, in this case, there is again a

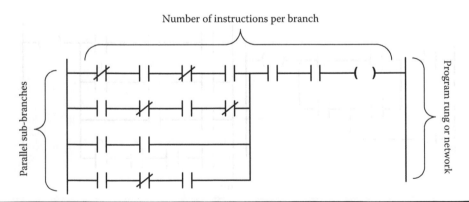

Figure 7.18 **There are limitations to the number of instructions per rung and to the number of sub-rungs.**

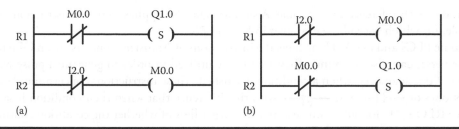

(a)   (b)

Figure 7.19 **(a) A two-rung program, whose result depends on the writing alignment of the rungs, while in (b) the rungs have an opposite order than in (a).**

simple activation of the auxiliary bit M0.0, but the output Q1.0 remains deactivated, as opposed to the result of the first program. It should be noted that the reset of output Q1.0 does not depend on I2.0 and M0.0, but on some other third variable, and hence it has been omitted from the programs of Figure 7.19. Therefore, it can be concluded that the order of rungs can directly affect the outcome of the program; however, since this is not the case in many problems, it does not mean that every time that the LAD language is utilized, special focus should be provided in the order of rungs. Instead, as a general concept, it is important to keep in mind that only in the cases of debugging the produced software may the order of rungs affect the automation program and be the source of unexpected results. For an automation program written in Boolean language, there are similar conclusions while, additionally, it should be considered that among similar activation instructions, the last one overrides the previous ones. For example, the following program is considered:

$$
\begin{aligned}
&\text{A} \quad &\text{I0.0} \\
&= \quad &\text{Q3.0} \\
&\vdots \\
&\text{A} \quad &\text{I2.0} \\
&= \quad &\text{Q3.0}
\end{aligned}
$$

where the output Q3.0 may be activated by input I0.0 and deactivated due to input I2.0. Thus, the final result will be Q3.0 = "0", because this is what is finally executed and therefore is saved last.

Additional instructions. Apart from the Boolean instructions OR and AND that have been included in Table 7.1, the instructions O (OR) and A (AND) also exist without reference to a variable, and therefore without the first sub-action of the instruction. This means that with the instruction "OR without a parameter", the CPU does not search in an area of the memory to find a desired value (state). For example, the instruction O without a variable is equivalent to creating a return path, as shown in Figure 7.20a. In Figure 7.20b, for comparison purposes, the OR

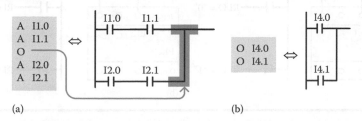

(a)   (b)

Figure 7.20 **The instruction OR without (a) or with (b) variable assignment.**

command is with reference to a variable. A similar logic also applies to the A instruction without a variable, for which we will see examples in Section 7.5.2.

In some PLCs and in LAD language, the instructions of "transient contacts" from OFF to ON, and vice versa, can also be found. They are mainly utilized in order to generate a pulse with the duration of one scan cycle when a condition is satisfied. These instructions concern the classical A and AN ones or ⊣ ⊢ and ⊣/⊢, with the difference that when their condition is satisfied, they give RLO = "1" for only a single scan cycle, regardless of whether the condition continues to be true. Figure 7.21 presents the programming symbols of two instructions in the case of LAD language. The instruction I0.0 ⊣↑⊢, if the input I0.0 is activated (OFF to ON), produces an RLO = "1" for one scan cycle. The instruction I0.0 ⊣↓⊢, if the input is disabled (ON to OFF), gives an RLO = "1" for one scan cycle. The above instructions are usually used in programming applications with sequential operations, mathematical operations, presetting of counters, etc.

Familiarity with the RLO. In Section 7.4.1 the concept of RLO has been described, as well as the process by which it is created by the CPU. This subsection will examine the specific RLO created by executing some instructions in the LAD and Boolean languages. Let's consider the PLC configuration in Figure 7.22a, where the input I0.1 is permanently disabled (switching contact S permanently open). Then the execution of the instructions ⊣ ⊢ and ⊣/⊢ will generate the RLOs that appear next to the instructions respectively. If the input I0.1 is activated and remains continuously activated (switching contact S is permanently closed) as shown in Figure 7.22b, then the exact same instructions will give different RLOs and, in particular, those that are displayed next to the instructions. The conclusion that can be drawn is that in programmable automation, if an input is activated or not (either closed contact or open contact), whatever RLO can be generated serves the logic of operation, provided that the appropriate AND or AND NOT instruction has been utilized.

The same conclusions and the same observations are valid if instead of the input we have as variable an auxiliary bit, as shown in Figure 7.23. The only difference is that instead of the closed

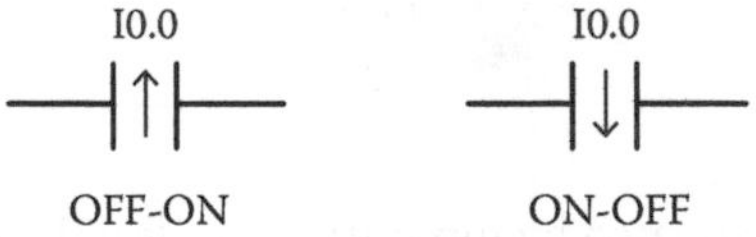

I0.0   I0.0

⊣↑⊢   ⊣↓⊢

OFF-ON   ON-OFF

Figure 7.21 Instructions in LAD language for the implementation of transitional contacts.

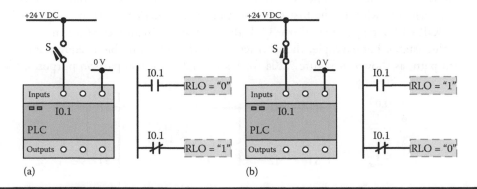

(a)   (b)

Figure 7.22 The RLO in relation to the non-energized (a) or energized (b) status of a digital input.

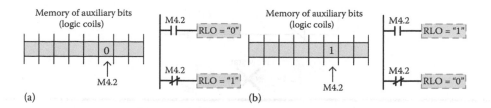

Figure 7.23   The RLO in relation to the status "0" (a) or "1" (b) of an auxiliary bit (logic coil).

or open input contact, the state "0" or "1" of the auxiliary bit is examined by the CPU in the corresponding memory location.

In a program with Boolean instructions, the creation of an RLO follows the same logical process. The only difference is the look of the programming where the RLO is created, because of a different language form. In Figure 7.24, a segment of a Boolean program is presented including for instructions A and AN among other instructions that precede or follow. Suppose that $P.RLO_i$ is the RLO created by executing all the preceding instructions. The execution of the instruction A I1.0 in conjunction with $P.RLO_i$ creates the new $P.RLO_{i+1}$, and the same procedure continues with the execution of the following instructions. The S.RLO in each instruction depends on the state of the input (whether or not it is activated) and each time affects the newly created P.RLO. A similar procedure is followed if, instead of the instructions A and AN, instructions O and ON or combinations of them are utilized.

Subsequently, let's consider the LAD program of one rung in Figure 7.25, where the execution of the program will give a different effect on each scan cycle. In particular, in odd scan cycles (first, third...) will give M1.0 = "1" and in even ones (second, fourth...) will give M1.0 = "0".

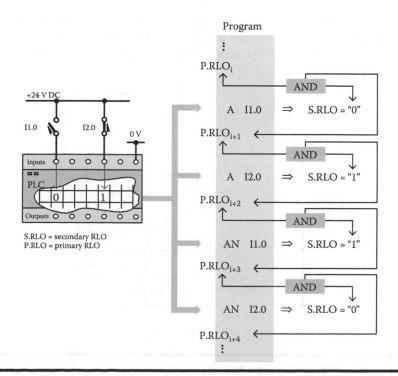

Figure 7.24   The procedure of RLO creation in a Boolean program.

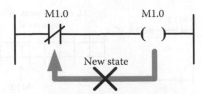

Figure 7.25 Program for RLO alternation per scan cycle.

This result is easily checked by looking at the generated RLO per scan cycle. With the utilization of this simple program, it is easy to explain why, in the case of PLCs, each program is executed only once in every scanning loop without feedback toward the instructions that have already been executed, and of the new states created in the activated or disabled variables. If such a situation did take place, then the program would lead the PLC to an endless loop of repetitions, from which it would never escape and therefore no control action would be applied.

The program of Figure 7.25 can also be used to measure the duration of a scan cycle when the PLC has no such diagnostic direct measurement. The auxiliary variable M1.0 has the behavior of a pulse train (1, 0, 1...) whose period has a two-cycle sweep time. The pulses of the variable M1.0 can be counted using a counter. By programming the counting from the counter of a large number of pulses of M1.0, and by combining the counter function with the measurement of corresponding time based on a timer, it is possible, with simple mathematical operations, to calculate the time duration of the PLC scan cycle for the specifically executed program.

7.5.2 Programming with Boolean Logic and Activation Instructions

Let's consider the classic automation circuit of Figure 7.26a, where the sensor S_1 controls the operation of the power relay C. It is preferable to operate a machine that is powered by a relay C when the NO switching contact of the sensors S_1 close. Before proceeding with the programming,

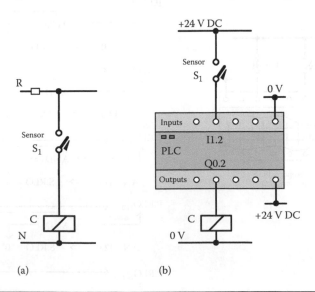

(a)

(b)

Figure 7.26 The relay C is energized when the contact S_1 closes. (a) Conventional circuit and (b) implementation of (a) with PLC.

it should be determined how the input and output devices are connected to the PLC, a step that should always be done initially regardless of the application. A PLC program has no meaning to be created, without or before the proper connection of the I/O devices and the corresponding definition of the addresses. These specifications, for the currently examined application, are presented in Figure 7.26b, where the NO switching contact S_1 is connected to the input I1.2 and the relay C to the output Q0.2. In this case, the required Boolean program is:

$$
\begin{array}{ll}
\text{A} & \text{I1.2} \\
= & \text{Q0.2}
\end{array}
$$

Subsequently, the classical automation circuit shown in Figure 7.27a will be considered, where the relay C is activated when the NC switching contact of sensor S_1 remains closed. For the implementation of the programmable automation and the connection of the I/O devices, as depicted in Figure 7.27b, the required Boolean program will be:

$$
\begin{array}{ll}
\text{A} & \text{I1.2} \\
= & \text{Q0.2}
\end{array}
$$

which has the same form as in the previous case, but with a different state of the switching contact. The explanation can be derived by examining the specifications of operation in each case and what they do in the PLC.

The automation circuit of Figure 7.28a permits a machine, powered by a relay C, to operate when the sensor contact S_1 is open (the machine stops working when the contact S_1 is closed). Such an operation is impossible to be implemented with a conventional automation circuit without the use of an auxiliary relay d. In contrast, in the programmable automation, this can easily be implemented by using the AN instruction. For the connection of the I/O devices and their addresses shown in Figure 7.28b, the required program in Boolean and LAD languages is the following:

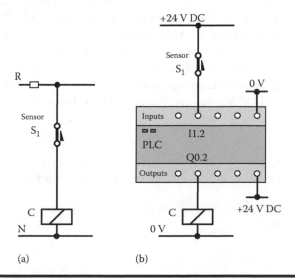

(a)　　　　　(b)

Figure 7.27　**The relay C is energized when the contact S_1 is closed. (a) Conventional circuit and (b) implementation of (a) with PLC.**

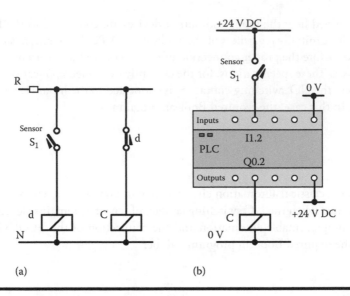

Figure 7.28 **The machine (C) operates when sensor contact S₁ is open. (a) Conventional circuit and (b) implementation of (a) with PLC.**

```
AN   I1.2
=    Q0.2
```

Let's now consider the classical automation of a START-STOP operation for a machine with a self-latch situation, as depicted in Figure 7.29. In this classical automation circuit, the STOP push button is typically a button with an NC contact, while the overall functionality cannot be implemented without the use of an auxiliary relay, as in the case of the previous programming example. Instead, in PLC programming, the STOP function can be implemented with both an NC contact button and an NO contact button. Thus, depending on the type of STOP button (NC or NO) that will be utilized, the program in the LAD language should be written accordingly in order to achieve the desired functionality. The two cases are shown in Figure 7.29, where for every way of connecting the I/O devices to the PLC, there is a corresponding program in the LAD language, while programming in the Boolean language is similar. The two programs in Boolean language for both the NC and NO STOP buttons are, respectively, the following:

```
A(
O    I1.4
O    Q0.2
)
A    I1.0
=    Q0.2
BE
```

```
A(
O    I1.4
O    Q0.2
)
AN   I1.0
=    Q0.2
BE
```

The same remarks are also valid for the START button, in the case that it is desired to be an NC contact button, where the need for developing a corresponding program is obvious.

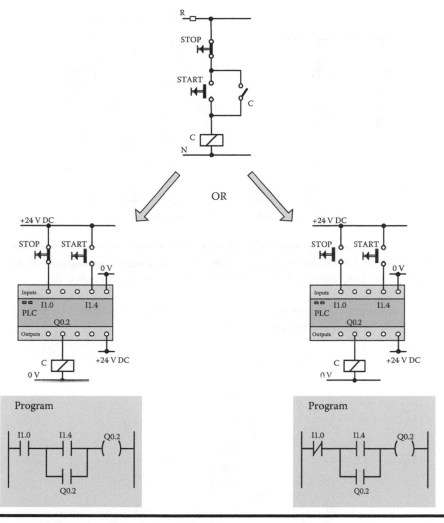

Figure 7.29 **The conventional START-STOP automation circuit for the LAD program for NC and NO STOP pushbutton.**

Subsequently, the automation circuit of Figure 7.30a will be considered, where the contained two branches have conflicting logic. This means that when the C_1 relay is activated, it is not possible to activate the C_2 and C_3 relays due to the sensor S_1. Thus, to activate the C_1 relay, the S_1 sensor should be activated, while in order to activate the C_2 and C_3 relays, the S_1 sensor should be deactivated, two conditions that cannot coexist. In particular, it is desired that the relay C_1 is activated when the switching contacts (in the corresponding branch) S_1 and S_3 are closed and the contact S_2 also remains closed. In the same way, the relays C_2 and C_3 need to be activated when the S_1 and S_3 contacts, in the second branch, remain closed and the contact S_2 is also closed. It is obvious that when the contact S_1 in the first branch is closed, then the corresponding contact S_1 in the second branch will open, since these are two different contacts of the same sensor (S_1). Figure 7.30b shows the connection of the I/O devices to the PLC in order to implement the circuit as a programmable automation. The indicated status of the switching contacts applies to the case of non-activated sensors. The C_2 and C_3 relays, connected in parallel to the conventional circuit, are

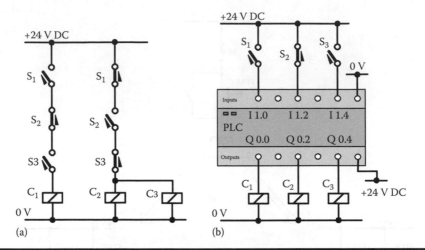

Figure 7.30 A conventional automation circuit (a) with a series of logic AND in both the branches and connection status of I/O devices for implementation in a PLC (b).

connected to two different outputs and their parallel operation is achieved by proper programming. The required Boolean program is as follows:

```
A     I1.0
A     I1.2
A     I1.4
=     Q0.0
AN    I1.0
AN    I1.2
AN    I1.4
=     Q0.2
=     Q0.4
BE
```

In Figure 7.31a a simple OR logic circuit is considered, including three-sensor switching contacts; S_1, S_2, and S_3. In this case, it is desired that the relay C_1 is activated when either the contact S_1 closes or the contact S_2 is closed or the contact S_3 closes. For a similar connection of the input devices to the PLC, as presented in Figure 7.31b, the required Boolean program is the following:

```
O     I1.0
O     I1.2
O     I1.4
=     Q0.2
BE
```

In the next programming example, the OR instruction, without a reference to a variable, will be utilized. This instruction has the same logic as the simple OR instruction, except that it does not refer to a specific variable, but to a complex logical expression from individual instructions that usually have an AND logic. In the circuit of Figure 7.32a, it is desired that the relay C_1 is activated

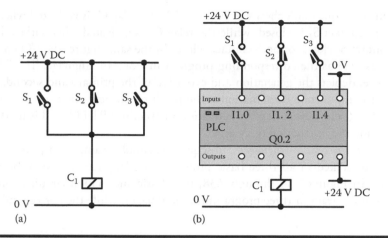

Figure 7.31 **A conventional automation circuit with parallel contacts (a) and connection status of I/O devices for implementation in a PLC (b).**

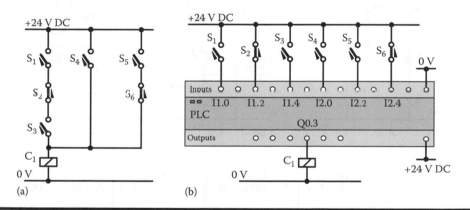

Figure 7.32 **A conventional automation circuit with parallel branches of seriously connected contacts (a) and the connection status of I/O devices for implementation in a PLC (b).**

when either the contacts S_1 and S_3 are closed and the contact S_2 remains closed; the contact S_4 is closed; or the contact S_5 is closed and the contact S_6 remains closed. For this example, the connection of the input devices is performed as presented in Figure 7.32b, while the required Boolean program is as follows:

A	I1.0		O		
A	I1.2		A	I2.2	
A	I1.4		A	I2.4	
O			=	Q0.3	
A	I2.0		BE		

After this first programming experience with some instructions expressing the Boolean logic and the simple activation instruction, the creation of the RLO in an entire small-sized program will be examined, since until now the RLO examination has been limited to individual instructions.

In the classic automation circuit shown in Figure 7.33, the relay C_1 is activated when the contact S_1 closes and the contact S_2 is closed, while the relay C_2 is activated when either the contact S_1 closes or the contact S_2 closes, or both contacts close. In the same figure, the I/O connections are also presented, as well as the corresponding program in Boolean language. The followed table, in the same figure, depicts the generation and evolution of the primary and secondary RLOs for all instances of the input state combinations. The result of the activation instructions obviously depends on the last (before executing the activation instruction) P.RLO, and is marked as logical "0" or "1" alongside the instruction (gray area).

Programming in three basic languages. Subsequently, simple examples of programmable automations will be presented in all three basic programming languages, namely Boolean (or IL), LAD, and FBD. In Figures 7.34 through 7.38, the classic automation circuit is presented with the corresponding program in three programming languages, while the corresponding electrical

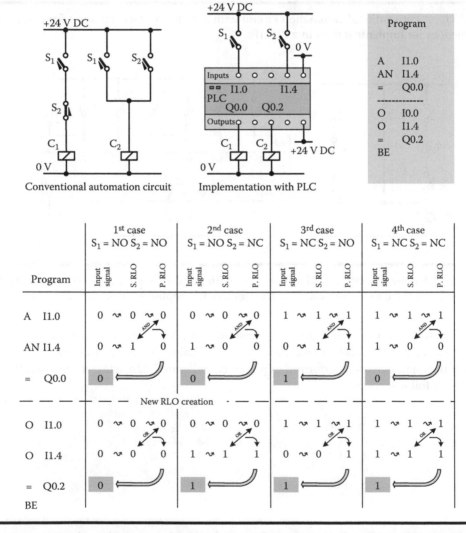

Figure 7.33 Creation of S.RLO and P.RLO in a Boolean program for all possible combinations of input statuses.

scheme showing the connections of the I/O devices to the PLC has been omitted. Furthermore, it should be clarified that in all these examples, the input switching contacts, connected to the PLC, have the same state as the one displayed in the classical automation circuit. In addition, in every conventional automation circuit, the corresponding addresses of the switching contacts and relays in the PLC are depicted for economy of notation. Figure 7.34 presents the AND logic of three switching sensor contacts, where their simultaneous activation energizes the relay.

In Figure 7.35, the OR logic of three switching sensor contacts is presented, as in a previous example.

In Figure 7.36, a combination of the AND logic before the OR logic of four switching contacts from an equal number of sensors is presented, with the expected goal of the activation of the relay Q3.1. The Boolean program uses the OR instruction without any reference to a variable. Obviously, there is another way of programming for achieving the same goal without using the previous instruction.

In Figure 7.37, a composite mixed circuit of four switching contacts from the same number of sensors is presented with the aim to activate the relay with any possible combination of states of the switching contacts.

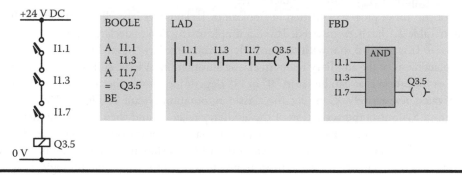

Figure 7.34   The AND logic in three basic languages.

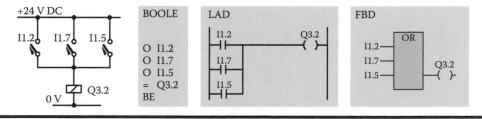

Figure 7.35   The OR logic in three basic languages.

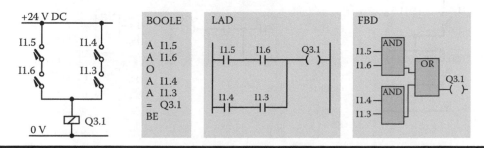

Figure 7.36   The AND logic in combination with OR logic in three basic languages.

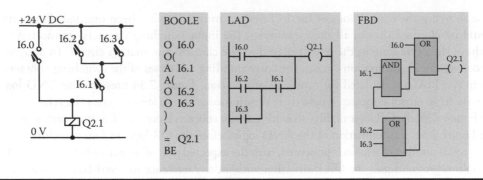

Figure 7.37 Mixed AND and OR logic in three basic languages.

Figure 7.38 shows a combination of an OR logic before an AND logic of four switching contacts from the same number of sensors. In this case, it is also desired to activate the relay with any possible combination of switching contact states.

In all of the above programs written in Boolean or IL language, the "BE" instruction that is not included in Table 2.1 has been inserted. This is a simple instruction to declare the end of the program.

Logic coils. It is recalled that with the term "logic coils" we have named the digital implementation or replacement of the auxiliary relays in classic automation. A logic coil is nothing more than a 1 bit memory location that can be set to "0" or "1" depending on whether the logic coil is switched off or activated respectively. Consider the classic automation circuit of Figure 7.39a, where two sensors S_1 and S_2 control the activation of the auxiliary relay d_0, and then d_0 in conjunction with a third sensor S_3 to control the operation of the power relay C_1. For the obvious purpose of operating the relay C_1 and connecting the I/O devices to the PLC, as shown in Figure 7.39b, the required program in Boolean language, under a direct representation of the classic circuit logic, is as follows:

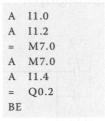

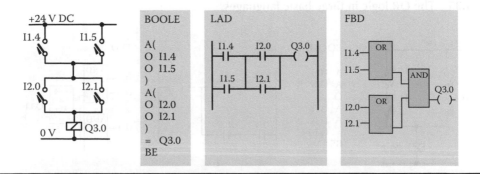

Figure 7.38 The OR logic in combination with AND logic in three basic languages.

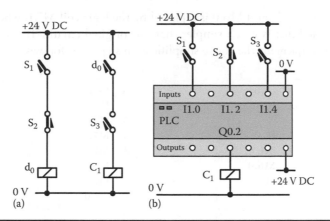

Figure 7.39 **The auxiliary relays of conventional automation are the internal digital elements (bits) in a PLC. (a) Automation circuit and (b) corresponding wiring for a PLC-based automation.**

Logic coils are very useful in programming and simplifying complex automation circuits. Parts of the composite automation circuit can be replaced at no cost by logic coils (as opposed to the cost of auxiliary relays in conventional automation) leading to a drastic simplification of the form of the circuit and hence of the overall PLC program. Let's consider the automation circuit of Figure 7.40a, which includes nine switching contacts as inputs to the PLC in a combination of branches of parallel and serial logic. Instead of programming the classical logic of the automation circuit faithfully, logic coils can be introduced as intermediate steps of this logic. In particular, the combination of the switching contacts I0.0, I0.1, and I2.0 is replaced by the logic coil M6.0, as shown in Figure 7.40b. In the next simplification step, the combination of the switching

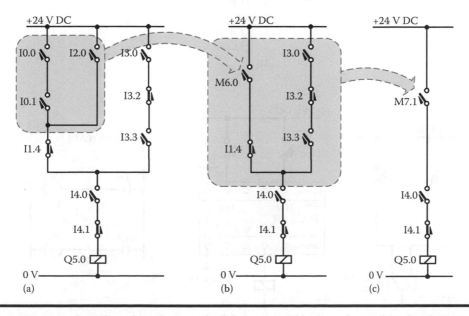

Figure 7.40 **Use of auxiliary bits (logic coils) for the simplification of complex circuits/programs. (a) Example automation circuit, (b) first simplification by using a logic coil, (c) second simplification by using another logic coil.**

contacts I1.4, I3.0, I3.2, I3.3, and M6.0 is replaced by the logic coil M7.1, whereby the circuit of Figure 7.40c is obtained that is much simpler than the original circuit. The required program in Boolean language to implement the above simplification steps is as follows:

```
A   I0.0              A   I3.0
A   I0.1              A   I3.2
O                    A   I3.3
A   I2.0             =   M7.1
=   M6.0             A   M7.1
A   M6.0             A   I4.0
A   I1.4             A   I4.1
O                    =   Q5.0
                     BE
```

Permanent activation of variables. As mentioned in the explanation of the basic instructions of Table 2.1, the instructions set and reset for permanent activation/deactivation are the digital implementation of the latch relay shown in Figure 7.41 (left). The latch relay has two coils, one for actuation (which is permanent once it happens, regardless of whether or not there is voltage at the ends of the coil) and a second coil to deactivate it, with the same characteristic of the residual state. Therefore, the latch relay requires two control signals, which are provided by the switching contacts of the sensors S_1 and S_2. In programmable automation, the relay C_1 no longer needs to be a latch relay (because of the higher cost) but is replaced by a common power relay and the "lock" function is implemented in the program using the S and R instructions. For the connection of the I/O devices to the PLC, shown in Figure 7.41 (middle), the required LAD program for permanently activating/deactivating a digital output (Q0.1) is shown in Figure 7.41 (right). The corresponding Boolean program is as follows:

```
A   I1.0
S   Q0.1
A   I1.4
R   Q0.1
BE
```

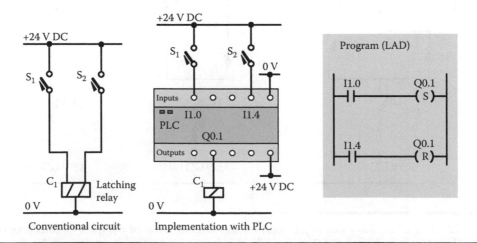

Figure 7.41   The instructions S and R implement the conventional latching relay.

In quite a similar way, the S and R instructions can permanently activate/deactivate a logic coil. Subsequently, let's consider the START-STOP classical automation circuit with a self-latch property, which can be utilized for the direct start and operation of an electric motor, as shown in Figure 7.42. The same figure also shows how the control buttons are connected to the PLC, as well as the Boolean language program, presented at the beginning of Section 7.5.2. The same operating logic can be achieved using the S and R instructions in the Boolean program that follows:

```
A    I1.3
S    Q0.2
AN   I1.0
R    Q0.2
BE
```

At this point the two Boolean programs can be directly compared to derive any difference. For those who are familiar with classical automation, it would be interesting to consider whether via the program based on the S and R instructions, the non-reoperation feature of the classic automation circuit is lost.

Figure 7.43 shows a complex machine, which is an industrial mixer of raw materials. The machine includes a central motor and a number of sensors that are necessary for its operation.

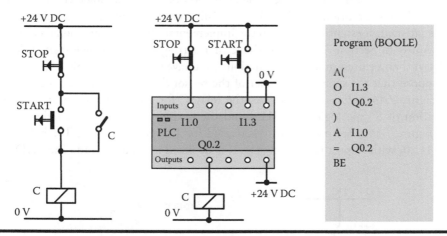

Figure 7.42   The conventional START-STOP circuit with self-holding logic can be implemented in a PLC via instructions S and R.

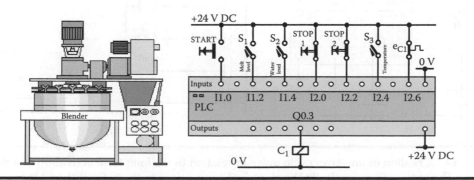

Figure 7.43   Operation of an industrial blender using instructions S and R.

The motor C_1 is switched on by pressing the START button, AND the material level sensor is activated, AND the water level sensor is also on. The machine stops operation if any of the two STOP push-buttons are pressed, OR if the temperature sensor is activated, OR if the thermal sensor of the motor is activated. All sensors and the thermal overload switch are shown in the deactivated state in the corresponding connection diagram in the PLC. The required Boolean language program to achieve the above function is the following one:

```
A    I1.0
A    I1.2
A    I1.4
S    Q0.3
ON   I2.0
ON   I2.2
O    I2.4
ON   I2.6
R    Q0.3
BE
```

Impulse generation. This term denotes the activation of a variable (logic coil or output) for a single scan cycle if a condition (RLO = "1") is satisfied regardless of the length of time that the condition is satisfied. It essentially refers to the implementation of the transitional contacts, mentioned previously in Section 7.5.2. The variable that is triggered for a single scan cycle (impulse generator) can then be used to perform functions that have to take place only once. As an example, let's suppose that the switching contact of the sensor S_1, shown in the classical automation circuit of Figure 7.44a, closes and activates the relay C for a sufficient time. It is desired that upon activation of S_1, the logic coil M11.0 and the output Q0.2 (indicator light) of the PLC (shown in Figure 7.44b) are momentarily activated for a single scan cycle. The behavior of the logic coil M11.0, with respect to the input I1.2, is depicted in Figure 7.44d. The LAD language

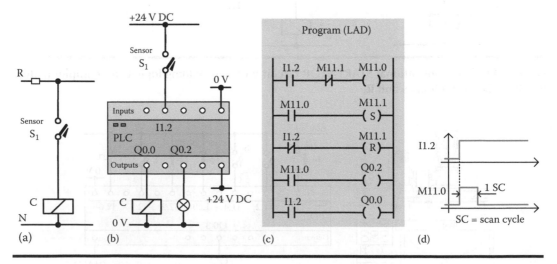

Figure 7.44 Creation of one-scan cycle pulse in relation to the long-term activation of a digital input. (a) The automation circuit, (b) the electrical connections to the PLC, (c) the LAD program, and (d) the timing diagram of I1.2 and M11.0.

program for executing the impulse pulse is shown in Figure 7.44c. In this case, the logic coil M11.1 is an auxiliary variable to achieve the required logic, while the corresponding Boolean language program is as follows:

A	I1.2		A	M11.0
AN	M11.1		=	Q0.2
=	M11.0		A	I1.2
A	M11.0		=	Q0.0
S	M11.1		BE	
AN	I1.2			
R	M11.1			

In order to produce a second impulse pulse, the input I1.2 should be deactivated, in order to reset M11.1, and then be reactivated.

Example: In an industrial manufacturing process, there are typically many machines (motors), each of which may present one or more operating faults, the appearance of which is provided in the automation program, and causes either a simple indication or some other action. Figure 7.45a shows a series of pumps that may exist in a pump station, each of which may exhibit a fault, such as a thermal overload drop, a flow stop, etc. In the examined case, it is desired that for any of the apparent fault triggers, a corresponding indication remains active (on) for as long as the fault lasts, while simultaneously triggering a corresponding logic coil momentary for one scan cycle. In addition, it is desired that the generated impulse of the logical coil corresponding to the fault permanently activates a common siren for all faults. The siren will stop by pressing a reset button. In the current example, only the case of two faults is considered, but the logic is expandable for any number of faults without a loss of generality. The connection of the I/O devices is shown in Fig. 7.45b, where the sensors S_1 and S_2 represent the two faults. Fault 1 is expressed by closing the contact S_1, while fault 2 is expressed by opening the contact S_2, as shown in the corresponding time diagrams of Figure 7.46. The required Boolean

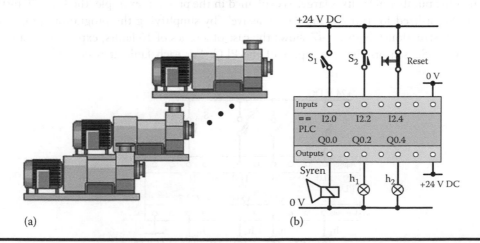

(a)   (b)

Figure 7.45 **A group of faults cause the activation of corresponding light indications and the sounding of a common siren. (a) A series of pumps, (b) the corresponding PLC wiring diagram for the case of faults from two pumps.**

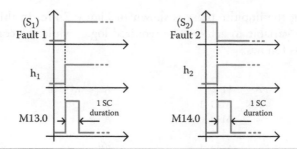

Figure 7.46   Creation of a one-scan cycle pulse for each fault.

language program that includes the creation of the two impulses and the logic of activating the output devices is as follows:

A	I2.0		O	M13.0
AN	M13.1		O	M14.0
=	M13.0		S	Q0.0
A	M13.0		AN	I2.4
S	M13.1		R	Q0.0
AN	I2.0		A	I2.0
R	M13.1		=	Q0.2
			AN	I2.2
AN	I2.2		=	Q0.4
AN	M14.1		BE	
=	M14.0			
A	M14.0			
S	M14.1			
A	I2.2			
R	M14.1			

When the number of faults is large, as explained in the previous example, the L and T instructions can be utilized for transferring data massively by simplifying the programming task (less number of instructions). Figure 7.47 shows the case of a series of 16 faults, expressed by an equal number of S_1–S_{16} sensors that are inputs of the PLC. For each fault, it is desired to activate a

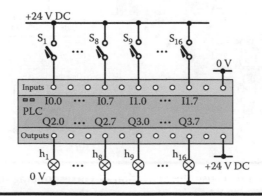

Figure 7.47   Successive inputs/outputs can be faced in groups as a byte or word.

corresponding indicator light for as long as this lasts. Similarly, it is desired to activate a logical coil for each fault, because a portion of the logic coil memory area is transmitted as information via a communication network to another PLC. The corresponding required programming of the PLC using the L and T instructions can be performed in the form of byte or word as shown in the following programs:

```
L    IW0                    L    IB0
T    MW100                  T    MB100
T    QW2                    T    QB2
BE                          L    IB1
                           T    MB101
                           T    QB3
                           BE
        WORD                       BYTE
```

At this point, it is suggested that the reader write the instructions needed to handle the faults one by one in order to have a full understanding of the required number of instructions and the programming shortening that has been achieved. A prerequisite for using the L and T instructions at word or byte level is that both inputs and outputs must have successive addresses.

7.5.3 Programming with Timers and Counters

As already mentioned in Section 7.4.3, each timer can implement various kinds of time functions, each one of which is determined by the corresponding activation instruction. Figure 7.48 shows the most basic time functions as a result of executing the same program:

```
A    I0.0
L    'T sec'
SX   T8
A    I1.0
R    T8
A    T8
=    Q4.0
BE
```

The only difference is the SX activation instruction that becomes:

- SP for the pulse timer
- SE for the self-holding pulse timer
- SD for the ON-delay timer
- SS for the self-holding ON-delay timer
- SF for the OFF-delay timer

and for the various cases of the input change (I0.0 or I1.0) that cause the set or reset of timers respectively. In these cases, every possible behavior of the output of a timer is clarified. The output of a timer, which has a real graphical form in the LAD language, will be presented subsequently. In Boolean language, it is the RLO of executing the instruction "A T8", which is a logical "1" when the timer has measured the time that has been set according to the time function that

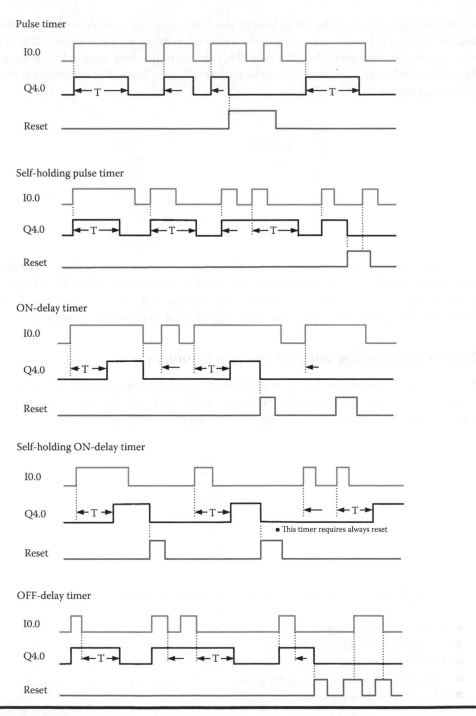

Figure 7.48 The basic possible time functions of a timer, each one selected by a corresponding instruction.

has also been defined via the corresponding activation. Therefore, the output Q4.0 through the instruction pair {A T8, = Q4.0} expresses nothing else but the behavior of the timer's output. Before each triggering instruction of a timer, the timer should be loaded with the time to be measured through the time instruction (L "T sec"). The way of declaring the number and type of time units in the program will not concern us further because it is a secondary issue and varies from manufacturer to manufacturer.

The *pulse timer* has an output that is activated and remains activated (ON) for a time T from the instant that the timer's input is activated with a positive rising edge (the input I0.0 changes from "0" to "1"). If the input of the timer is deactivated before the time T expires, then the output of the timer is deactivated at the same time instant. The timer output is also deactivated as soon as the reset input (I1.0) is activated. In this case, the output of the timer cannot be activated again unless a new positive rising edge pulse in the input is preceded.

The *self-holding pulse timer* has a behavior similar to the simple pulse timer, except that if the input of the timer (I0.0) is switched off before time T expires, the output remains energized until the T time measurement has been completed, thus this timer has the term "self-holding". If one or more positive rising pulse edges occur to the timer's input, while the timer output is activated, the T-time measurement is updated to each applied pulse.

In the case of the *ON-delay timer*, its output is activated after a time T has elapsed from the moment of the positive rising pulse edge at the input (I0.0), and remains activated for as long as the input is activated (ON). If the activation of the input stops before the T time expires, then obviously the timer's output is not activated, and the restarting of the time T measurement requires a new positive rising pulse edge to occur at the input. Activation of the reset input (I1.0) immediately deactivates the output and inhibits the T time measurement.

The *self-holding ON-delay timer* has a similar behavior to the previous timer, except that the time T countdown continues even if the timer input (I0.0) is deactivated. Therefore, the output of the timer will necessarily be activated after time T, and the only way to deactivate this is to use the reset input (I1.0). Successive deactivations and activations of the setting input cause the T time measurement to be updated.

The *OFF-delay timer* activates its output at the same time when the input is activated and it remains activated for as long as the setting input is activated. If the timer's input is deactivated, then the deactivation of the output will be delayed for a period of T time (thus the term "OFF-delay"). On successive activations and deactivations of the input, there is also a renewal of the time T measurement.

In the LAD language, the graphic symbol of the ON-delay timer instruction is shown in Figure 7.49, while the graphic symbol is similar for all the other types of timers. In particular, the graphic symbol of a T_{xx} ON-delay (SD) timer activation instruction accepts five connections or definitions. The input S is the setting input (logic "0" or "1") of the timer. The numerical value

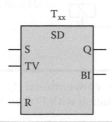

Figure 7.49 **Graphical symbol of a timer in LAD language including instructions for activation, resetting, outputting, time setting, etc.**

of the time and the time unit that T_{xx} will measure are declared in the TV input. The input R is linked to or denoted by the variable that is required to perform the reset action of T_{xx}. At the Q output of T_{xx}, it is connected or declared the variable that needs to be activated when T_{xx} terminated the measurement of the defined time. Finally, the output BI is provided by the CPU, and the remaining time to be measured in integer form becomes zero when Q = "1".

In the conventional automation circuit shown in Figure 7.50a, the usual industrial need for starting up a machine (relay C) with a time delay is implemented. The initiation of the machine with a delay time T is performed with respect to an event expressed by a discrete signal. When the S_1 sensor is activated and its switching contact closes, the timer T is activated, which measures time T, closes the switching contact T after the elapse of the time T, and finally the relay C is activated. In Figure 7.50b, the required I/O connections for the implementation of the above operation to a PLC are presented. In this case, the reset button allows the reset of the current time measurement. Whether the timer will restart the time measurement depends on the state of the sensor S_1 and the time function ON-delay of Figure 7.48. The required program in the LAD language is shown in Figure 7.50c, where the time T has been defined as 30 s. From the graphical symbol of the timer T12, the output BI has been omitted, while a rung has been added to the program to activate the logic coil M3.7 with a delay as well at the output. The corresponding Boolean language program is the following:

```
A    I1.0
L    '30 sec'
SD   T12
A    I1.3
R    T12
A    T12
=    Q0.2
=    M3.7
BE
```

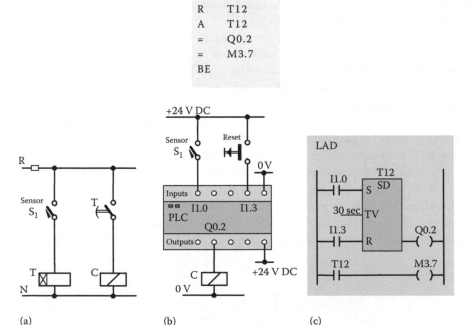

(a)   (b)   (c)

Figure 7.50 Implementation of the conventional ON-delay starting circuit in a PLC. (a) Conventional circuit, (b) implementation with PLC and (c) program.

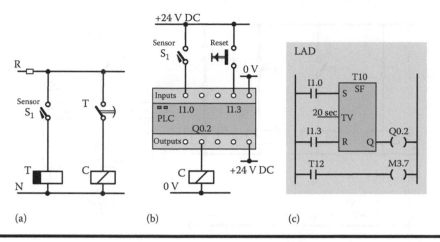

Figure 7.51 Implementation of the conventional OFF-delay stopping circuit in a PLC. (a) Conventional circuit, (b) implementation with PLC and (c) program.

In a similar way, a timer can be programmed to implement any other time function of Figure 7.48. As an example, Figure 7.51 shows the classic automation circuit, the I/O connections to the PLC and the LAD program for the operation of a machine (relay C) with a time delay for stopping. When the sensor contact S_1 closes, the OFF-delay timer is activated, which immediately closes the delay contact T, and therefore the relay C is energized. When, at some time instant, the contact of S_1 becomes open, the timer is immediately deactivated, but the timer's switching contact T opens with a time delay T and hence the relay C will be deactivated with the same delay (20 s in the program). The corresponding Boolean language program is the following:

```
A    I1.0
L    '20 sec'
SF   T10
A    I1.3
R    T10
A    T10
=    Q0.2
=    M3.7
BE
```

Example: For a machine (relay C) it is desired, from the moment that an operating condition will be satisfied, to operate for time T_2, to stop for time T_1 and for this to be repeated automatically until the operating condition ceases. The required time function—ON for T_2, OFF for T_1—is implemented with the conventional automation circuit of Figure 7.52a. The described time function follows the auxiliary relay d, as shown in Figure 7.52b, while the sensor S_1 plays the role of the operating condition. For the connection of the I/O devices shown

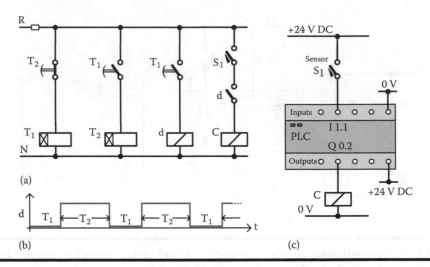

Figure 7.52 **Operation of a machine (C) with two time constants, T1 and T2 for OFF and ON statuses, correspondingly. (a) Conventional circuit, (b) time response (ON–OFF) of relay d and (c) implementation with PLC.**

in Figure 7.52c, the required Boolean language program, for $T_1 = 20$ s and $T_2 = 80$ s is the following:

```
AN   T2
L    '20 sec'
SD   T1
A    T1
L    '80 sec'
SD   T2
A    T1
=    M3.4
A    I1.1
A    M3.4
=    Q0.2
BE
```

Counters. Counting as is a very common function in all industrial environments. There is no industrial production process (especially consumer products from beer or water bottles, to medicines boxes, to all kinds of food packaging) that does not require enumeration of objects. A series of features associated with the operation of a counter is the ability to place it at an initial value, to increase or decrease its content, to define the step of its content change and to reset its current value. Figure 7.53a shows a conveyor belt for carrying luggage that is counted by the photocell (PC). The input devices that affect the counting function of the PLC are shown in Figure 7.53b. With the set button, the counter is set at an initial value of 100. With the reset button the contents of the counter are reset for any reason and at any time. It is desired when 100 pieces of luggage are counted that the relay C_1 is activated. The required Boolean language program is as follows:

```
A    I0.5          A    I0.3
CD   C10           R    C10
A    I0.0          AN   C10
L    '100'         =    Q2.3
S    C10           BE
```

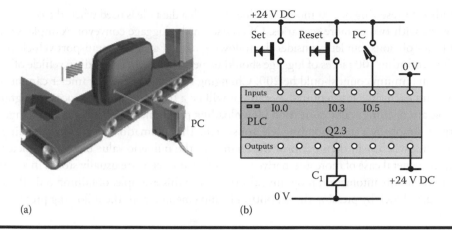

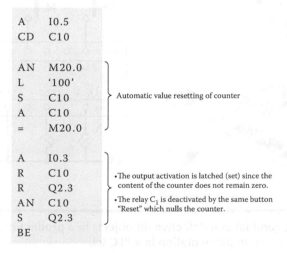

(a)   (b)

Figure 7.53   (a) Counting luggage on a belt conveyor set-up (b) PLC wiring of the related automation components.

In the description of the counting instructions of Section 7.4.4, the behavior of a counter's output with respect to its content was explained. The instruction "AN C10", from which the activation of output Q2.3 is dependent, is based exactly on the condition that the counter's output is logic "1" if its content is C10≠0.

In the previous example, the presetting of the counter to the value 100 was performed manually (by pressing the set button). This means that in order to operate the installation continuously, someone should press the button set each time that 100 luggage have passed. Subsequently, two different ways for counter presetting at a value will be considered. The first way will automatically achieve the resetting of the counter to its initial value, while the second one will be indirect through a data block. More specifically, automatic resetting of the counter means that when the counter of the luggage in the previous example has counted 100 objects and therefore its content is zero, then it will obtain the value of 100 again to perform the subsequent counting. The Boolean language program, where the automatic resetting is achieved, has a small number of additional instructions in relation to the previous one (manual presetting), and also does not use the input I0.0 for set. The instructions of the program are as follows:

A	I0.5
CD	C10
AN	M20.0
L	'100'
S	C10
A	C10
=	M20.0
A	I0.3
R	C10
R	Q2.3
AN	C10
S	Q2.3
BE	

Automatic value resetting of counter

• The output activation is latched (set) since the content of the counter does not remain zero.

• The relay C_1 is deactivated by the same button "Reset" which nulls the counter.

The indirect presetting of a counter to a value through a data file is used when the controlled process can work with two or more scenarios. Of course, in the luggage conveyor example, something similar it is not obvious, but let's consider the following case. If a loading transport vehicle of type A is used, then counting 100 pieces of luggage should be performed when a loading vehicle of type B is used, so the maximum count should be 200. Changing the value of this parameter cannot be done in the program each time it is required, because it will be a time-consuming process (programming device, program opening, value correction, downloading program, etc.). Instead, this change should be performed simply by a corresponding selector switch. The automation program, depending on the position of the switch, will then perform the change in the numeric value of the parameter. These two, or in the general case of more alternative values of a parameter, are usually stored in a data block from where the main automation program calls them. For this example, obtaining only the value of 200 from a data block by pressing the set button is implemented with the following program:

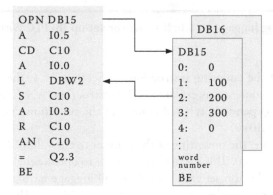

Comparison of numerical values. Figure 7.54a shows a conveyor belt transferring caps of switches from the production department to the final packaging section. Photocell 1 (PC_1) counts all the caps passing in front of it, whereas photocell 2 (PC_2) detects if there is a cap of different size. When

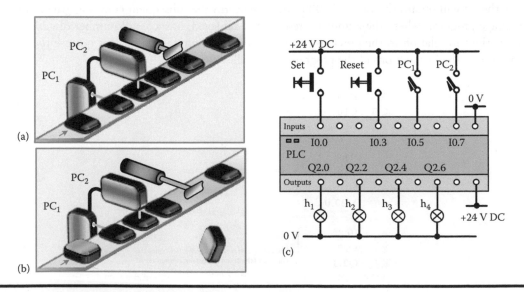

Figure 7.54 Counting good (a) and defective (b) objects in a production line, and the connection status of I/O devices for implementation in a PLC (c).

PC_2 detects a defective cap, it causes the compressed air piston to extend and push the defective cap into a bin of waste products, as shown in Figure 7.54b. This part of automation will not concern us. What is desired is a counter to increase its content with pulses from PC_1, and to decrease it with pulses from PC_2, so that the final content of the counter expresses the well produced caps. Additionally, we want four indicator lights (h1, h2, h3, and h4) to be activated when the content of the counter is C0 ≤ 200, C0 = 400, C0 ≥ 600, and 750 ≤ C0 ≤ 800, respectively. For the I/O connections shown in Figure 7.54c, the required Boolean language program is the following:

A	I0.5	L	C0	L	'600'	≤		
CU	C0	L	'200'	≥		=	M30.1	
A	I0.7	≤		=	Q2.4	A	M30.0	
CD	C0	=	Q2.0	L	C0	A	M30.1	
A	I0.0	L	C0	L	'750'	=	Q2.6	
L	'0'	L	'400'	≥		BE		
S	C0	=!		=	M30.0			
A	I0.3	=	Q2.2	L	C0			
R	C0	L	C0	L	'800'			

With the set button, we set the counter to zero value or any other value is desired, and with the reset button we can reset its content at any phase of operation.

Comparison of times. Let's consider an example where a number of machines are desired to start their operation with the instantaneous push of a common pushbutton with a time difference of 5 s from each other. One solution would be to use an equal number of timers, each one programmed in the corresponding ON-delay time. Instead of this solution, a single timer and comparison instructions will be utilized. In this implementation, the L Tx instruction is utilized, since it permits the remaining time to be measured to load in the accumulator, thus Tx is going to measure from the time instant that the instruction is executed. Figure 7.55a shows the connections of the I/O devices for three machines (C_1, C_2, and C_3); however, the program can be expanded for any number of machines. The time diagram of starting up the three machines and the corresponding times to be used in the programming are presented in Figure 7.55b, including the start of a fourth hypothetical machine. One or more machines may be added in the same way with a corresponding increase of the timer setting time. The required Boolean language program is the following:

A	I0.0	L	T3	S	Q2.6	
L	'18 sec'	L	'13'	AN	I0.3	
SE	T3	≤		R	Q2.0	
L	T3	S	Q2.3	R	Q2.3	
L	'18'	L	T3	R	Q2.6	
≤		L	'8'	BE		
S	Q2.0	≤				

A self-holding pulse timer is used because pressing the start button is momentary. With the reset button, all the machines will stop at the same time.

7.5.4 Using Structural Programming

The purpose and benefits of structural programming were discussed in Section 7.3, where the basic structural components of such an automation program have also been described (e.g., the three types of POUs). Here, how to implement structural programming with specific examples is presented. Figure 7.56 shows the I/O connections in the PLC for three direct-starting motors

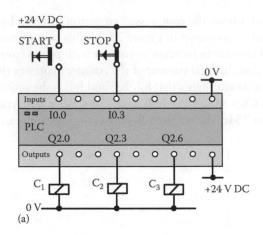

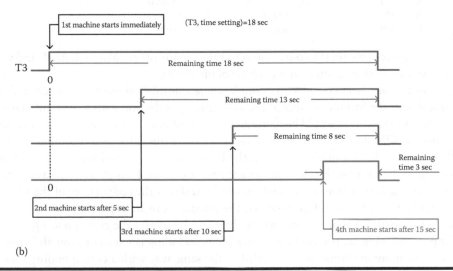

Figure 7.55 **Start of machines with time difference between each other, (a) PLC electrical wirings, (b) timing diagram of the overall operation.**

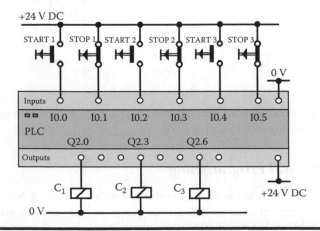

Figure 7.56 **Connection status of I/O devices in a PLC for START-STOP operation of three motors.**

(C_1, C_2, and C_3) with their corresponding START-STOP control buttons. Because START-STOP automation is exactly the same for the three machines, a simple FC can be used instead of a single unified program. The first step is the FC programming that includes the declaration of variables, the FC naming, the introduction of comments, and the automation program implemented by the FC. These actions will have the result that follows:

FC3

Variable declaration			
Name	Declaration	Type	Comment
SRT	INPUT	Boole	Starting motor
STP	INPUT	Boole	Stopping motor
MRT	OUTPUT	Boole	Power relay of motor

Program

```
A(
O    SRT
O    MRT
)
A    STP
=    MRT
BE
```

The second step is to write the general program that will call the FC3 in series for the three machines, and which will have the following form:

Main program

```
CALL    FC3
    SRT:=    I0.0
    STP:=    I0.1
    MRT:=    Q2.0

CALL    FC3
    SRT:=    I0.2
    STP:=    I0.3
    MRT:=    Q2.3
```

```
CALL    FC3
    SRT:=    I0.4
    STP:=    I0.5
    MRT:=    Q2.6
BE
```

Of course, the number of instructions included in the FC3 is small due to the simplicity of the example. The use of an FC is greatly facilitated when the repetitive operation requires a large number of instructions and repetitions. For example, let's assume that FC3 has 80 instructions and 25 motors. Then the number of the required instructions and thus the size of a single unified program without using FCs can be easily imagined.

Both the main program and the FC3, as well as any other POU, are programmed, as expected, in the PLC software environment that will be utilized. Since the main program and FC3 are declared, programmed, and stored, the software environment is the one that will establish their interface without the need for any additional programming action by the user. In Figure 7.57, the I/O devices of two motors with inversion of rotation (from the overall 10 available, whose operation will be programmed

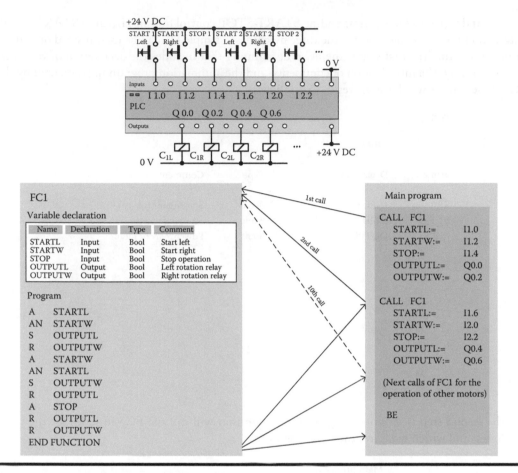

Figure 7.57   Programming operation of forward-reverse motors using FC.

using an FC) are shown. The same figure shows the FC1 that implements the inversion function of motors as a subroutine, the main program, and the instruction calls among them.

The programming of an FB that is directly linked to one or more data blocks is also addressed in a similar way. Figure 7.58 shows a production process consisting of three conveyor belts (M_1, M_2, and M_3) for transferring different objects, three processing machines (W_1, W_2, and W_3), a central conveyor belt (M_4) and a finishing, assembly, and packaging station (S). The four conveyor belts have a similar function (e.g., ON, OFF, counting, etc.), but because they carry different objects, certain technical parameters and data are different for each conveyor. This may be, for example, the transport speed, the number limits of objects that each conveyor transfers, etc. Therefore, the operation of the conveyor belts can be controlled by a single FB, which will be associated with four data blocks (DB1,..., DB4) in which the corresponding parameter values for each conveyor belt will be stored. The processing machines W_1, W_2, and W_3 have a completely similar function and the only element in which they differ also belongs in the START-STOP logic function. For example, if W_1, W_2, and W_3 are coloring machines, only the paint color they apply can vary, and therefore the output that will be activated in each one. It is understood that each machine can paint with three different colors. If these are labeling machines, they can only differ in the type of label and so on; many similar examples can be assumed. For this reason, the processing machines W_1, W_2, and W_3

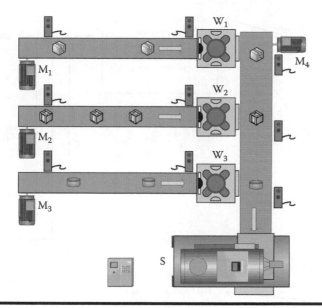

Figure 7.58 **Industrial station for transferring and processing different products.**

can be programmed with a simple FC. Finally, the station S cooperates with the rest of the equipment, but has its own separate automation logic that can be programmed into the main program. At this point, the automation logic of each production unit will not be considered in detail, except for the presentation of the overall organization of the structural programming and the definition of the calling hierarchy of the various program blocks. Figure 7.59 shows the structural form of the overall automation program of the industrial process. The components of the program are FB1 for the conveyor belts, which receive data from four different data blocks; and FC1 for processing machines and the main program, including the central station automation.

7.5.5 Programming Mathematical Operations

Programming a mathematical operation in an automation application does not present any difficulty. Regardless of the type of mathematical operation, it is only required to "load" the two numbers (among them the operation will be applied) in two corresponding registers, and then invoke the corresponding instruction to execute the mathematical operation. In most cases, the instruction of the arithmetic operation is accompanied by the declaration of the type of numbers such as real, integer, etc. The only point that someone needs to be aware of is the issue of the sequence in which registers are loaded. The mathematical operation is made with the first number from the content of the register 2 (i.e., the first one loaded first) and the second number from the content of the register 1 (i.e., the second one loaded last). When the mathematical operation refers to a single number (e.g., ABS), then it will obviously be performed on the content of the register 1. Figure 7.60 is illustrative of how the subtraction of two integers is performed. The result of the mathematical operation is always present in register 1, so that by using the T instruction (remember that the transfer instruction always acts on register 1) the result can be transferred to another location in the memory of the PLC. In the subtraction example, the result is transferred to the MW 100, while the other three basic numerical operations are dealt in the same way. Examples of automation programs with a need for mathematical operations are described in Section 7.5.6. Finally, it should be noted that currently

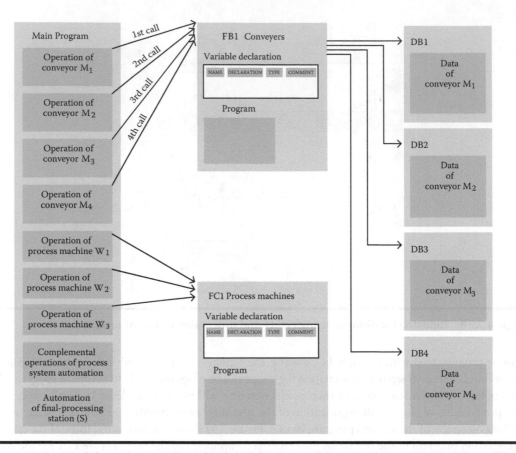

Figure 7.59 **Modular programming of industrial process shown in Figure 7.58 and the calling sequence of program blocks.**

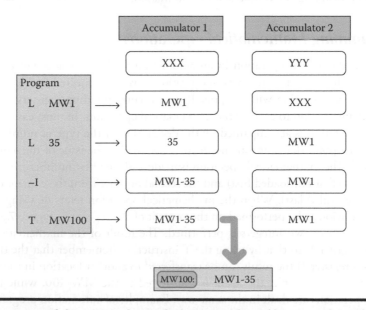

Figure 7.60 **Contents of the accumulators during execution of integer subtraction.**

all the large PLCs dedicated for industrial applications have a large set of algebraic, arithmetic, and trigonometric operations available, and their utilization is performed in a similar way.

7.5.6 Applications of PLC Programming

In this section, some practical automation layouts are presented, which are useful both for acquiring more experience in programmable automation and for their application to similar electromechanical automation projects. At this stage, it is assumed that the reader has understood the basic programming principles of PLCs and has acquired enough experience in programming the logic of simple automation layouts, thus there will not be a detailed description of how each application is programmed. Instead, the problem is briefly described, clearly listing the operating specifications, and giving the automation program directly.

Reversing Motor Operation. The reversing motor, which is controlled by two power relays, can be rotated in both directions, as shown in Figure 7.61a. When the motor is mechanically combined with a lead screw, the rotary motion is converted to a linear one, such as the one moving left or right on the worktable in the layout. The desired direction of rotation is selected via the two START buttons, respectively. The limit switches LS_1 and LS_2, located at the two extreme motion limits, ensure that if we forget to stop the movement of the worktable, then it will stop automatically, and the mechanism of the layout will not be destroyed. Figure 7.61b shows the classic circuit of automation and the operation of the two relays with a self-latch logic, while there is also the electrical latch of one branch from the other. Figure 7.61c shows the I/O devices and the way of connecting them to the PLC, which will implement the programmed automation. The only issue

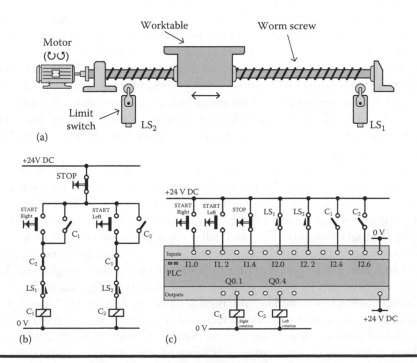

Figure 7.61   Linear movement of a worktable (a), conventional circuit for automatic stopping at the left and right limits of motion (b), and the connection status of I/O devices for implementation in a PLC (c).

that requires an explanation is the fact that there are two switching contacts of the relays C_1 and C_2, connected as inputs in the PLC, a need that does not arise from the classic automation circuit. Their usefulness will be explained after the presentation of the corresponding automation programmed in Boolean language that follows:

A	I1.4			
A(			Alternative	
			instructions	
O	I1.0			
O	I2.4	}	O	Q0.1
)				
A	I2.0			
AN	I2.6	}	AN	Q0.4
=	Q0.1			
A	I1.4			
A(				
O	I1.2			
O	I2.6	}	O	Q0.4
)				
A	I2.2			
AN	I2.4	}	AN	Q0.1
=	Q0.4			
BE				

The role of the C_1 and C_2 switching contacts, which are digital inputs of the PLC, is to provide additional information if the C_1 and C_2 relays are really energized. If the inputs I2.4 and I2.6 are not available, then instead of the corresponding instructions in the program, we would have the alternative instructions shown (in a box) next to the first ones and refer to the output variables Q0.4 and Q0.1. These variables, however, are internal elements of the PLC and, in particular, memory locations that have the information whether these outputs of the PLC are activated or not. However, this information (which is output activated) does not ensure that the relay is really energized (e.g., the relay coil was burned). Therefore, in cases where it is important to know that a relay has been activated, then a relay's switching contact should be an input to the PLC and thus be a kind of relay state sensor.

Taking one step further, the activation of a power relay does not ensure that the machine is working properly (e.g., the power cable between the power relay and the motor was cut off). If it is just important to know that the motor operates normally, then some kind of sensor should be used to detect the motor rotation and the motor's normal operation. The sensor output switching contact will be a digital input to the PLC and will transfer this required information.

Star-delta automation. Each star-delta motor starts with its windings connected in a star configuration (Y) and, after a short period of time through a change in the wirings caused by the automation circuit, operates with its windings connected in triangle configuration (Δ). Because these topics are generally well-known and have also been described in Section 4.3.2, the description will be limited to a brief mention of the functional requirements. The automation program should, after pressing the START button momentarily, activate the relay C_3 (see Figure 7.62a) that performs the star node. Then, if the relay C_3 (precondition) is activated, the relay C_1 is also activated. After the elapse of time T, the relay C_3 switches off and, without deactivating C_1, the relay C_2 is activated to perform the triangle connection of windings. This operation is accomplished with the classic automation circuit shown in Figure 7.62b. The switching contact C_1 with time delay is a contact of

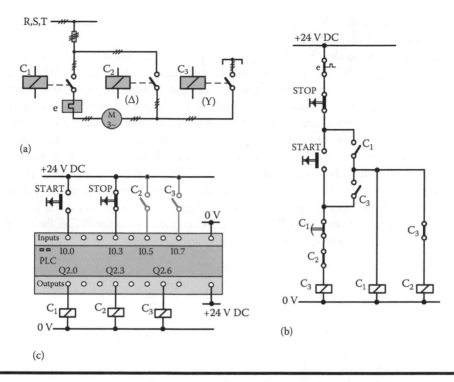

Figure 7.62 The conventional Y/Δ automation circuit and its implementation in a PLC. (a) Power circuit of star-delta (Y/Δ) motor, (b) conventional Y/Δ automation circuit and (c) implementation of Y/Δ automation with PLC.

an ON-delay pneumatic timer that is mechanically coupled on the relay C_1. For the connections of the I/O devices shown in Figure 7.62c, the required Boolean automation program is as follows:

A	I0.3		A	M30.0
A	I0.0		AN	T7
O			AN	Q2.3
A	I0.3		=	Q2.6
A	Q2.0		A	M40.0
A	Q2.6		L	'8 sec'
=	M30.0		SD	T7
A	I0.3		=	Q2.0
A	Q2.0		A	M40.0
O			AN	Q2.6
A	M30.0		=	Q2.3
A	Q2.6		BE	
=	M40.0			

The thermal overload switch of the motor has been omitted from the I/O connection diagram and from the automation program because its handling is standard and straightforward. Also, the switching contacts C_2 and C_3, which constitute the digital inputs I0.5 and I0.7 (to be used in the next example of parking fullness check), have not been taken into account in the automation program. Both the classic automation circuit and the above automation program function without any problem and implement the Y/Δ operation. However, in a few cases it is possible to experience an operating fault and a three-phase short circuit if for any reason, mechanical or other, the relay C_3 is delayed in deactivation with respect to C_2 at the time of change. This operation can be easily observed by inspecting the power circuit of Figure 7.62a, so if this does occur, then the star node is transferred from the ends of the motor windings to the three phases of the network, thus causing the three-phase short circuit. Although the probability that this fault exists is small, in a conventional automation circuit it is not taken into account, and the automation is implemented with one timer. In the case of programmable automation, where the PLCs have a plurality of timers, two timers can be used, one to measure the time of change from the star to delta configuration and the second to enter a delay of 2–3 s between the deactivation of the C_3 relay and the activation of C_2. In this way, any possibility of a three-phase fault is eliminated. Figure 7.63 shows the classic automation circuit that implements the above logic with two timers. The timer T_1 introduces the delay between the relays Y and Δ, while the timer T_2 determines the time instance of change from Y to Δ. For the same I/O connections to the PLC of the Figure 7.62c, the required Boolean language program will be in the following form:

A	I0.3		A	M20.0		=	Q2.3
A(			L	'10 sec'		A	M20.0
O	I0.0		SD	T2		A(	
O	M20.0		A	M20.0		O	I0.7
)			AN	T2		O	I0.5
=	M20.0		AN	I0.5		)	
A	M20.0		=	Q2.6		=	Q2.0
A	T2		A	M20.0		BE	
L	'2 sec'		A	T1			
SD	T1		AN	I0.7			

Parking fullness check. In Figure 7.64, there is a parking lot whose entry and exit of cars are electronically controlled by the photocells PC_1 and PC_2, respectively. In this case, it is desired for the counter to "monitor" the number of cars inside the parking lot so that the PLC will notify drivers whether or not the lot is full. In particular, if the number of cars within the lot exceeds 950, it is desired to activate a warning light (h1), and if the number of cars reaches the lot limit, a siren will sound to prevent other cars from entering. With the set button, the counter is set at an initial value

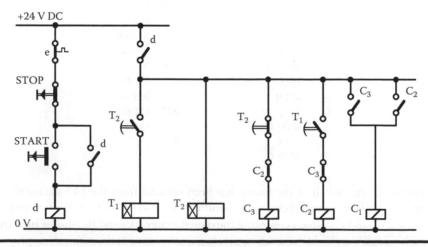

Figure 7.63   Conventional Y/Δ automation circuit with two timers for implementation in a PLC.

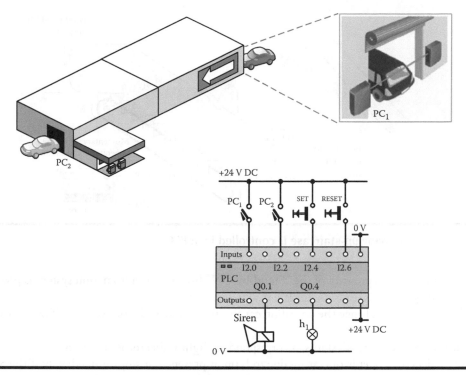

Figure 7.64   The PLC controls the fullness of the parking lot.

(defined in the program). It is also assumed that the initial number of cars at the beginning of the PLC operation is 350. With the reset button, the content of the counter can reset if required for some other reason, while the connections of the I/O devices to the PLC are shown in the same figure. The required Boolean language program has the following form:

A	I2.0		L	C7
CU	C7		L	'950'
A	I2.2		>	
CD	C7		=	Q0.4
A	I2.4		L	C7
L	'350'		L	'1000'
S	C7		≥	
A	I2.6		=	Q0.1
R	C7		BE	

Electric ladder control. Figure 7.65 shows an escalator that can operate only when a person is approaching (and not permanently, as happens sometimes). The detection of the approaching person at the up-escalator is performed via a photocell. At the entrance and exit of the ladder there are, at appropriate points, two emergency STOP buttons for use by the public, so if any accident occurs during the operation, anyone can stop the movement of the ladder. In the space of the escalator machinery, there is a control panel with buttons and indicators for controlling the electric ladder during maintenance periods. Specifically, the operating specifications of the above installation are as follows:

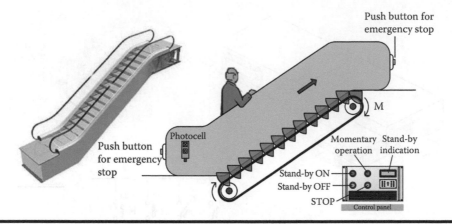

Figure 7.65   The moving up-staircase is controlled by a PLC.

1. By instantaneous pressing of the "Stand-by ON" button, the automation system is powered, and the escalator is ready to operate if someone approaches it.
2. The readiness and hence the operation capability is removed by pressing the "Stand-by OFF" button.
3. For the "Stand-by ON" status, a corresponding light indicator is activated.
4. If the electric ladder is operating due to human presence, it stops immediately if the STOP button on the control panel is pressed, if one of the two emergency buttons at the entrance and exit of the ladder is pressed, or if the thermal overload switch of the motor is activated.
5. If the escalator is in "Stand-by ON" status and a person is approaching, the activation of the photocell will cause the ladder to operate for 40 s, after which it will automatically stop.
6. Successive arrivals of people will cause a reset to the 40-s operating time.
7. During the maintenance period, it is necessary to operate the ladder manually. This is accomplished with the "momentary operation" button, which means that the ladder operates as long as the button is pressed, regardless of whether the system is in "Stand-by ON" status or not.

The escalator operation, based on the above specifications, is accomplished with the classic automation circuit shown in Figure 7.66. Two auxiliary relays and an OFF-delay timer are used. Relay C is the power relay that powers the motor of the escalator. Figure 7.67 shows the connections of the I/O devices in the PLC and the corresponding programs in both LAD and FBD languages. In Boolean language, the program will have the following instructions:

A(			O	I2.0		A	T1
O	I1.0		O	I2.2		=	M5.0
O	M3.0		O	I2.4		O(	
)			)			A	M5.0
AN	M4.0		A	M3.0		A	M3.0
=	M3.0		=	M4.0		)	
=	Q0.4		A	I2.6		O	I1.6
A(			A	M3.0		=	Q0.1
O	I1.2		L	'40 sec'		BE	
O	I1.4		SF	T1			

The three programs do not constitute a perfectly direct translation of the classic automation circuit. For example, the logic coil M5.0 does not exist as an auxiliary relay in the classic automation circuit.

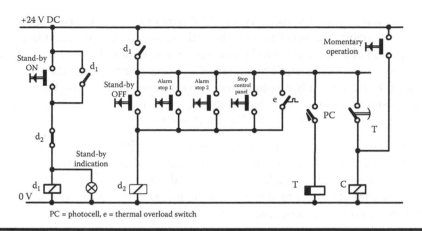

Figure 7.66 Conventional automation circuit of the moving up-staircase shown in Figure 7.65.

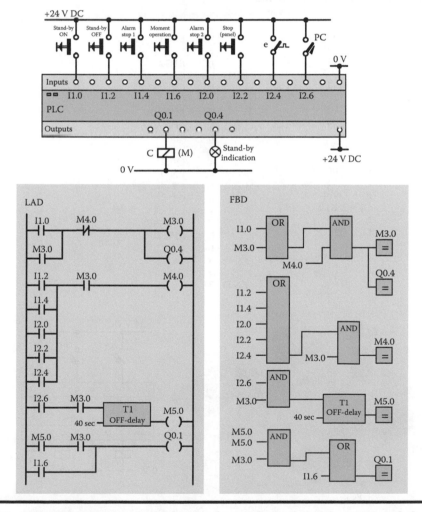

Figure 7.67 Connection status of I/O devices for implementation in a PLC and the required programs in LAD (left) and FBD (right) languages.

Counting and numerical calculations. Figure 7.68 shows two conveyor belts that move similar products from one point to another in an industrial manufacturing process. On each conveyor belt, the transported objects are counted by a corresponding photocell (PC_1 and PC_2). We will be dealing with only a part of the whole operation of the conveyor belts and therefore only part of their automation logic. Thus, it is supposed that the two conveyor belts have been put into operation based on some conditions and a corresponding set of instructions that do not need to be analyzed further in this example. Specifically, it is desired:

1. When the number of objects on a conveyor belt exceeds 500, the conveyor stops.
2. When the difference between the numbers of transferred objects on the two conveyor belts exceeds 200, a light indicator h has to be activated without stopping the conveyor belts.
3. The current sum of the counted objects on both conveyors needs to be stored into a data block (DB) as a numeric value renewed in each scan cycle.

The required Boolean language program for this section of the entire operation and automation of the two conveyors is the following:

```
    ⋮                           >
S    Q0.0                       JC   BIG1
S    Q0.4                       <
A    I1.0                       JC   BIG2
CU   C1              BIG1: L    C1
A    I1.3                  L    C2
CU   C2                    −
L    C1                    T    MW100
L    '500'                 L    MW100
≥                          L    '200'
R    Q0.0                  >
L    C2                    =    Q0.2
L    '500'                 JU   END
≥                   BIG2: L    C2
R    Q0.4                  L    C1
OPN  DB12                  −
L    C1                    T    MW200
L    C2                    L    MW200
+                          L    '200'
T    DBW3                  >
L    C1                    =    Q0.2
L    C2              END:  BE
```

```
DB12

0:
1:
2:
3: "C1+C2"
4:
```

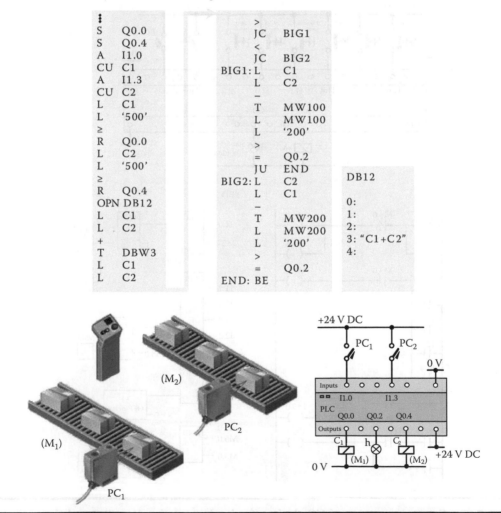

+24 V DC
PC_1 PC_2
0 V
Inputs
I1.0   I1.3
PLC
Q0.0   Q0.2   Q0.4
Outputs
C_1 h C_2
(M_1)   (M_2)   +24 V DC
0 V

(M_2)

PC_2

(M_1)

PC_1

Figure 7.68 Belt conveyors for transferring products and the PLC controlling their operation.

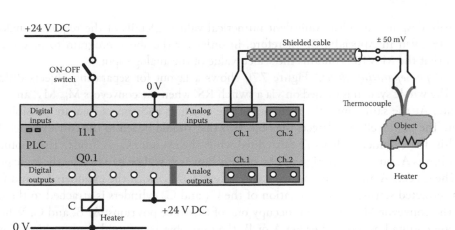

Figure 7.69   Temperature regulation via ON-OFF control performed by the PLC.

Temperature control. Figure 7.69 shows a temperature sensor (thermocouple), which measures the temperature of a body. The output of the sensor is an analog voltage (±50 mV) and is connected to a corresponding analog input (Ch. 1) of the PLC. Based on the measurement made by the sensor, it is desired to control the body temperature by activating and deactivating the heater C, (ON-OFF control). In particular, it is desired to keep the temperature at 300 °C with an acceptable variation of 2%, and therefore to keep it between the limit values of 297 °C and 303 °C. This means that the heater will operate (ON) for t ≤ 297 °C and will stop (OFF) for t ≥ 303 °C. In order to proceed, it is important to know how to read (in which memory address) the analog input value in the particular PLC being utilized, and where it is stored. For this example, it is assumed that the analog input value (Ch. 1) is stored in the MW 100 memory word and the layout is turned on with an ON-OFF switch (input I1.1). The Boolean language program will include the following instructions:

L	MW100		A	I1.1
L	'297'		AN	M0.3
≤			A(	
=	M0.2		O	M0.2
L	MW100		O	Q0.1
L	'303'		)	
≥			=	Q0.1
=	M0.3		BE	

At this point it is necessary to clarify the following issue. In the above program, it has been implicitly assumed that by executing the instruction L MW100, a numeric value, which expresses the temperature in degrees Celsius, is loaded into the register. This is not really the case, since the temperature sensor, in reality, applies to the analog input voltage, and the numerical value of this voltage is stored in the memory. Therefore, the comparison that follows in the program refers to dissimilar things. In order for the comparison to be correct, it is necessary to make the so-called "analog value scaling". The scaling procedure means converting the voltage values to equivalent temperature values, or vice versa, and thus to have comparable sizes. As a rule of thumb, the first solution is preferred because humans understand the engineering units (values) of physical size more easily despite any other equivalent numerical values. Details of the required analog value scaling are given in Appendix B. Therefore, in order for the above program to be correct, it is assumed that the MW100 contains the scaled value of the analog input.

Layout for separating objects. Figure 7.70 shows a layout for separating objects differing in height. The whole system is turned on via a switch RS, when the conveyors M_0, M_A, and M_B start to operate. At the same time, the contents of the counters and timers that are possibly utilized are reset. The photocell PC_2 detects the objects B that should be led to the conveyor belt B (M_B). To do this, the cylinder C_B has to be extended while the cylinder C_A is inside. The photocell PC_1 detects objects A, but it is also triggered by objects B, which will be handled by the logic programming. The objects A are driven on the conveyor belt A (M_A) with the cylinders C_A and C_B to be both in retracted status. The combination of the C_A and C_B cylinders is attached to the movable part of the conveyor M_0, which can occupy one of the three positions: A, B, and C. When 1000 objects are counted, whether they are A or B, the next objects are guided to position C with the C_A and C_B cylinder pistons both in extraction status, while both the conveyor belts M_A and M_B stop. This continues until a reset in the operation of the layout is performed by the switch RS, whereby the conveyor belt M_0 also stops. Furthermore, it is desired that, if for any reason, it does not reach an object A or B (in front of photocells) for a time interval of 10 min, a light indicator

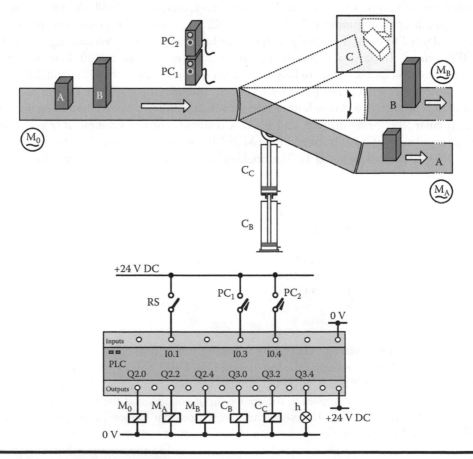

Figure 7.70 **An electro-pneumatic arrangement for separation of objects with different heights and the PLC controlling its operation.**

(h) is activated. The cylinders C_A and C_B are of double action at this point, and the automation will not deal with the logic of the pneumatic circuit. Thus, it will simply be considered that when, for example, the C_B relay is activated then the corresponding cylinder is extracted, the C_B is deactivated and the cylinder is retracted. According to the I/O connections in the PLC and the addresses listed, the required Boolean language program is as follows:

A	I0.1	L	'10 min'	≥	
AN	M11.1	SS	T1	S	Q3.0
=	M11.0	A	I0.4	S	Q3.2
A	M11.0	SS	T1	R	Q2.2
S	M11.1	R	Q3.2	R	Q2.4
AN	I0.0	S	Q3.0	A	T1
R	M11.1	CU	C1	=	Q3.4
A	M11.0	A	I0.3	AN	I0.1
R	C1	AN	I0.4	R	Q2.0
R	T1	SS	T1	R	Q2.2
A	I0.1	R	Q3.0	R	Q2.4
S	Q2.0	R	Q3.2	R	Q3.0
S	Q2.2	CU	C1	R	Q3.2
S	Q2.4	L	C1	R	T1
A	M11.0	L	'1000'	BE	

The program originally creates an impulse with duration of a scan cycle because the switch RS causes the input I0.1 to remain continuously active. Therefore, the resetting of C1 and T1 could not depend on the input I0.1. With impulse of M11.0, the resetting is done once.

Assembly station. A composite assembly machine consisting of a rotating table T and the two conveyor belts M_1 and M_2 is controlled by a PLC, as shown in Figure 7.71. The conveyor belts M_1

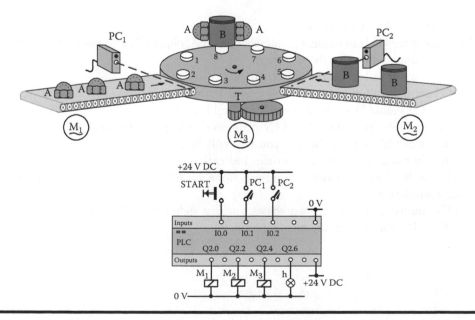

Figure 7.71 Assembly station of devices ABA, each consisting of two objects A and one object B, controlled by a PLC.

and M_2 feed the assembly table with type A and B components, respectively. The assembly table has eight work positions, while each assembled device in each work position consists of a type B component and two components of type A. The operation of the assembly station should follow the following specifications:

1. The table rotates stepwise and takes in a series of eight components A, eight components B, and then eight components A again.
2. The stepwise rotation of the table is realized by a mechanism (it does not matter how) that rotates the table one position at each rising pulse received by the motor M_3, which is the respective relay for the power supply (output Q2.4).
3. The assembly-collection procedure starts with an instantaneous START signal and finishes after the complete assembly of eight devices that make up one "assembly cycle".
4. After the START signal, the conveyor belts M_1 and M_2 start to operate in the order defined by the specification 1. Any conveyor belt that does not feed the table with accessories remains stationary.
5. When the machine assembles 400 devices, the light indicator h should be activated, an act that means that an "assembly series" has been completed.
6. Objects are conveyed randomly onto the conveyor belts and are detected by the corresponding photocells. The time from activating a photocell until the component is placed at the corresponding table position is considered negligible, and the method of placement is mechanically automatic.

Before proceeding with the automation programming, it is necessary to predefine certain parameters and features of the program logic, which facilitates programming tasks and prevents errors. The larger and more complex the application, the more necessary this step is. For this application, the following remarks should be highlighted:

- There is no need to use a timer.
- There are three kinds of counting, namely objects A, objects B, and assembly cycles.
- The C1 counter will count the objects A.
- The C2 counter will count the objects B.
- The C3 counter will measure assembly cycles. For the completion of an assembly series (400 devices), C3 should be set to measure 50 assembly cycles.
- In each assembly cycle, the C1 and C2 counters should be automatically reset.
- In each assembly series, the C3 counter will be reset with the next pressing of the START button, in order to keep the indication light on until that moment, which would not be the case if the zeroing was done automatically at the end of an assembly series completion.
- The C1 counter will count in two phases, the first eight objects A in the first phase, and the next 8 (9–16) objects A in the second phase.

After taking these remarks into account, the required Boolean language program is the following:

A	I0.0	⎫ Start of assembly cycle
S	M0.0	⎭
A	I0.1	⎫ Counting of objects 'A'
CU	C1	⎭
A	I0.2	⎫ Counting of objects 'B'
CU	C2	⎭
L	C1	⎫
L	'8'	⎬ A<8
<		
=	M1.1	⎭
=!		⎫ A=8
=	M8.1	⎭
L	C1	⎫
L	'16'	⎬ A=16
=!		
=	M16.1	⎭
L	C2	⎫
L	'8'	⎬ B<8
<		
=	M1.2	⎭
=!		⎫ B=8
=	M8.2	⎭
A	M0.0	⎫ If "assembly cycle is
A	M1.1	⎬ ON" AND if A<8 then
=	Q2.0	⎭ conveyor M₂ operates
A	M0.0	⎫ If "assembly cycle is ON"
AN	M1.1	⎬ AND if A≥8 AND B<8,
A	M1.2	⎪ then conveyor M₂ operates
=	Q2.2	⎭

A	M0.0	⎫
A	M8.2	⎪ If "assembly cycle is ON"
AN	M1.1	⎬ AND if B=8 AND A≥ 8
AN	M16.1	⎪ AND A≠16, then
=	Q2.0	⎭ conveyor M₁ operates
A	Q2.0	⎫
A	I0.1	⎪ If M₁ operates and an
O		⎪ object A arrives
		OR
A	Q2.2	⎬ If M₂ operates and an
A	I0.2	⎪ object B arrives,
=	Q2.4	⎭ then motor M₃ operates
A	M16.1	⎫ Counting of
CU	C3	⎭ assembly cycles
A	M16.1	⎫
R	C1	⎪ Resetting for starting
R	C2	⎬ a new assembly cycle
R	M0.0	⎭
L	C3	⎫
L	'50'	⎪ If an assembly series is
=!		⎬ complete, then h=ON
=	Q2.6	⎭
A	I0.0	⎫ Resetting for starting a
A	Q2.6	⎬ new assembly series
R	C3	⎭
BE		

Some of the "logical latches" in the operation of the conveyors may be redundant and not the least possible, but this is something that should not be considered as problematic. In an automation program, and especially in those of large applications, in addition to the logic extraction and writing of the corresponding instructions, tests of the system operation should be carried out and in extended operation time. For example, in the above program, it has not been considered how the layout will behave if an operator presses the START button continuously for both cases where objects are coming on two conveyor belts. At the very least, thorough testing should be performed in an automation program simulation environment. However, these issues are more concerned with the implementation of automation projects, and the acquisition of relevant experience and is beyond the scope of this book.

Swimming pool control. The initial filling of the swimming pool shown in Figure 7.72 starts by switching on the ON-OFF rotary switch and then the following procedure takes place:

- From the zero level to the S_1 level, only the water pump operates.
- From the S_1 level to the S_2 level, both the water pump and the chlorine pump operate.
- From the S_2 level to the S_3 level, only the water pump operates again.

A chlorine meter measures the residual chlorine in the tank and its analog output (0–10 V) is applied to an analog input (Ch. 1) of the PLC. Lets suppose that the numerical value of the analog input is stored in the MW100 memory word and expresses ml of chlorine per cubic meter of water (ml/m³). After the initial filling of the pool, the water pump operates whenever it is necessary to keep the level at S_3, and the chlorine pump whenever it is required to keep the measurement of residual chlorine between 0.5 and 1.5 ml/m³. The light indicator h shows all the possible logical errors of the S_1, S_2, and S_3 level sensors that are likely to occur.

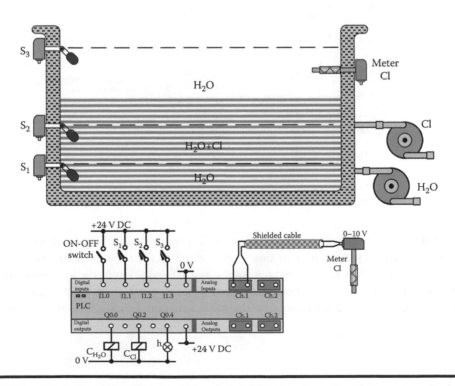

Figure 7.72   Control of water level and dissolved chlorine of a swimming pool performed by a PLC.

According to the I/O connections of the PLC and the addresses listed, the required Boolean language program is as follows:

A	I1.0	R	Q0.0	L	MW100
AN	I1.1	R	Q0.2	L	'1.5'
S	Q0.0	A	I1.3	>	
A	I1.3	AN	I1.2	=	M4.0
R	Q0.0	O		A	I1.3
AN	I1.3	A	I1.3	A	M3.0
A	I1.0	AN	I1.0	=	Q0.2
=	Q0.0	O		A	M3.0
A	I1.0	A	I1.2	=	QX.X
A	I1.1	AN	I1.1	A	M4.0
AN	I1.2	=	Q0.4	=	QY.Y
S	Q0.2	L	MW100	BE	
A	I1.2	L	'0.5'		
R	Q0.2	<			
AN	I1.0	=	M3.0		

Outputs QX.X and QY.Y are hypothetical that energize indications "Low Cl" and "High Cl" correspondingly.

Drilling machine automation. In Figure 7.73 a drilling machine is shown, for holes opening on objects A, which are manually placed by an operator. The machine has a movable cover to protect the operator from lubricating liquids, insertion of the hands, etc., during the operation of the drilling head. In particular, the machine includes the following equipment:

C_1 = Cylinder that lowers or raises the chuck.
C_2 = Cylinder that lowers or raises the protective cover.
S_0 = Sensor detecting the lower position of the piston C_1, that is, when the drilling has finished.
S_1 = Sensor detecting the presence of an object in the drilling position.
S_2 = Sensor detecting the lower position of the protective cover.

The stand-by control panel is the two-button control box with which the machine is set to a stand-by status, which means that it is ready for operation. Also, there is a light indicator expressing the "Stand-by ON" status.

The operation control panel is the two-button control box in a double function to start or stop the drilling procedure.

M = The motor of the drilling head rotation.

The following specifications are valid for the operation of the machine:

1. The initial position of the two pistons is "up".
2. The machine can operate only if it is in "Stand-by ON" status, i.e., if the "Stand-by ON" button has been pressed and the readiness indicator lights up.
3. Drilling can only be started if an object has been placed and has been detected by sensor S_1.
4. If there is an object in the drilling position and the two buttons "start drilling" are pressed at the same time (requires both hands of the operator for safety reasons), then the protective

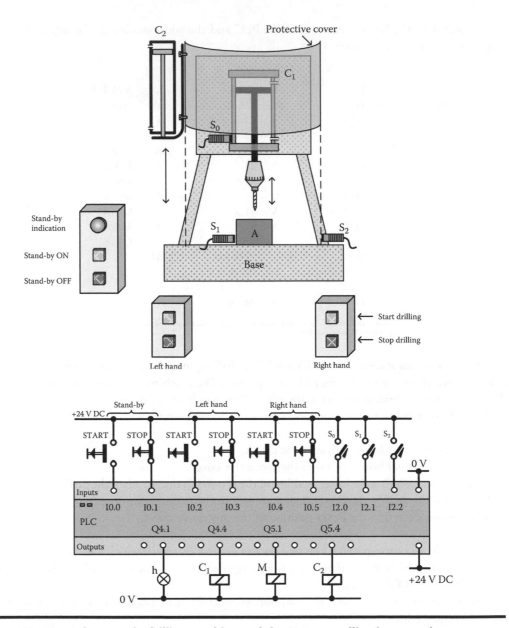

344 ■ *Introduction to Industrial Automation*

Figure 7.73 A semiautomatic drilling machine and the PLC controlling its operation.

cover (activation of the relay C_2) starts to lower. If the two buttons are released, the cover returns to its original position.

5. When the protective cover is brought to its final position, it is detected by sensor S_2 and then the piston C_1 (activation of the relay C_1) starts to lower, while simultaneously the drilling head (M) starts its rotation. In this phase, we do not have to press the two "start drilling" buttons because the protective cover has already been lowered.

6. Piston C_1 and the drilling head press down on the object until drilling is complete. This is established by the sensor S_0 activation.

7. As soon as S_0 is activated, the drilling head and the protective cover return simultaneously to their initial position. Then the machine is ready to accept a new item.

8. If at any time, either the two "stop drilling" buttons are pressed simultaneously or the "Stand-by OFF" button is pressed alone, the process stops and the two pistons return to the "up" position.

9. For the connection diagram of I/O devices in the PLC and the listed addresses, the required Boolean language program is as follows:

A	I0.1		A	I0.4
A(			O	
O	I0.3		A	Q5.4
O	I0.5		A	I2.2
)			)	
A(			AN	I2.0
O	I0.0		=	Q5.4
O	Q4.1		A	Q5.4
)			A	I2.2
=	Q4.1		=	Q5.1
A	Q4.1		A	Q5.1
A	I2.1		=	Q4.4
A(			BE	
A	I0.2			

The corresponding program in LAD language is shown in Figure 7.74.

Bowling automation. In Figure 7.75 an automated runway of the well-known bowling game operated and controlled by a small PLC is shown. Of course, it is not an industrial application but since it is familiar to anybody and presents interesting automation features it can be considered as a small automation project. In each game (coin throw) the player has the right to do 12 double attempts. Any double attempt means that the player can roll the

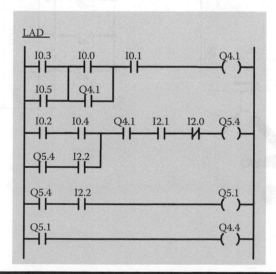

Figure 7.74   Program in LAD language for operation of the drilling machine shown in Figure 7.73.

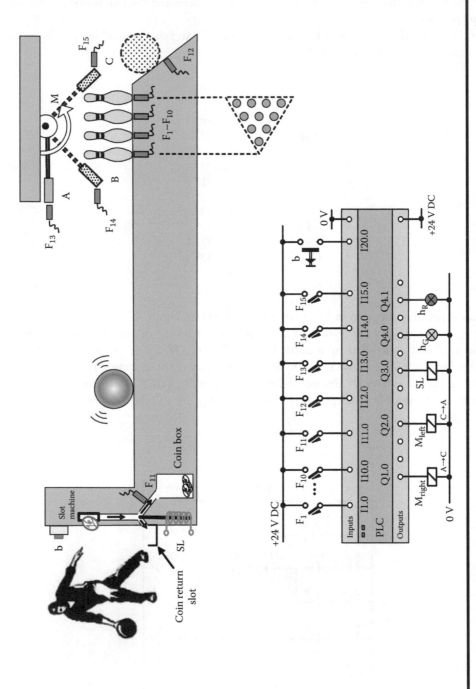

Figure 7.75 The electromechanical bowling equipment (in simplified form) is controlled by a PLC.

ball twice in his effort to knock down all the bowling pins. If during the first try the player knocks down all the pins, the attempt ends. If after the first try, there are any pins still standing, the player has a second chance to knock down the remainder of the pins. The operating specifications are as follows:

1. Bowling works with a coin slot mechanism. In order for the coin box to accept a coin, three preconditions should exist: [to press the button b] AND, [the 10 pins to be in their place] AND, [the pin bracket is in position B]. This situation is noted as a normal resting state.
2. To drive the coin into the coin box (according to 1), the solenoid (SL) should be activated. When the SL is not activated, the coin that the player is putting in the box comes out in the coin return slot.
3. As soon as the coin passes in front of the sensor F_{11}, the solenoid SL is deactivated because the slot does not accept coins during the game and returns them to the player. The mechanism of the pin bracket comes in position A, allowing the player to roll the ball. The light indicator h_R is turned on, which means that the game is in progress, i.e., the player has not completed the 12 double attempts. Simultaneously, the light indicator h_G turns off (see specification 7).
4. All the pins are detected by a corresponding proximity switch (F_1-F_{10}). After the first attempt, the number of knocked down pins is stored, and if all of them were not knocked over, the system expects the second attempt of the player. If all the pins were knocked over, the actions listed below after the second attempt should follow. After the second attempt and after the ball passes in front of the F_{12} sensor, the number of knocked down pins is also stored. The mechanism (M) rotates the pin bracket from point A to point C, pushing the upright or fallen pins towards the mechanism for their repositioning (how this mechanism operates is not included in this current automation study). After a time delay of 3 s, the pin bracket (M) returns to point A, and at the same time, 10 pins are repositioned by the corresponding mechanism.
5. The score of each player's attempt consists of the total number of pins that were knocked down in both attempts, which are stored in the PLC.
6. The player's overall score is the sum of his individual scores over 12 double attempts, which also has to be recorded in the PLC.
7. When 12 double attempts have been made, the system should return to its initial resting state, that is, the mechanism (M) rotates the pin bracket from point A to where it blocks point B, hence the continuation of the game until the next coin insertion in the coin slot machine. The h_G indicator lights up, thus indicating that the game is over and that the bowling system can accept a new coin, which deactivates the h_R indication at the same time.

The most difficult issue in this project, from the programming point of view, will be the counting of the knocked down pins by a single counter, under the real assumption that it is impossible for all the pins to be knocked over at different scan-cycles of the PLC.

Problems

7.1. Write a program in LAD language so that a machine can operate or stop permanently via two push buttons START and STOP respectively, but also to operate momentary via a third push button JOG (i.e., to operate as long as the JOG button is pressed).

7.2. In programmable automation, it is possible to use an NO or NC button for stopping the operation of a machine. Suppose that, during machine operation, the wire connecting the STOP button to the power source is broken due to a fault action as shown in the figure. In such a case, which of the two configurations is the most appropriate and why?

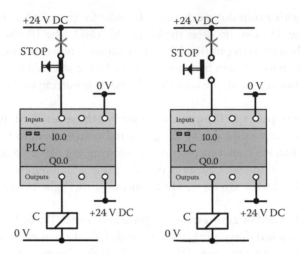

7.3. An On-Delay timer can be programmed to be self-activated when a measured time is elapsed. Then, the timer output (T2) will present the response shown in the figure. Combine the timer operation with a digital output (Q0.0) in order to have an alternation (ON-OFF) of Q0.0 with time period double of that presented by the T2 pulse train. Write the required program in LAD language.

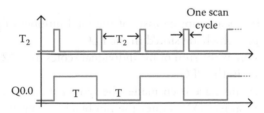

7.4. A digital input (I2.0) of a PLC is a pulse train with a frequency T. Write a program in LAD language so that a digital output (Q4.0) has a pulse train with frequency 2T, as shown in the figure.

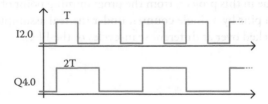

7.5. Translate the conventional automation circuit, shown in the left-hand figure, in a LAD program for the I/O connections in PLC shown also in the right-hand figure.

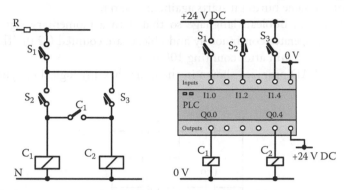

7.6. The example for counting and arithmetic operations described in Section 7.5.6 was addressed using a Boolean program. Program the same automation logic in LAD language.

7.7. The arithmetic value stored in memory word MW200 of a PLC expresses a temperature measurement in °C. Write a program to convert this value in °F and store it in a different memory word.

7.8. The ventilation of a road tunnel is achieved with the continuous operation of four ventilators. In the control room of the tunnel, there is suitable signaling provided by a PLC so that the responsible operator is informed of the ventilation status according to the following plan:

   a. When three or four ventilators are in operation, the green traffic light is ON and the cars freely enter the tunnel.

   b. When only two ventilators are in operation, the yellow traffic light is flashing (green light is OFF) and the operator has to be on standby to take necessary action.

   c. When only one ventilator is in operation or all have stopped, the red traffic light is ON and the operator has to intervene by limiting the traffic outside the tunnel.

   Detection of whether or not ventilators are operating is achieved by four corresponding sensors which are digital inputs to the PLC. After defining the kind of input contacts in relation to ventilators' operation and their I/O addresses, write the required program that has the minimum number of instructions in any language you wish.

7.9. Complete the RLO which is created by the CPU of the PLC during the execution of the instructions of the program in the empty rectangles (dash line) shown in the figure at the first scan cycle and at the first scan cycle after the time T2 (60 sec) has elapsed. The input devices have the statuses shown in the PLC figure.

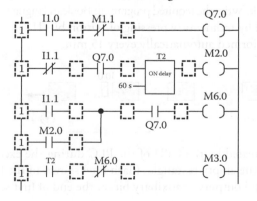
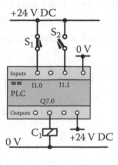

7.10. Write a program in Boolean language so that at the first press of a push button a machine starts to operate, at the second press of the same pushbutton the machine stops, and at the third press of the same button it starts again, and so on.

7.11. Write a program in Boolean language so that with a momentary signal Start (I0.0) a machine (Q4.0) operates continuously and objects are counted (I2.0). The machine must stop automatically 5 min after counting 100 objects.

7.12. Examine if the LAD program shown in the figure has any logical or syntax error. Justify your answer.

```
        I0.0            Q2.0
       ──┤ ├──         ─( S )─
                        Q3.0
                       ─(   )─

        I1.0            Q2.0
       ──┤ ├──         ─( R )─
                        Q3.0
                       ─(   )─
```

7.13. Convert the logic-flow diagram of the figure to a LAD program equivalent from a logical point of view, where A = Q1.0, B = I2.0 and C = I3.0.

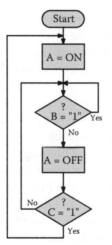

7.14. In an industrial production line, objects are transferred along a conveyor belt which operates continuously as shown in the figure. It is desired to count objects passing in front of photocell (I0.0) in time intervals of 1 min. It is also desired to store the last three measurements (objects/min) in memory locations or in a data block. Write the required program in Boolean language for two cases:
   a. Each measurement is caused by momentary pressing of button b0 (I1.1).
   b. Three measurements are performed automatically every 15 min.

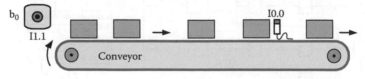

7.15. Complete the RLO which is created by the CPU of the PLC during the execution of the instructions of the program in the empty rectangles (dash line) shown in the figure. Please also complete the status of digital outputs or auxiliary bits at the end of first scan cycle and at the end of the first scan cycle after time T2 (60 sec) has elapsed. The input devices (two sensors) are in their normal status.

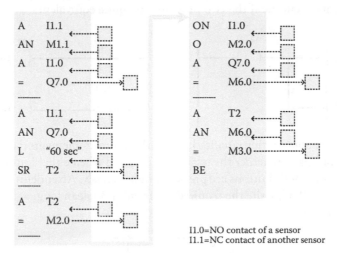

I1.0=NO contact of a sensor
I1.1=NC contact of another sensor

7.16. A machine operates with the aid of a direct starting motor in the START/STOP procedure of which it is desired to insert an intermediate state for "confirmation". In particular, the following specifications have to be satisfied:

a. If the motor does not operate (C = OFF) and the START/STOP button is pressed, then the message "Surely you want to Start?" is activated. By pressing the START/STOP button a second time, the motor starts to operate (C = ON) and the message turns off, or by pressing the button CANCEL the message turns off and the motor remains in its stop state.

b. If the motor operates (C = ON) and the START/STOP button is pressed, then the message "Surely you want to Stop?" is activated. By pressing the START/STOP button a second time, the motor stops to operate (C = OFF) and the message turns off, or by pressing the button CANCEL the message turns off and the motor remains in its operation state. Write the required program in any language according to the previous specifications and for the I/O connections shown in the figure.

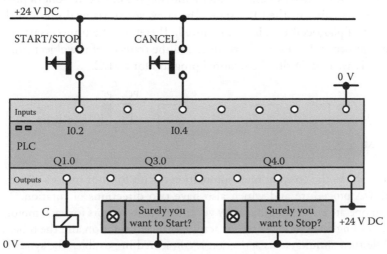

7.17. A digital output (Q2.0) of a PLC must be energized when either one of two digital inputs I1.0 and I1.1 or both, have been activated. Two different programmers wrote the following LAD programs. Examine if the two programs are quite equivalent.

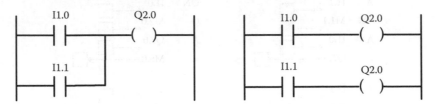

7.18. The conventional automation circuit, shown in the figure, implements the EXCLUSIVE-OR function according to whether the relay C is energized when either button A or button B is pressed, but not if both buttons are pressed simultaneously. After defining the I/O connection status in the PLC, write the required program in Boolean language for achievement of the same function.

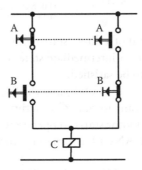

7.19. The figure shows an object transport conveyor and its manual takeover. The conveyor belt M is desired to operate as follows:
   a. Objects are coming accidentally. If only one object comes (with detection by photocell PC_1) the conveyor stops until a worker removes it, and then the conveyor starts again.
   b. If two or more objects come near each to other, then the conveyor stops (after detection of an object from PC_1), but after removing the object in front of photocell PC_1 the conveyor remains in the OFF state. The conveyor starts again only if the next object in front of photocell PC_2 is also removed. Therefore, the unnecessary operation of the conveyor occurs for a very short time, and the transfer of the object from position PC_2 to PC_1 is avoided. Write the required program for a PLC.

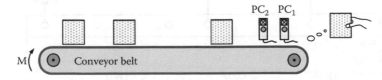

7.20. The figure shows the conventional automation circuit for four motors which start with the classical Y/Δ procedure and which also have two directions of rotation. The motors are desired to operate either automatically via sensors (two sensors for each motor), or manually via the handling buttons. One sensor activates the left rotation and the other the right rotation, while their simultaneous activation is considered impossible. The symbol Y/Δ ↔ inside the boxes expresses the required automation for starting such Y/Δ motors. Since, the multiplicity of motors and the uniformity of operations are suitable for using modular programming, write the required program using any kind of program organization units.

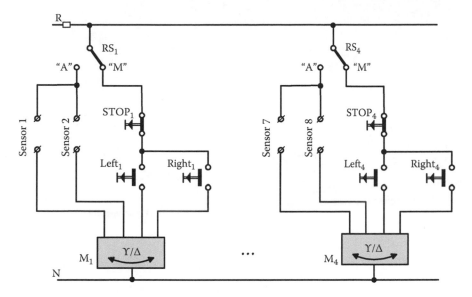

Chapter 8

Industrial Networks of PLCs

PLCs, which are the leading, stand-alone control tools of industrial processes today, can be interconnected with each other to exchange data, information, and control signals. The current form of automation of most modernized industries is based on integrated control systems implemented by the communicative interconnection of computers and PLCs. The interconnected systems of PLCs offer to an extensive industrial process, flexibility, transparency, and effective coordination of all its parts. The utilization of PLC networks has a direct huge economic benefit, as has been proven in practice in terms of quality improvement, reduction of production costs, flexibility, and reliability of industrial production.

It is an undisputable fact that it is impossible for an industrial process to take place in a limited space or in a single building. Typically, an entire industrial process is implemented in a number of buildings (production departments) and therefore a need is created for effective coordination of these various sections of production. Initially, during the first decades of the industrial age, a lot of time was wasted and labor costs were high mainly due to the need to send "messages" along all sub-processes in order to achieve better process operation and exploitation. The first step towards automation is the introduction of central control stations for the production process. However, the management and control of an industrial process is effective only if information can be received from every point in the network. This means that for every "bit" of information sent to the central control station, a corresponding wire is needed. Therefore, for an increased volume of transmitted information, an increased amount of wires or, equivalently, a significant cabling task is needed. Hence, it is understandable that in order to install a central station for monitoring and controlling an industrial process, many, large bundles of wires are needed (peer-to-peer connections), which leads to a significant increase in the cost of the automation task. At this time, the idea of using communication networks in the industrial world has matured as a physical step after the parallel developments in the area of computer networks. With the help of communication networks, all the bundles of cables are replaced by a single cable, which contains only two or four conductors instead of the thousands that we had before. In this approach, digital devices of any kind, participating in the operation of an industrial process, are connected via a network, and can communicate and exchange data with each other and with a central monitoring station. The data received by a digital device can relate the information required for its own control decisions, or the automation commands of a coordinator PLC for specific control actions directly enforceable to the actuators, etc. In conclusion, it can be said that one of the purposes of using communication networks in an industrial environment is to reduce the enormous task of wiring. The essential result is a significant reduction in installation, pipeline, and cable channel costs, as well as maintenance costs of wiring. But the financial benefit is not the only profit that can be achieved from the application of computer networks in industrial automations. With the use of communication networks in an industry, we have the ability to receive and process any data desired in a central control station. This, in addition to many others, leads to faster and more accurate production planning, as well as faster optimization and tuning of controlled processes based on reference values received from the central control station. With the use of communication networks, increased operational reliability is also achieved. In the networked automation approach, errors are detected at an early stage, while the source of errors can be easily identified using the existing data of a control process. With the adoption of communication networks, individual processes can be controled independently and take place in parallel. This has, as a consequence, the whole industrial process to be controlled in a decentralized manner and at high speed from a central control station providing coordination, monitoring, and recording of data. The communication networks developed for the use in an industrial environment are called "industrial networks". They interconnect primarily programmable logic controllers, as shown in Figure 8.1, and secondly personal computers or any other form of digital communication-enabled devices, such as a robot controller.

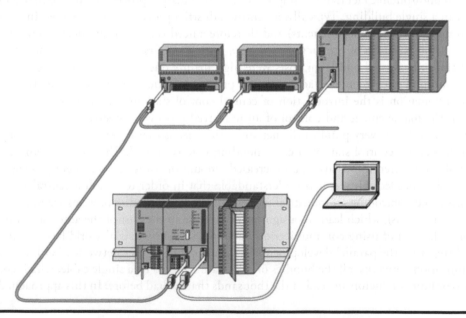

Figure 8.1   Interconnection of various PLCs and remote I/O units through the Profibus network (Siemens).

In the field of industrial networks, a lot of related products have been developed over the years by European and US companies, and organizations of automation. For only indicating the range and the variety of the current industrial networks, the names of the most famous industrial networks existing in the market today are listed below:

AS-I	Hart	Interbus	Modbus
Profibus	CC-Link	LonWorks	CANbus
Industrial Ethernet	Device Net	SERCOS	EtherCAT
DIN MESSbus	P-NET	SDS	ControlNet
Foundation Fieldbus	DH+/DH-485	Profinet	EthernetIP
Ethernet Power Link			

In addition to those, the networks LIN-bus and FlexRay are used mainly in motorcars, and the network Instabus is used in building automation. More details on the specific functionalities and characteristics of these networks can be found on the Internet in the form of online manuals, as well as from the corresponding manufacturers, thus these networks will not be further analyzed in detail.

8.1 Topology of a Network

The term "topology" refers to the way in which the PLCs of a network are joined together or, in other words, to the shape of the graph formed by interconnected lines. The topologies of networks are derived in two basic ways from their node interconnection, point-to-point connection and multipoint interconnection. The first consists of a circuit that connects two PLCs of the network without the interference of any intermediate node, while the second consists of a communication line that is shared between more than two PLCs in the network. Generally, industrial network topologies have various forms regarding the natural method of their node interconnection. The most prevalent topologies are the star, the ring, and the bus and these are analyzed in the following subsections.

8.1.1 Star Topology

In networks with a star topology, the connection between several PLCs is made through a central PLC, as shown in Figure 8.2. The central PLC accepts messages from the PLCs' senders and forwards them to the destination PLCs, except if the messages are intended for the receiving PLC. The star topology of an industrial network has the advantage that the linking of any two PLCs requires two communication lines at most. Thus, the time needed from sending until receiving a message is relatively small. On the other hand, the fact that the host PLC participates in the procedure of dispatching any sent message, can lead to situations of overcrowding and, consequently, to delays of the messages' distribution. The extension of a star network is fairly easy, because the only nodes involved in this implementation are the new and central PLCs. To achieve the network's extension, (a) a new communication line should be installed, (b) the two PLCs should be connected in the line and (c) the network routing tables that exist on the remaining PLCs should

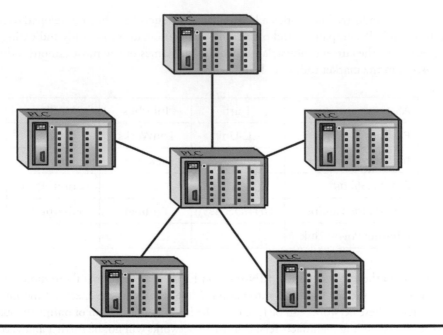

Figure 8.2   Star topology of a PLC network.

be updated for the addition of the new node. The networks with a star topology have relatively low reliability. Any damage to the central PLC will result in the shutdown of the entire network, while damage to any other PLC does not have a serious impact on the functioning to the network.

8.1.2 Ring Topology

In networks with a ring topology, all PLCs are connected in a point-to-point configuration, thus forming a continuous ring, as shown in Figure 8.3. Every message transmitted from a PLC sender contains the address of the destination PLC and is circulated from node to node in the direction

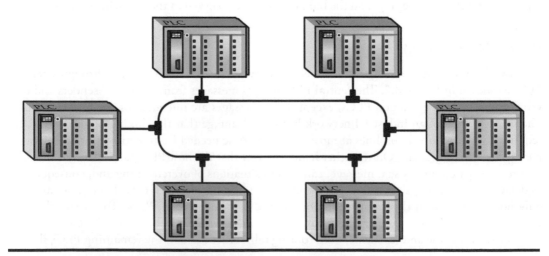

Figure 8.3   Ring topology of a PLC network.

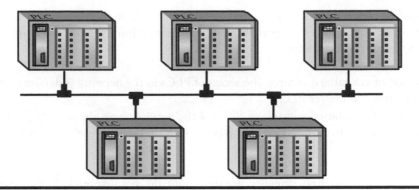

Figure 8.4   Bus topology of a PLC network.

of a ring, until it is received from the destination PLC that recognizes its address and is relayed to the PLC sender. The PLC receiver, before relaying the message, can notice that it has been read, and therefore when the message is returned to the PLC sender, it understands that the message has been received correctly. All the PLCs of this network usually have the same rights to send and receive messages, and therefore the control of the information flow in the network is distributed. Ring topology networks have a high reliability. In case of a failure of one PLC, the messages are driven to the adjacent PLC, and thus the network ring remains unbroken. Obviously, if a PLC is disabled, all other PLCs continue to communicate normally. The extension of a ring topology network is performed relatively easily, because only the PLCs that are adjacent to the new node are affected.

8.1.3 Bus Topology

In networks with a bus topology, all PLCs are connected to a common communication line that is a multi-point or multi-drop physical medium with open ends for transmitting data-messages, as shown in Figure 8.4. This common line is called a "bus" and is the only channel of communication between the PLCs of the network. Only one PLC can send a message through the common communication channel at any time. The message with the address of the destination node is transmitted along the bus until received by the PLC that recognizes its address.

The reliability of bus-type networks is quite satisfactory. The only case that may cease the operation of the entire network is when the common communication channel becomes totally disabled, something that is very rare. Any possible damage to a PLC has no effect on the functioning of the other PLCs, except that they cannot communicate with it. In general, the extension of a bus-topology network is excellent and straightforward by the addition of new PLCs. This addition is performed very easily at any point of the network, without requiring multiple connections and without hindering the operation of the network. Due to the above advantages, bus topology is prevalent and widespread in the area of industrial networks.

8.2 Communication Protocols

The term "protocol" in the field of communication networks generally refers to the rules according which a node can have access to the network bus. Access rules essentially constitute the control methods for the information flow in the network, which ensure that there will not be collisions in the channel of communication that could lead the network to instability, and that they will not interrupt or corrupt small or large amounts of transmitted information. The control of nodes' access may be central or distributed.

- In the case of centralized control, the access of PLCs to the network is determined by a central PLC called a "master PLC".
- In the case of distributed control, all PLCs of the network have the same rights to send and receive messages. Therefore, they don't have to wait in order to receive rights from a master PLC.

Access methods of the first category are deterministic, because each PLC communicates with another in a given time period. Access methods of the second category may be deterministic or stochastic. In the case of stochastic methods, the time a PLC communicates with another PLC is not definite, and is generally dependent on different access methods that also define the access times.

8.2.1 Master/Slave Method of Access

According to the Master/Slave protocol, a PLC in a communication network is defined as the master PLC, and thus all the remaining PLCs become slaves, as shown in Figure 8.5. The PLC defined as the master station controls the bus traffic entirely. The master PLC with a predefined order gives the control of the bus to the rest of the PLCs, and thus the ability to communicate to each slave PLC. The slave PLC acquires the exclusive use of the network for some definite time and distributes the information that is required by the corresponding application to the other PLC/PLCs in the network.

8.2.2 Carrier Sense with Collision Detection Method of Access

The full name of this access method is "carrier sense multiple access with collision detection" (CSMA/CD protocol). The term "carrier sense" derives from the fact that each PLC can sense

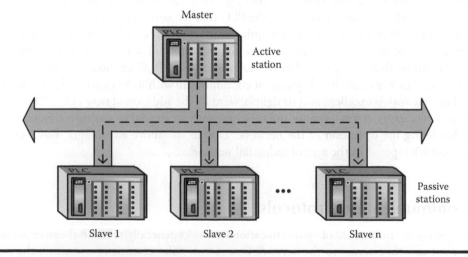

Figure 8.5 A network with one master and multiple slave PLCs.

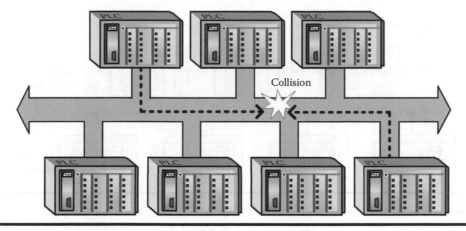

Figure 8.6   Communication collision in a network with carrier sense.

the bus (check its availability for transmitting information), while the term "multiple access" is defined from the characteristic that all the PLCs in the network have the privilege to access the bus. Furthermore, the term "collision detection" is defined from the protocol's ability to detect communication collisions that may occur, application of methods for avoiding them, and guarantee of the proper operation of the network. According to this access method, when a PLC needs to transmit a data packet, first it "hears" the bus, which means that it senses whether the line is idle and therefore available to be used. If the bus is busy, the PLC waits until it is released, otherwise it immediately transmits the intended information. In the case that two or more PLCs start to transmit at the same time in the free bus, then a collision will occur, as shown in Figure 8.6. In this situation, after the detection of the collision from the PLCs in the same network, all the collided PLCs stop their transmission, wait for the elapse of a random period of time (standby), and after that, repeat the whole process again from the beginning until they successfully achieve the transmission of their data packets. In this approach, the main disadvantage of this method is that theoretically there is no upper limit to the time that a PLC should wait in order to send a message, and thus it is characterized as a stochastic method with the characteristic that this back-off time can be increased widely until the collapse of the network. This drawback is the main reason that the method is not used when real-time control is needed, which normally happens in industrial processes.

8.2.3 Token Passing Method of Access

According to this method, a PLC gets the token, which is equivalent to the authorization of a data transmission, and transmits the data to the desired PLC. After the end of the transmission, the token passes to the next PLC, and the previous procedure is followed. The token is a certain number of bits, also called a "coded packet", which circulates throughout the network. Depending on the network's topology, the token circulation could pass in a ring, a bus approach, or a combination of these access methods. Figure 8.7 shows the application of an access method where the token passes initially in a bus topology. In this configuration, the token passes only by some selected PLCs in the network with a predetermined *a priori* sequence, creating a logical ring in this way. Each PLC of the logical ring knows which PLC is before and after it. When the logical ring is created for the first time, the PLC that initially gets the token transmits the data and, after

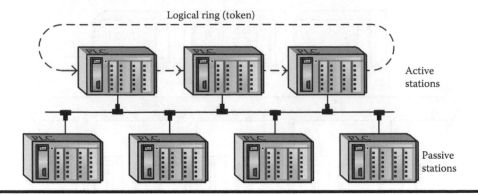

Figure 8.7 Token bus network with a logical ring.

the end of the transmission, passes the token to the next PLC in the sequence. Subsequently, the process is repeated cyclically along the logical ring, while the station that has the token is the only one that can transmit data on the bus. Finally, it should be noted that in industrial networks, the hybrid method of multiple master PLCs is commonly encountered, which combines the features of the token passing method on a bus and master/slave method.

8.3 Implementation of Industrial Networks

8.3.1 Data Transmission Media

In general, all the physical media for the communication of data transfers for classical public communication networks can also be utilized in the case of the industrial networks. The most common data transmission media (hardware) are presented next.

Twisted-pair cable. The simplest medium for data transmission in a communication network is a twisted-pair cable. It consists of two insulated copper conductors twisted around one another. The twisting of wires is made in order to avoid the electromagnetic interactions between two or more pairs when they are near. When several wire-pairs go parallel and at long distances, then these are grouped in bundles with an external protective casing. The wire pairs at these bundles would create an electromagnetic interference between them if they were not twisted. The signal transmission with twisted-pair cables is quite satisfactory (amplification is not needed) for distances of a few kilometers. For very large distances between two network nodes, repeaters are required in order to communicate. Analog and digital signals can both be transmitted via a twisted-pair cable. In this case, the bandwidth depends on the thickness of the wires and the transmission distance, while generally some Mbits/sec for distances of a few kilometers can be achieved. In the twisted-pair case, the main advantages are the low cost and the ease of installation, while the main disadvantages are their sensitivity to noise and the relatively slow speed of transmission.

Coaxial cable. The coaxial cable consists of a stiff copper wire (core) that is surrounded by an insulating material. The insulating material is covered by a woven copper braid or a metallic foil that acts both as the second wire in the circuit and as a shield for the inner conductor. The external conductor (grid) is covered by a protective plastic cover. There are two types of coaxial cables, the 50 Ω cable used for digital transmission (base band) and the 75 Ω cable used for analog transmission (wide band). The coaxial cable is characterized by a high bandwidth and a satisfactory tolerance to noise. The bandwidth varies depending on the length of the cable. For coaxial cables with a length of 1 km, the data rate can be as high as 10 Mbps. For smaller or larger distances the data rate is correspondingly higher or lower. Coaxial cables are used by many companies in commercial industrial networks for the interconnection of PLCs.

The 75 Ω coaxial cable through which analog signal transmission is performed, is called a broadband cable, and is used between others in cable television systems for transmitting video, voice, and data. It is characterized by a high bandwidth and a long transmission distance. In wideband transmission systems, the total available bandwidth is divided into a series of non-overlapping frequency channels, usually in 6 MHz channels, used for television broadcast. Each channel could transmit analog TV signals, audible signals, and digital data independently from other channels. Video channels, after being digitized, could be compressed to better fit the channel width. For transmitting digital signals in an analog network, every digital connected device should be equipped with electronic circuits able to convert the outbound train of bits into an analog signal and the incoming analog signal into a train of bits. In conclusion, the 75 Ω coaxial cable is suitable for all the types of digital audio, digital video, and data signals.

Fiber-optic cable. Digital data transmission via an optical fiber can now be considered as a classic broadband trend. Many manufacturers of PLCs have adopted optical fibers as the transmission medium in the respective networks on the market. It is known that a light pulse can be used to represent the binary digit "1", while the absence of a pulse can represent the binary digit "0". To perform an optical transmission, a light source, a transmission medium, and a light detector are required. The transmission medium is a very thin filament of glass or fused silica. The light source is either an LED or a laser diode that both emit pulses of light when voltage is applied. The detector is a photodiode that produces an electrical signal in response to an incident of light. By connecting an LED or a laser diode at one end of a fiber optic cable and a photodiode at the other end, a one-way data transmission system can be created, which is able to accept an electrical signal to convert and transmit light pulses subsequently, and then, at the other end or at the signal exit, reconvert it into a similar electrical signal as the input one.

8.3.2 The ISO/OSI Model

In the field of communication networks, there has been a complete incompatibility in the various hardware and software manufacturers, as well as between the communication protocols used in these networks. Thus, at the beginning of the communication network era, it was impossible to connect two computers or PLCs from different manufacturers on a common network. The International Organization for Standardization (ISO) developed the Open Systems Interconnection model (OSI) in order to describe the communication process in the field of networks. The model has a key feature, that of a hierarchical seven-layer stratification. Each layer is directly linked with the previous one and has a definite relation with the rest. Each layer takes care of a very specific job, and then offers its services onto the next higher layer. The interface is so flexible that each manufacturer has the possibility to create its own communication protocol consistent with the OSI model. The ISO/OSI model is a reference tool for data communications, and its seven layers are ranked from the lowest physical layer, data link, network, transport, and presentation to the highest application layer. In these layers, all the specifications of the communication networks are defined, from electrical details of bit transmission, the voltage that would represent the binary digit "1" and bit "0", the time length of each bit-pulse, up to the applications concerning the user. Finally, the ISO/OSI model defines terms that networking professionals can use to compare basic functional relationships of different networks.

8.3.3 Network Devices

Due to various factors, industrial networks (as well as networks in general) have a limited capability in terms of both the number of interconnected stations as well as the maximum length of interconnecting network cables. Therefore, problems are often created when it is necessary to extend a network of PLCs or to interconnect an existing network with another one. To overcome these limitations, four network devices are used, namely the repeater, the bridge, the router, and the gateway. With the help of these devices, it becomes possible to extend a PLC network, either in terms of the number of connected stations, or in terms of geographic area and range. Also, with the help of these devices, the communication between heterogeneous networks of PLCs is possible.

Repeater. The repeater is used only in homogeneous networks, i.e., networks that follow the same exact communication protocol. The repeater just regenerates and forwards bits from one sub-network to another, making two sub-networks look like one. Great distances are covered by concatenating two or more network segments via a repeater. Repeaters do not need software to operate, since they just copy and regenerate bits blindly, without any further processing on the repeated data packets.

Bridge. The bridge performs the connection of two networks which have different data link layers (data frames with different media access control addresses) but the same network layer as in the OSI model. The data frames from one network segment, reaching the bridge, are analyzed. After checking and filtering the destination address, they are forwarded to the second network segment. Bridges, unlike repeaters, use software for their operation. They are programmed for exchanging certain data between subnets and generally for carrying out the necessary modifications during their operation. As an industrial example, by using two CAN@net II/Generic bridges by the HMS/IXXAT Company, a CAN-Ethernet-CAN* bridging can be implemented. This bridge allows the exchange of CAN messages between two separate CAN networks via the Internet, where filter tables can be defined.

Router. A router can connect two dissimilar networks. In terms of the OSI model layers, a router connects two networks that differ in the three lower levels. This means that the router is used when two networks have the same transfer layer but different network layers of the OSI model. The router, unlike the bridge, may have its own address in the network. Some routers are also used as industrial secure devices protecting control networks from critical assets.

Gateway. The gateway is the most general type of device for interconnecting heterogeneous networks. It has the ability to interlink completely different networks in terms of communication protocols, even networks that are different in all the seven layers of the OSI model. The gateway is also used for the interconnection of networks that do not follow the OSI model. Gateways are generally more complex than bridges or routers. Figure 8.8 shows the use of the above devices

---

* Controller area network (CAN) is a well-known industrial network.

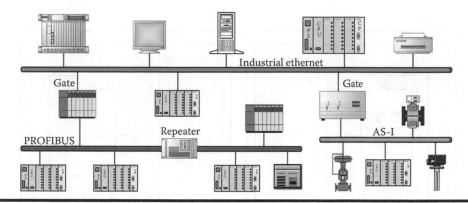

Figure 8.8 Three different networks (corresponding to machine, production control, and supervisory levels) are interconnected through network devices.

to create an integrated interconnection of different or similar networks, called a "multi-network communication structure".

8.3.4 The Communication Task of PLCs

The need for data exchange between PLCs and other digital devices or controllers in an automated industrial process has led all manufacturers to supply their controllers with additional communication capabilities. The hardware for the communication capability of a PLC appears in two forms, as is described subsequently. In compact type PLCs, the communication hardware and software are embedded in the module of their CPU, with the communication port as the only visible part. In modular type PLCs, the communication hardware is available as separate modules of communication of various types and features, which have already been mentioned in Section 6.7. Regardless of the form of the hardware, the communication process of a PLC fits together with the classical automation program inside the scanning cycle of the PLC, as shown in Figure 8.9, where two PLCs communicate over a network. It is obvious that this communication process refers to the read/write operations of the variables to be transferred over the network and not to the communication task of the network operation that is performed by the communication processor-module. The transmitted data over the network are stored in the buffers of the communication modules according to rules defined by the communication protocol of each network. As shown in Figure 8.9, the PLC in each scan cycle updates its memory with data that have arrived over the network, executes the corresponding instructions based on the updated values in the PLC's memory, and updates the buffer of the communication module with variable values obtained from the execution of the automation program that needs to be transmitted through the network. These steps are combined suitably with the rest of the classical steps of the automation program execution, including the reading of inputs and writing of outputs in the PLC, which have been described in detail in Section 6.1.

The addition of the communication task on the rest of the computing work for the scanning cycle increases the duration of the latter. On the other hand, the communication network has its own response times, which are not related to the duration of the scanning cycle. For the connected system of two PLCs of Figure 8.9, it is defined as a response time R_T; the time that elapses from the moment that an input is activated in the PLC1 until the moment that an output will be activated in the PLC2. Obviously, in the automation logic program of the PLC2, the output status depends

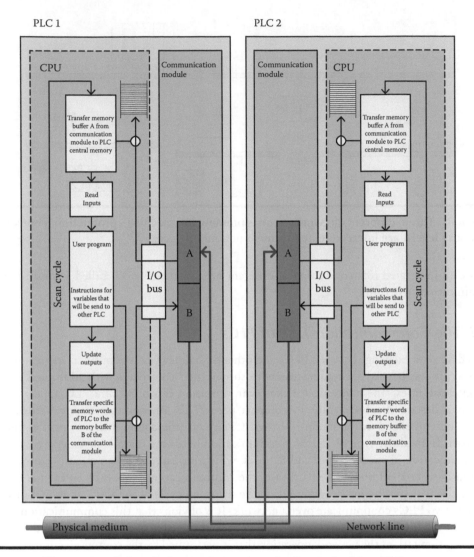

Figure 8.9 **The scan cycle in relation to the communication task of two PLCs.**

on the status of the PLC1 input. The response time or the total communication time is calculated by the equation,

$$R_T = I_T + 2ST_1 + PT_1 + A_T + T_T + PT_2 + 2ST_2 + O_T$$

where

I_T = Input delay time, which is the time that elapses from the moment that an input contact closes until the moment that this 1-bit information is available for reading in the PLC1.

ST_1 = The scan cycle time of PLC1.

ST_2 = The scan cycle time of PLC2.

PT_1 = The processing time in the PLC for the preparation of data to be transmitted.

PT_2 = The processing time in the PLC2 for receiving arrived data and preparation in order to be available for the automation program executed in the PLC2.

A_T = Waiting time of a PLC to access the physical network.

T_T = Data transmission time for the given data transfer rate (baud rate) of the network.

O_T = Output delay time, the time that elapses from the moment that 1-bit information is available in the output module until the moment that the corresponding output is energized in the PLC2.

The numeric values for all the above parameters make sense only in the case that these refer to a specific network and with a certain PLC. To give a sense of the time correlation between a PLC and a corresponding network, it is reported that, in many cases, the access cycle time of the network may be faster than the scanning cycle of a PLC that is connected to the network. For example, in a Profibus network with the smallest possible speed (selected and programmed based on the length of the network) and a usual number of nodes, it is possible to have a 2–3 network operation cycles per scan cycle of a PLC.

8.3.5 The Actuator-Sensor Interface (AS-I) Network

In an industrial process, all actuators and sensors should be connected to the programmable automation system. The industrial network AS-I that first appeared in 1993 and was reissued in 2000 supports communication at this level, known as machine level. Figure 8.10 shows a typical configuration of the industrial network AS-I. At the machine level, where there is a large number of sensors and actuators, the installation and use of the AS-I network allows us to avoid the numerous wiring that is replaced with a two-conductor cable for power and data transfer simultaneously, as shown in Figure 8.11a. The special cable of the AS-I network is depicted in Figure 8.11b, and is known as the yellow cable, and the applied technique of connection without removing the insulation make the wiring work simple, while providing greater security in the data transmission. In the same figure, it can also be observed that nowadays it is possible to interconnect even simple push buttons, something that was impossible as a non-supported feature a few years ago. The topologies that can be configured to implement the AS-I network are the bus, star, ring, and

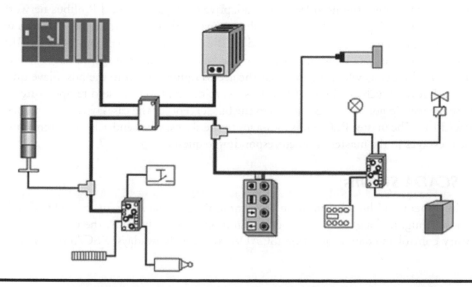

Figure 8.10 **Interconnection of PLCs, sensors, and actuators through the AS-I network (Siemens).**

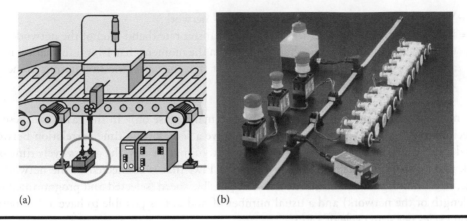

Figure 8.11   **The AS-I network can interconnect sensors and actuators (a), as also simple devices (buttons, switches, etc.) (b) at the machine level.**

tree. Furthermore, the AS-I network has the ability to communicate easily with other industrial networks like DeviceNet, Profibus, etc.

8.3.6 *The Profibus Network*

Except for interconnection of sensors and actuators, there is a big need for interconnecting controllers of all kinds that exist at the level of control of a production process, such as robot controllers, industrial PCs, PLCs, CNC machine controllers, programming devices of PLCs, etc. At this level of a computer integrated industrial environment, strong communication abilities are required, such as a large volume of data, real-time high speed, a large number of nodes, and a large range. One of the networks with such characteristics is the Profibus network for a wide range of industrial applications in manufacturing, process, and building automation. A PLC can communicate with machines from different manufacturers via a Profibus network that is a vendor-independent fieldbus standard without the need for special adaptive settings. A typical Profibus network configuration is shown in Figure 8.12. In the Profibus network, there is separation of PLCs on master/slave units. The token passing procedure is used for the communication of master PLCs. The master PLCs designate the data communication on the bus, and can transmit messages without the need for external requests when the PLC has the access rights (token) to the bus. Slave units may be peripheral devices such as PLCs, I/O devices, servo drives, inverters, and temperature controllers. These devices do not have direct access to the bus, except for the cases where it is requested by the master PLC. The master PLC sends messages to the slaves, who send acknowledgment or send response messages to the master upon corresponding requests.

8.3.7 *SCADA Systems*

The establishment and broad utilization of communication networks in the field of industrial automation brought, along with the economic and operational benefits, the development of the supervisory control and data acquisition (SCADA) systems. Nowadays, a SCADA system is used

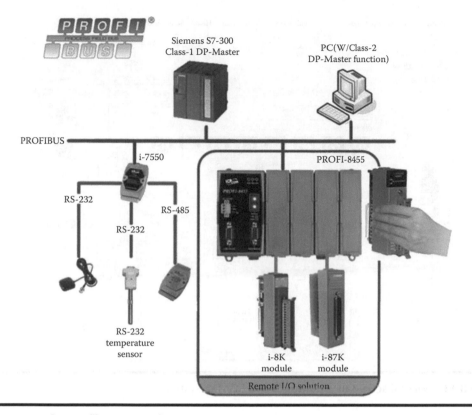

Figure 8.12   The Profibus network.

to monitor and control an entire production system or plant and consists of a series of operations, the collection of information, the transfer of it to a central station through the network, the carrying out of the necessary data analysis, the indication of data in a number of virtual "screens", and finally the possible control actions. It is essentially a large software package-tool, which collects all the required information through the network from the controlled system, so that it can implement a number of operational functions while visualizing it to the user in a friendly and realistic approach. The most significant feature of SCADA systems is the ability for downloading and storing data, graph production, the graphical display of the process with simulated motion of the machines, the announcing of alarm situations, the presentation of real-time values in active fields, the recording and printing of normal or exceptional events, the statistical processing of data, the status monitoring of the overall communication network, the communication with the external databases, and many other features that provide the end user with a direct overview of the controlled industrial process. It is obvious that the control actions applied by a SCADA system can be automatic or manual. This means that an operator can start or stop machines through the SCADA system while Figure 8.13 shows an example of an industrial process depicted in a SCADA software environment.

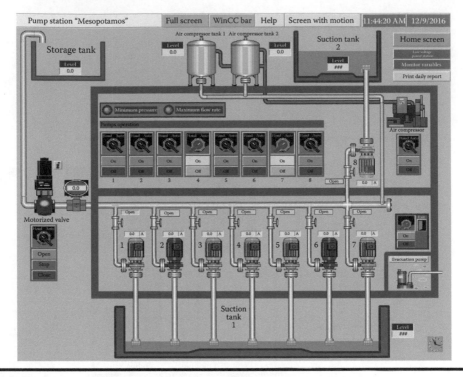

Figure 8.13   Virtual screen of a pump station SCADA system.

Review Questions

8.1. What is the purpose and what are the advantages of using communication networks in industry? Comment upon the existence of many industrial networks in the international market.

8.2. In addition to PLCs and cable, what others devices can be used in an industrial network and for what purpose?

8.3. At the lowest level of communication of production machines, the so-called "field level", what is the most significant characteristic that the industrial network must present and for what reason?

8.4. Why do we need industrial networks in an industry? Write down as many technical impacts of the operation of an industrial manufacturing or process line as possible.

8.5. A new PLC has to be connected on a preexisting industrial network. What hardware and software are required in order to perform the connection? What kind of programming and which units are required in order to achieve the operative connection of the new PLC with the other interconnected PLCs to exchange information and control commands?

8.6. Describe the access method of a PLC in an industrial network called "token passing in bus topology".

8.7. In your opinion, which are the two most basic network topologies that are used in industrial networks? Please describe their advantages and disadvantages.

8.8. What is a "SCADA system"? What does it include, how does it operate, and for what purpose? Which operational functions does it offer the user?

8.9. If a system is geographically distributed (like a town water system, a gas or oil pipeline, etc.), how do you think that its communication integration is achieved?

8.10. In an industrial process, there are 100 smart sensors with embedded microprocessors and communication ports, and 150 pure analog sensors all of which must be interconnected. In which manner will you interconnect the 250 sensors through a suitable industrial network?

8.11. Using Internet databases, find what is called a "computer-integrated manufacturing" (CIM) model. What kind of networks and technical characteristics are suitable at each one of the three levels of the CIM model?

8.12. The figure shows a possible configuration of an industrial Ethernet network consisting of a number of the same kind of subnetworks. Consider any kind of information that can be derived from the figure and write comments for any topic such as communication architecture, topology, devices, physical medium, etc.

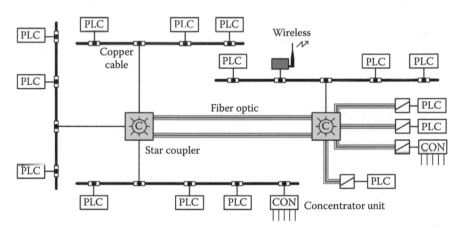

Chapter 9

PID Control in the Industry

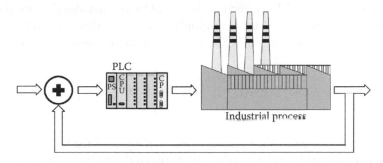

In this chapter, the basic topics of PID control are described, while focusing particularly on those properties of PID control that are mainly related to an application in industrial processes. The theoretical aspects of PID control are subject to automatic control systems and are taught in the corresponding textbooks, thus no further analysis of this subject will be performed here, as it is assumed that the reader already has a basic background on control theory for understanding the operation of PID from a mathematical approach. Thus, in this chapter, specific reference will be made only to the essential elements of the PID control theory with a focus on the application of PID and the proper corresponding tuning of the overall control scheme. More analytically, this chapter will present how PID controllers operate and why they have prevailed in the field of industrial control for almost 70 years now, without having appeared as something more effective than their digital implementation.

9.1 PID Control

A feedback controller is designed to produce an "output", which acts correctively in one process, in order to lead a measured process variable to the desired value, known as the set point. In Figure 9.1, a typical feedback control loop is shown, where the blocks represent the dynamics of the whole system (controller and controlled process) and the arrows represent the flow of information either in the form of electrical signals or in the form of digital data. All the feedback controllers determine their output by taking into account the error between the desired and the measured actual value

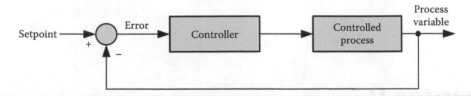

Figure 9.1 The controller determines its output based on the difference between the desired value and the actual value.

of the controlled variable. For example, a home thermostat is a simple ON-OFF controller that activates the heating system when the difference (error) between the actual and the desired room temperature value exceeds a threshold. A PID (proportional, integral, derivative) controller implements the same function as a thermostat, but determines the output with a more complex control algorithm. In particular, it takes the current value of the error in the series, the integral of the error in the latest time period, and the current value of the derivative of the error into account, in order to determine not only the size of the correction that should apply, but also the time duration of the corrective action. These three quantities are multiplied by three different gains (P, I, and D) with their sum as the final controller's output (CO(t)) according to the following equation:

$$CO(t) = P \cdot e(t) + I \left(\int_0^t e(\tau) d\tau \right) + D \left(\frac{d}{dt} e(t) \right) \tag{9.1}$$

In Equation 9.1, P is the gain of the analog term, I is the gain of the integration term, D is the gain of the differentiation term, and e(t) is the error between the desired value (set point) SP(t) and the actual value of the variable PV(t) at time t,

$$e(t) = SP(t) - PV(t) \tag{9.2}$$

If the current error is large and unchanged for some time or is changing rapidly, the PID controller will try to make one large correction of the system behavior by producing a respectively large output. Conversely, if the process variable is very close to the desired value for some time, the PID controller will remain idle. Of course, the issue of the proper operation of a PID controller is not as simple as the last paragraph probably implies. The difficult task is that of tuning the PID controller, which consists of selecting the values of gains P, I, and D so that the sum of the three terms in Equation 9.1 give an output that will drive the process variable with stability to changes that will eliminate the error.

How well the PID controller will be tuned depends on how the controlled process responds to the corrective actions of the controller. Processes that respond instantaneously and predictably don't require a feedback signal. For example, a car's lighting system rapidly reaches the desired output value (light) when the driver presses the corresponding switch without needing corrections of the real value from a controller. On the other hand, the fixed-economic speed controller (cruise control) of a car cannot accelerate very quickly to the desired speed that the driver chooses. Due to friction and inertia of the car, there is always a delay between the time that the controller activates the throttle and the desired speed of the car. Generally, a PID controller should be adjusted by taking these delays or the physical parameters of the controlled system into account.

PID Control in Practice. For an investigation of the PID's operation in real life, let's consider a wastewater treatment system that is a very slow chemical process where, as is widely known, the cleaning of an average wastewater quantity lasts several hours and generally responds slowly to the action of the controller. If there is a sudden error (e.g., change of set point), the PID controller's reaction will be determined mainly by the differentiation term in Equation 9.1. This will cause the controller to start an explosive corrective action when the error will change value from zero. The proportional term will then affect the control signal in order to maintain the output of the controller until the error reaches zero.

Meanwhile, the integration term will also begin to contribute to the output of the controller as the error accumulates over time. After a period of time, the integration term will prevail on the output signal, because the error will slowly diminish in the sewage treatment process. Even after clearing the error, the controller will continue to produce one output based on past errors that have accumulated in the controller's integrator. Then, the process variable will surpass the desired value, creating an error with an opposite sign from the previous one. If the integration gain (I) does not have a large value, the next error will be smaller than the initial one, and the integration term will begin to diminish as negative errors will be added to the previously positive term. This operation can be repeated a few times until the current error and the accumulated error are eliminated. Meanwhile, the differentiation term will continue to add its portion in the output of the controller, based on the derivative of the varying error signal. The proportional term will also contribute positively or negatively to the output signal of the controller, depending on the error.

In the case that a fast process responds quickly to the action of a PID controller, the integration term will not have a significant contribution to the output of the controller because the errors will have a very short duration. On the other hand, the differentiation term will tend to get large values because of rapid changes in the error and absence of long delays.

It is clear, from the above detailed description of the behavior of a PID controller that the effect of each term in Equation 9.1 on the value of the output of the controller depends on the behavior-response of the controlled process. For the sewage treatment process, a large value of the differentiation gain D could be desired in order to accelerate the action of the controller. An equally large value of the D gain for a quick process could cause an unwanted fluctuation of the output of the controller, as any change in the error will be amplified by the action of the differentiation term. In conclusion, the optimal choice of the three gains P, I, and D for a specific application is the essence of PID controller tuning.

PID Controller Tuning Techniques. There are three principal techniques to configure the parameters of a PID controller that are going to be mentioned briefly without expanding from a theoretical analysis. Therefore, the reader should refer to the corresponding theory of automatic control systems for a detailed description of the first two of them, since the third technique is based on engineering experience.

The first technique is based on a mathematical model of the process that associates the value of the process variable PV (t) with the rate of its variation and a number of previous values of the output of the controller, for example the equation,

$$PV(t) = K \cdot CO(t-d) - T\left(\frac{d}{dt}PV(t)\right) \tag{9.3}$$

Equation 9.3 refers to a process with a gain K, a time constant T, and a dead time d. The gain K of the process represents the size of the controller's action on the process variable. A large value of K means that the process converts the small controller actions in major changes of the process variable, while the time constant T represents the time delay of the process. For the sewage treatment process, the time constant T will have a large value. The dead time d refers to another kind of delay that is found in all processes when the sensor used for measuring the controlled variable is located at a distance from the actuator that implements the action of the controller. The time required for the actuator's action to affect the process is the dead time. During this time interval, the process variable does not react to the action of the actuator. Only after the dead time, the process variable starts to react substantially and then it begins the measurement of the time delay T.

The second technique, known as the Ziegler-Nichols method (which first appeared in 1942) is the most popular because of its simplicity and applicability in any process that can be modeled in the form of Equation 9.3. The technique consists of three tuning rules for the PID controller, which convert the parameters of Equation 9.3 into values for the gains P, I, and D of the controller and are expressed by the equations,

$$P = \frac{1.2T}{Kd}, \quad I = \frac{0.6T}{Kd^2}, \quad D \frac{0.6T}{K}, \quad (9.4)$$

The Ziegler-Nichols method also proposed a practical method for the experimental estimation of the values of the parameters K, T, and d of a process.

The third technique is empirical and based on the iterative procedure "trial and error" by trying out a set of three values for the constants of the PID controller and observing the behavior of the error. Depending on the behavior of the corresponding error, the gains of the PID controller can be further tuned by their direct increase or reduction. Experienced control engineers know very well according to the controlled process and after some test steps, how much is required to increase or reduce any constant of the PID controller in order to improve its behavior.

9.2 PID Control in PLCs

In the case of PLCs, PID control is implemented in two ways: In the first case, the PID algorithm is integrated in the programming software of the PLC (software controller) and is called as a subroutine, which means as an FB that has been developed by the manufacturer of the PLC. The PID algorithm exists in the relevant library, so that it only requires the declaration of variables and the parameters of the controlled process. The values of these parameters, which are necessary for the operation of the PID FB, are stored in a corresponding DB. In some PLCs, the PID algorithm is integrated in the CPU, and the programming environment provides the user with a menu in order to communicate with the controller and to set the required parameters. The physical magnitude that controls a PID FB is taken as a variable from one of the analog inputs of the PLC, and subsequently, the PID FB regulates the value of an analog output of the PLC. Depending on the capabilities of the PID algorithm developed as software from the manufacturer of the PLC, the user can select any kind of a control configuration from the P, PI, PD, and PID controllers that is suitable for the controlled process while, depending on the form of the selected PID controller, the necessary gains need to be tuned. For example, in the case of a PI controller, the D gain is automatically zeroed and the engineer needs to define the P and I gains only.

Concerning the declaration of parameters for a PID FB controller, the reader, who will encounter such an application in practice, should be prepared for a much larger number of parameters

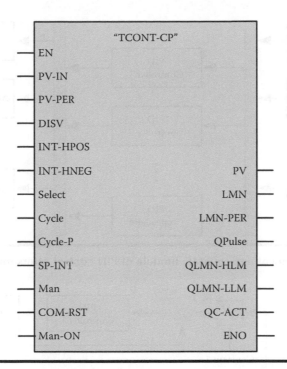

Figure 9.2 Example of parameters to be assigned in a PID FB program (Siemens).

than the three described in classical PID control theory. For example, Figure 9.2 shows the graphic symbol of the PID FB's call in FBD language of Siemens Step7 software, with more than 20 parameters that may be declared in accordance with the desired function of the PID controller.

After programming a PID FB, it will run along with the rest of the automation program within the time horizon of the scan cycle. Depending on the controlled process, this may result in an applied PID control that is not satisfactory because of the duration of the scan cycle. The latter has, of course, always a short duration that is satisfactory for the execution of the ON-OFF control, but which may not be sufficient for the action of the PID control, especially for fast processes. In this case, a second method of implementing a PID controller is followed.

This second approach consists of supplying a separate PID control module that was mentioned in Section 6.7, which has its own autonomous microprocessor and its own independent analog inputs and outputs. In this way, the PID control module carries out the function of control, regardless of the scan cycle of the CPU of the PLC, but communicates with the PLC to exchange the values of the parameters and variables. In addition, the independent module of the PID control has much more features than the built-in FB to the software algorithm. Furthermore, this module has the ability to implement more than one PID control loop, depending on the number of pairs of analog inputs and outputs. Figure 9.3 shows the block diagram of four PID control loops that can implement a PID control module with four analog inputs and outputs. The bars between the controllers and the analog inputs and outputs represent the ability of each PID controller to be combined with any (one or more) from the analog inputs and outputs. The four PID controllers can operate independently of each other or combined in different ways, as shown in Figures 9.4 and 9.5. The combined or individual operation of controllers is independent of the order of execution of operations on each controller. This is specified by the manufacturer, and is sequential from controller 1 to controller 4. This means that the conversion of analogue input 1 and the PID controller algorithm 1 will be performed first, and then the

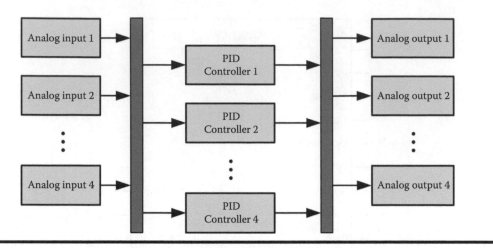

Figure 9.3 **An independent and separate module of PID control offers multiple control loops.**

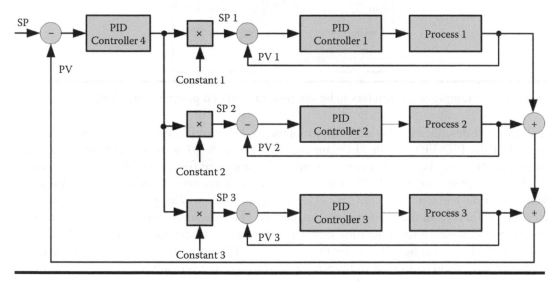

Figure 9.4 **Combination of PID controllers for controlling three interrelated sub-processes (Siemens, FM 355 PID Module).**

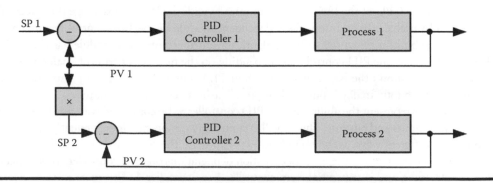

Figure 9.5 **Sequential control of two processes with two PID control loops.**

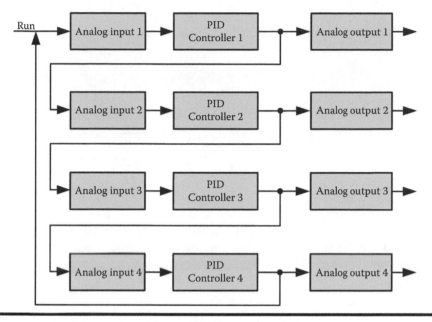

Figure 9.6 **Sequential execution of PID control in four controllers with a direct update of analog outputs per execution (Siemens, FM 355 PID Module).**

controller 2 operation will follow, and so on, as shown in Figure 9.6. The result of the PID algorithm on each controller is transferred directly to the corresponding analog output without waiting for the operation of the next controller. If a controller is not used by the user, then its function is bypassed and therefore does not contribute to the total time of the operation cycle of a PID control module. In this case, special virtual panels provide the user with a form of menu, the ability to easily define the number and type of controller terms (P, PI, PD, and PID), the combination of inner loops to introduce the values of the PID controller's gain and, in general, to organize the control task of the process, as happens in the FM-355 closed-loop control module of Siemens, for example.

In some PLCs and in their corresponding software for a PID FB, a method for self-regulation (auto-tuning) of the PID controller's gains has been integrated. This facilitates the work of the control engineer, provided that the basic characteristics of the controlled process and especially the slow or fast response of the process under control are known *a priori*. One method for the self-regulation of the PID controller that is widely used from control engineers is the so-called technique of "feedback with relay" by Aström and Hägglund (1984). Over the last 25 years, this technique has been used by a large number of manufacturers, as well by developing proper FBs. The central idea of the technique is the supply of a small and sustained oscillation of the controlled process, which is generally stable. Based on the period of oscillations and the size of variations that are observed in the process variable, the fundamental frequency and the gain of the controlled process can be calculated. Based on these two values, the auto-tuning algorithm is able to calculate the values of the gains P, I, and D of the controller, which obviously would not be the optimum but would still be very close. Finally, it should be noted that special attention should be followed in this method not to destroy or create fatigue in the mechanical parts of the controlled process by the application of the oscillating inputs.

Chapter 10

Industrial Applications

An automated production system is one in which a process is performed by a number of machines without the direct participation of a human worker. The whole automated system is built from smaller automated units suitably coordinated. Such smaller industrial applications of automation are presented in this chapter, most of which have the form of a project for development rather than the scope of a tutorial problem. Some industrial applications met outside a factory environment are also included. If the reader can create proper industrial automation solutions for all of all these projects, it means that he/she is ready to work as professional engineer in the field of industrial automation.

10.1 Cyclic Operation of Traffic Lights

Figure 10.1 shows the time function of traffic light operation at an intersection for two modes of traffic regulation: normal and night operation. The selection of the operation mode is made manually through a rotational switch, but may also be automatic by a sensor signal or a remote signal received from a traffic control center. Write the required program for the operation of the traffic lights.

10.2 Conveyor System for an Assortment of Objects by Pairs

In the conveyor layout of Figure 10.2, the arrangement of objects in pairs is performed. Each pair is received by the conveyor M_3 and should contain one black (B) and one white (W) object, labeled either as BW or WB. The conveyor M_1 brings objects with a random color. The conveyor M_2 is a specific buffer of four positions, each of which can be placed in a central line M_1M_3 by the corresponding activation of relays C_1–C_4 that drive four pneumatic cylinders (e.g., the activation of relay C_3 brings position 3 in the central line). Each position of the buffer M_2 accepts only two objects. The movable metallic plate binds the forward movement of a single object. After the pairing of a BW or WB is formed, the metallic plate goes down for 5 s by the activation of relay C_5 and the pair is transferred to M_3. Of course, there is a mechanical coupling of the M_3 motion with each position of M_2 for the transferring of the pairs. When the position of M_2 in central line

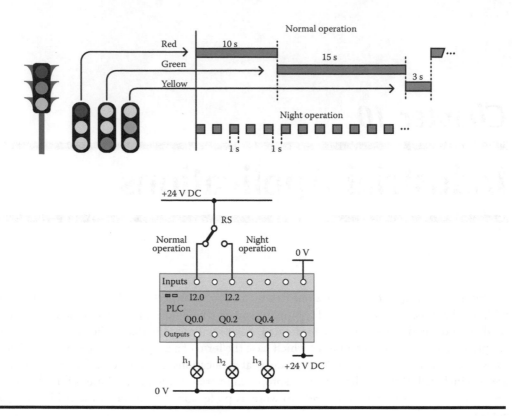

Figure 10.1   Cyclic operation of traffic lights.

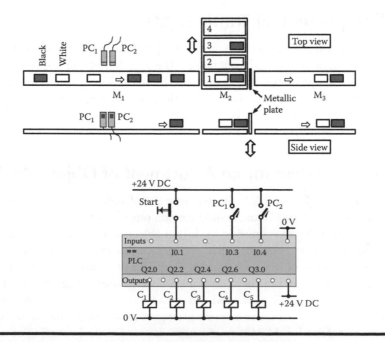

Figure 10.2   Conveyor system for an assortment of objects by pairs.

contains an object, and the next coming object has the same color, then the buffer M_2 must be moved to an empty position or to a position with an object of different color. The arrival of objects is random but suppose that it is impossible to find more than four objects of the same color. The conveyor system starts to operate after a momentary signal from the push of a START button and stops similarly through the push of a STOP button. Write the required program for a PLC.

10.3 Packaging System of Different Balls

Figure 10.3 shows a packaging system of balls in packets of six different items. There are 12 feeding lines, each one with different design. The balls come from all lines in a random way and at relatively slow rate. The packaging system is controlled by a PLC and operates according to the following specifications:

- With the momentary signal START, the machine produces 15 packets of 6 items (90 balls) and waits for the next start signal.
- Each ball, coming from the corresponding feeding pipe, falls on an inclined plane only if the solenoid SL_i is energized, and is detected by the photo-switch F_i. After detection, the solenoid is deactivated in order to block the falling of the same second ball, since the packet must contain different balls (i = 1–12).
- Each ball falling on the plane rolls towards to point A and is detected by the photo-switch F_0. The completion of six balls is verified by the permanent activation of photo-switch F_0.
- Just after six balls fall, the other balls stop falling, and after the completion of six balls, the wrapping is started at point A. The wrapping is performed by rotating the geared motor M by 360°, and the end of rotation is detected by the proximity switch S_0.
- If during the production of 90 balls, the number of balls from the same feeding line is greater or equal to 21, then this kind of ball is excluded from the next two sets of 90 balls in production and is reverted normally in the following set of 90.

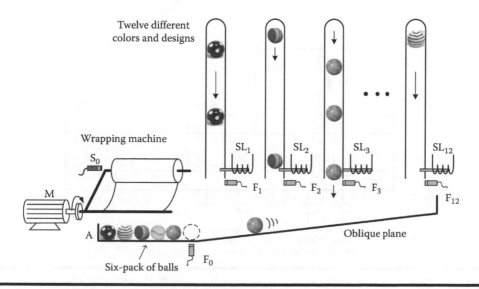

Figure 10.3 Packaging system of different balls.

After making a diagram of the I/O devices connection to a PLC, the required program may be written in Boolean or Instruction List language.

10.4 Conveyor System for Transferring Granular Material with Weight Control

Figure 10.4 shows a conveyor system for transferring granular material from a silo to a process machine (not shown in this figure). The silo has an electrically driven outlet door, which is open when the output Q4.0 is energized. The DC motor moves the conveyor belt with three speeds through the connection of the convenient resistors to the motor coils. A weight

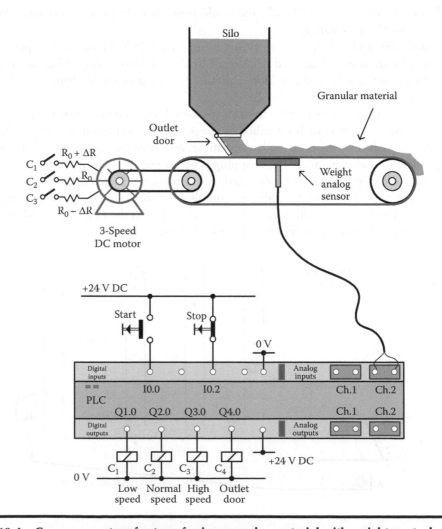

Figure 10.4   Conveyor system for transferring granular material with weight control.

transducer measures the weight of the granular material in the indicated position. It is desired that the conveyor belt operates at normal speed when the measured weight is equal to a pre-defined value of 1 Kg.

The normal speed is achieved by the activation of output Q2.0 (C_2). Due to irregular quantities of the material falling from the silo, the measured weight may vary. When the measured weight is less than 1 Kg, the conveyor speed decreases to "low" by energizing the output Q1.0 (C_1), which adds resistance to the motor coils. If the measured weight is greater than 1 Kg, the conveyor speed increases to "high" by energizing the output Q3.0 (C_3), which subtracts resistance from the motor coils. Obviously, only one of the three outputs connecting the resistors to the motor coils must be energized each time. The conveyor system starts to operate by momentary pressing the START button and then the conveyor belt turns on, and the outlet door opens. By pressing the STOP button at any phase of the operation, the conveyor stops and the door closes immediately. A program for the PLC implementation may regulate the speed of the conveyor belt automatically according to the above specifications.

10.5 The Food Industry: A Machine for Production of Tzatziki Salad

In the food industry, the machine for production of Greek tzatziki (a kind of green salad made from cucumber, yogurt, and garlic) consists of:

- Three silos containing the raw materials
- Three electronic weighing scales with rotating platforms
- A conical tank vessel for mixing materials
- A drum blender
- A dosing pump
- The conveyor belt for packaging of the produced salad, as shown in Figure 10.5a.

Technical characteristics of the equipment.

1. The outlet doors of silos open for feeding the weighing scales via the activation of the pneumatic cylinders C_1, C_2, and C_3, which are extended at the rest state. Each cylinder is retracted by the activation of the corresponding PLC output.
2. Each raw material falls on the corresponding weighing scale with a slaw rate of 100 gr/s. The electronic weighing scales have an SPST (NO) output, which is energized at each Kg of weight according to the graph shown in Figure 10.5b.
3. The existent raw material on each scale after weighing, falls in the conical tank vessel via the activation of pneumatic cylinders C_4, C_5, and C_6, which at rest state are also extended.
4. The drum blender and the dosing pump have direct starting motors. At the bottom of the tank vessel there is a level sensor for viscous material with an SPST output that is NO when there is no material in the tank vessel.
5. The conveyor belt operates with a direct starting motor (M_1) and brings the plastic containers to be filled at specific positions.

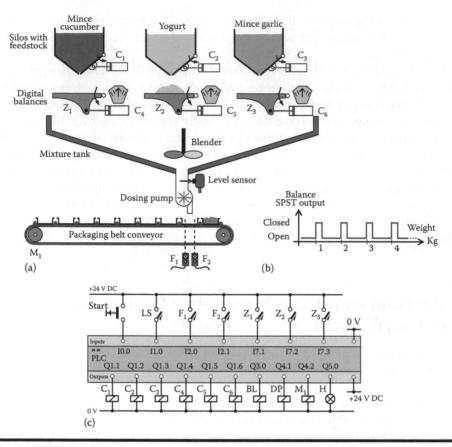

Figure 10.5   Machine for production of tzatziki salad (a), digital balance output (b), I/O device connections in PLC (c).

Operation specifications. The production procedure includes three phases: weighing the raw materials, the feeding and mixing phase, and the packaging phase, which are performed sequentially according to the following specifications:

1. The production process starts with a momentary signal START from the push of a button. Then, the outlet doors of the silos open for feeding weighing scales until the required quantity of material is completed at each scale. After the completion of a material quantity, the corresponding outlet door closes.

2. When the three required quantities have all been completed on the weighing scales, the cylinders C_4, C_5, and C_6 are retracted by activation of the corresponding outputs and cause them to fall in the tank vessel. The three cylinders remain retracted for 5 s in order to be sure that the viscous yogurt has fallen.

3. At the end of the time interval of 5 s, the rotation of the blender starts and keeps on for 12 min.

4. After the end of the mixing procedure, the dosing pump and the conveyor belt start to operate simultaneously. The pump and the conveyor are synchronized by construction during the overall operation.

5. Two photoelectric switches (with SPDT outputs) detect the presence of the plastic container below the outlet of the dosing pump. The two photoelectric switches are necessary due to the small gap between the two successive containers. When there is no plastic container below the pump outlet, the pump stops and operates only the conveyor until it brings the container below the pump outlet. Then the dosing pump starts to operate again.
6. The applied recipe requires 80 Kg yogurt, 20 Kg minced cucumber, and 5 Kg garlic pulp.
7. The filled containers must be counted. The applied recipe corresponds to approximately 300 filled containers. If for any reason there are fewer than 300 filled containers, then the indication lamp H turns on.
8. After the emptying of the tank vessel, all parts of the system stop or reset and the machine is ready for the next START signal.

It is recommended to write the required program in Boolean or IL language and use modular programming.

10.6 Retentive Reciprocating Movement of a Worktable

Retentive operation in a PLC means that in the case of a power interruption, its energized internal elements (M, T, C, and O) preserve their state that existed before the interruption. When the power supply is retrieved, the program of the PLC runs based on the conserved states of these internal elements. In other words, the PLC operation is continuous from the point that it had stopped. Suppose that the memory bits M10.0 to M10.7 have retentive property. The mechanism shown in Figure 10.6 has a worktable that performs a continuous reciprocating movement which (up-down) starts and stops by the same button alternately. The LAD program of the figure only performs the reciprocating logic without the exiting input and does not satisfy a retentive

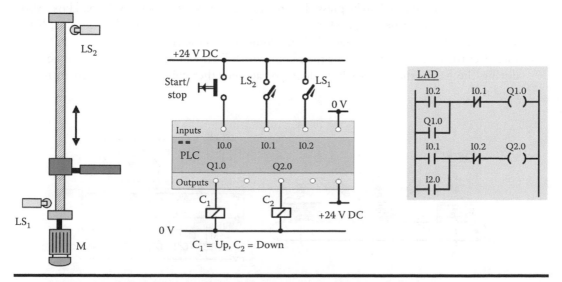

Figure 10.6 Retentive reciprocating movement of a workable.

operation. Write a program for a PLC to achieve the continuous reciprocating movement of the worktable, and present the retentive property, either by modifying the LAD program suitably or writing a new one.

10.7 Wooden Plate Stacking, Painting and Transferring Process

Figure 10.7 shows an arrangement for stacking, painting, and transporting wooden plates. After pressing an ON button, the feeder FD of wooden plates starts to operate. This means that a stack of 12 plates is created at point A, with each plate being counted by the photo switch F_1. Then the feeder stops and the conveyor belt M operates until it brings the stack from point A to the location B, with the last detected by F_2. Then the spray painting nozzles (PN_1 and PN_2) are energized for a time interval of 18 s. At the end of the time interval of 18 s, the painting nozzles are deactivated and the conveyor belt transfers the stack to location C, the last detected by F_3. By pressing the OFF button at any time instant, the operation of the arrangement stops immediately. Write the required program for a PLC.

10.8 An Automated Billiard Table Controlled by a PLC

This application is outside the industrial world, but includes operations very common in industrial processes, such as multi-input counting, signaling, and multiple phases of operation and detection. Figure 10.8 shows an automated billiard table controlled by a small PLC. The specifications of operation are the following:

1. During the game, balls fall from the four corners of the table and are detected by four photo switches F_1, F_2, F_3, and F_4, while all are gravity-driven to the ball box. Of course, how many balls are going to fall at each corner is a quite random event.
2. The table operates with a coin slot machine. In order for a coin to be accepted by the coin box, it is necessary for button b to be pressed, and all nine balls to be present in the ball box. Also, the solenoid SL must be energized in order for the coin to be driven to the coin box. If for any reason the SL is not energized, the coin comes out in the coin return slot.
3. Just after the detection of a coin by the photo switch F_5, the SL must be deactivated because during the game, the slot machine does not accept other coins but returns them to the player.

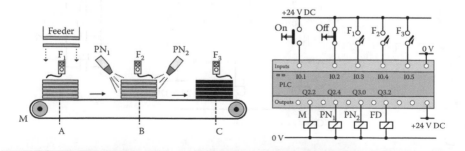

Figure 10.7   Wooden plate stacking, painting, and transferring process.

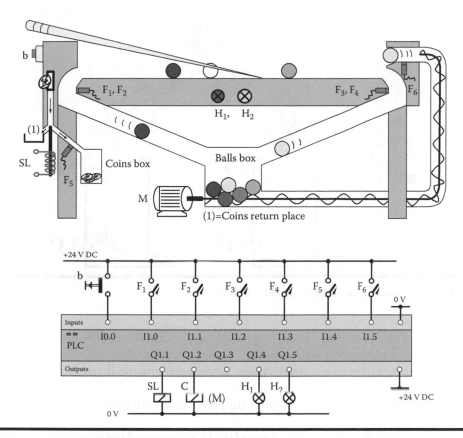

Figure 10.8   Automated billiard table controlled by a PLC.

4. With the same conditions in Step 3, two more actions must take place:
   a.   The motor M starts to operate, the balls are detected and counted by the photo switch F_6 and sent to the table. The motor M stops when the F_6 detects nine balls.
   b.   The red indicating lamp H_1 turns on, which means that the game is in progress.
5. When nine balls have fallen in the ball box, the red lamp turns off and the green lamp H_2 turns on, which means a new game can be started.

The required program has to be written for the I/O addressing shown in the figure.

10.9 Automated Filling of Two Milk Tanks

In a milk factory, a PLC automatically fills two milk tanks. As shown in Figure 10.9, each milk tank has two floater switches, up and down, which provide the signals "fill" and "empty" respectively. The specifications of operation are the following:

1. The automatic filling of the tanks is energized/de-energized by pressing the ON and OFF buttons respectively. The indication lamp H_0 shows the operation state.
2. Only one milk tank may be filled at a time. If the two tanks are simultaneously empty, then the first tank fills first.

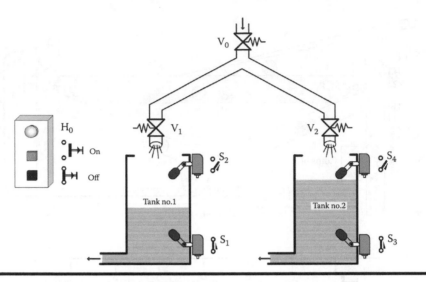

Figure 10.9   Two milk tanks with automatic filling.

3. When the PLC receives an "empty" signal (S_1 or S_3), then the electrovalve V_1 or V_2 opens immediately and the central electrovalve V_0 opens after 5 s. Then, when the signal "fill" (S_2 or S_4) is received, the electrovalve V_0 closes immediately and the electrovalve V_1 or V_2 closes after 10 s.
4. All electrovalves close immediately after the press of the OFF button.

Design a diagram with the connection statuses of the I/O devices and the corresponding addresses, and subsequently, write the required PLC program.

10.10 Modular Programming for a Set of Processing and Repairing Stations

A production line includes six processing stations and one repairing station, each one of which performs its own process or assembly as the produced device passes through them. Each station operates according to its automation program P_i if this is called (executed) by the PLC. The sensor S_2 detects possible defects of the produced device. In such a case, the device is forwarded to the repairing station that operates according to the program P_{123}, while stations 1 to 6 stop operating. When the repairing procedure has finished and the repaired device is ready to reinsert in the central production line (something that is expressed by the activation of sensor S_4), then the operation of stations 1, 2, and 3 must be bypassed. This means that only stations 4, 5, and 6 must start their operation or, in other words, the repaired device does not need processing in stations 1, 2, and 3. Sensor S_1 signals the arrival of a device for processing and assembly. Sensor S_3 signals the end of a device processing. The operation of the conveyor system is not examined here, as also it is considered that a new device does not come before the end of the previous device processing. It is ideal to develop a modular program for the stations' operation, called organization block, including all sub-programs P_i, i = 1–6 and P_{123} (Figure 10.10).

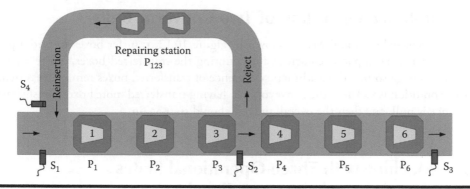

Figure 10.10   A group of processing and repairing stations.

10.11 Traffic Light Control of a Complex Intersection

At an intersection between three towns, as shown in Figure 10.11, the traffic lights are controlled by a PLC. Their operation is based exclusively on timers, i.e., there are no traffic sensors. The green left light flashes every second for 10 seconds. All the other lights operate according to the bar diagram. The cyclic operation of the lights starts/stops through the rotary switch RS_{0-1}. It is recommended that the required PLC program be written in Boolean or LAD language.

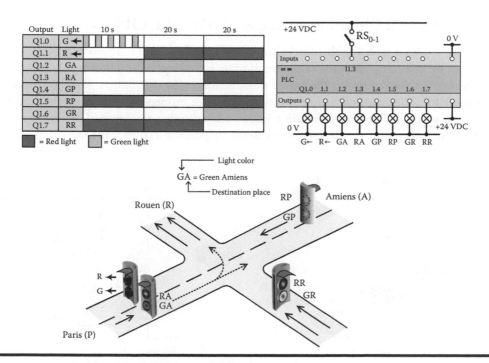

Figure 10.11   Traffic light control of a complex crossroad.

10.12 Combined Operation of Two Conveyor Belts

The two conveyor belts M_1 and M_2, as shown in Figure 10.12, transfer boxes toward a packaging machine. Each has a photocell sensor for counting the transferred boxes. Write a program in Boolean language so that the arithmetic difference of transferred boxes remains less than ten. When this condition is violated, the conveyor belt, having transferred more boxes, must stop until the difference is null, and then the overall process should start again.

10.13 A Machine with Three Operational States and an Acknowledgement Signal

A complex machine has three operational states: stoppage, preheating, and operation. If the machine is in stoppage state, the pressing of a START-STOP button energizes the message "Preheating-Wait" and the output of heating resistors. By pressing the START-STOP button for a second time and if the time interval of three minutes has passed, the motor C is energized, while the message indication and the resistors are de-energized. If during the preheating phase the cancel button is pressed, the message indication and resistors are de-energized. If during this operation the cancel button is pressed, nothing happens, while the START-STOP button is pressed the motor stops immediately. Write a PLC program to satisfy the desired operation specifications (Figure 10.13).

10.14 Chemical Cleaning Process of Metallic Objects

Figure 10.14 shows a tank for chemical cleaning of metallic objects. The basket with metallic objects must be immersed inside the tank several times according to the following plan. By instantaneously pressing the button b_0, the basket must be immersed three times into the tank and to remain there for 20 seconds each time. Between the three immersions, the basket returns to its initial (up) position and goes down again without delay through the forward/reverse motor M_1. After the third immersion, the basket returns to the up position and remains there until the next

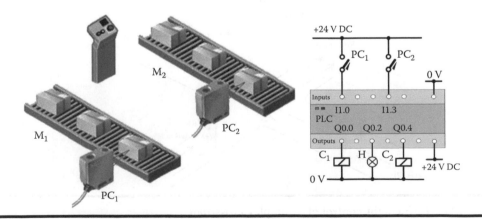

Figure 10.12 Combined operation of two conveyor belts.

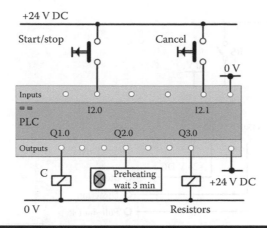

Figure 10.13   I/O device connections to PLC for the application of Section 10.13.

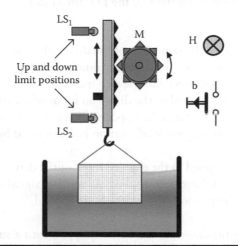

Figure 10.14   Tank-elevator system for chemical cleaning of metallic objects.

pressing of the button b_0. The indication light H_0 is turned on during the cleaning procedure. After making a diagram of the connection of the I/O devices to a PLC, write the required program in the LAD language.

10.15 Driving a Step Motor Through a PLC

Step motors are electric motors of specific construction for high-precision positioning applications in complex machines. In particular, step motors are brushless DC motors that rotate in discrete steps usually of 1.8° and in both directions, while they can operate in full or half step mode. Step motors require an electronic circuit called a controller or driver to energize the motor phases in a timely sequence to make the motor turn. As shown in Figure 10.15, the stepper driver accepts the following four control signals:

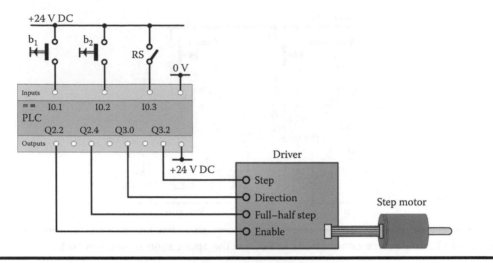

Figure 10.15   The step motor is controlled by the PLC through a driver.

Step. For each pulse applied to the step input of the driver, the motor turns one step ahead. For a series of pulses the motor turns by an equal number of steps. The frequency of the pulse waveform defines the rotation speed of the step motor.

Direction. A logic high signal applied to the direction input of the driver determines one direction, while a logic low signal causes the opposite direction of rotation.

Full half step. The full or half step mode of rotation is determined by a logic high or low signal respectively.

Enable. A logic high signal applied to the enabled input of the driver enables the motor to rotate if the previous three signals have been applied. A logic low signal disables the step motor that cannot rotate, even if the previous three signals are applied.

The above signals can be produced from a PLC equipped with a suitable digital output module. Write a program in Boolean language so that the following operations are achieved:

1. With the press of button b_1 and the RS switch open, the motor rotates left with a full step.
2. With the press of button b_2 and the RS switch open, the motor rotates right with a full step.
3. With the press of button b_1 and the RS switch closed, the motor rotates left with a half step.
4. With the press of button b_2 and the RS switch closed, the motor rotates right with a half step.

The desired speed of rotation corresponds to a pulse frequency of 25 Hz.

10.16 Stacking Machine of Light Objects

Suppose that a step motor is used in a stacking machine of light objects. The controller of the step motor has to save the number of objects that have already been stacked each time in its memory, in order to stack the next object. It is desired to develop such a controller for the motions in a PLC. Using the program written for the problem presented in Section 10.15,

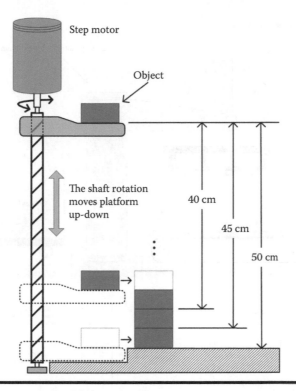

Figure 10.16 Stacking machine of light objects.

write a complementary one so that the PLC accomplishes the sequence of the step motor motions shown in Figure 10.16 and specified as follows:

- With the first pressing of b_0, the platform goes down 50 cm and returns to its upper location.
- With the second pressing of b_0, the platform goes down 45 cm and returns to its upper location.
- With the third pressing of b_0, the platform goes down 40 cm and returns to its upper location.
- With the tenth pressing of b_0 the platform goes down 5 cm and returns to its upper location.
- With the eleventh pressing of b_0 the platform, as in first pressing, goes down 50 cm and returns to its upper location.

Suppose that the rotation of motor axis by 1° causes a linear displacement of the platform of 1 mm. The b_0 is an NO button connected to an additive input I1.0 (not shown) of the PLC, while the outputs are as shown in Figure 10.15.

10.17 A Simple Robotic Arm for Pickup and Placement of Light Objects

A simple robotic station for receiving and stacking light objects is shown in Figure 10.17. For the needs of this industrial automation, suppose that all the movements of the robot are achieved with step motors. The whole system is controlled by a PLC, the I/O connections of which are shown in

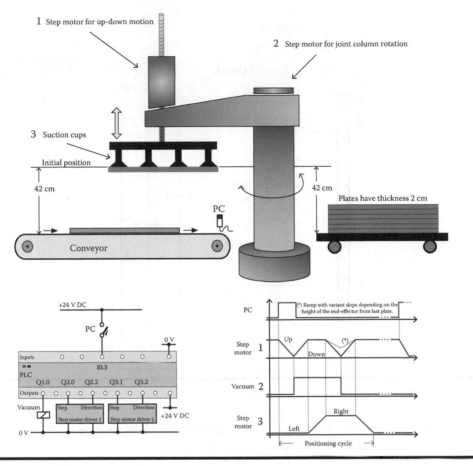

Figure 10.17 Robotic arm for pickup and placement of light objects.

the figure. Using the programs developed in the previous applications in Sections 10.15 and 10.16, write a PLC program so that each arrived object (signal from the photocell PC) is grasped by the robot and placed on the pallet, creating a stack of 20 objects. Particularly at each signal PC, the diagram of the robot movements shown in the figure must be accomplished, as well as the vacuum creation for suction cup feeding.

10.18 Heat Treatment Process in a Chamber Furnace

In heat treatment processes, a gas-fired chamber furnace is usually required. The sliding door of such an industrial furnace, shown in Figure 10.18, is controlled by a PLC. The furnace door, at its open and closed states, activates the proximity switches PS_1 and PS_2 respectively, while the normal state is considered the closed one.

Opening door. The door opens by the momentary pressing of button b_0 through the forward/reverse motor (M) operation. When the sensor PS_1 is activated, the door is open and the motor M stops.

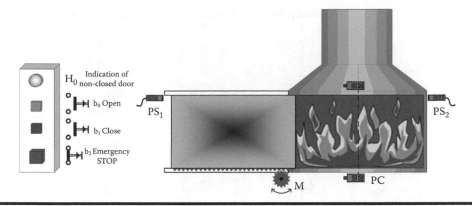

Figure 10.18   Chamber furnace for heat treatment.

Closing door. The door closes automatically after 30 s. The door can also close manually before the completion time of 30 s, by momentary pressing of button b_1. The indication light H_0 turns on either during the door movement or by being in the open state.

Emergency actions. If during the door closes, the photocell PC is activated, the door stops immediately, while the door continues to close if the photocell is deactivated. The movement of the door, either opening or closing, is interrupted permanently by the pressing of emergency button b_2.

After making a diagram of the I/O device connections to a PLC, write the required program.

10.19 Working Time Monitoring of a Machine under a Three-Shift Schedule

An industrial plant operates in a three-shift schedule. A complex machine of the production procedure operates intermittently due to the kind of process performed. Its operation also depends on the mood of the machine operator for its working time. Therefore, it is desired to know (e.g., weekly) the total hours of the machine's operation. For this reason, there are three control panels, one for each shift work, and three hour-meters that record the hours of the machine operation per shift work, as shown in Figure 10.19. Each operator can start up the machine only from the control panel corresponding to his shift work if:

a. The operator turns on the switch with a key
b. The corresponding START button is pressed

Then the machine operates and the hour meter records the time of operation. Any handling from the other two control panels must cause no action. The machine operation stops either by the operator by pressing the STOP button, or by a sensor detecting a critical temperature. After defining the I/O devices and their connections to a PLC, write the required program. Examine also, if it is possible to have the same operation with only one pair of START-STOP buttons common for the three shift works.

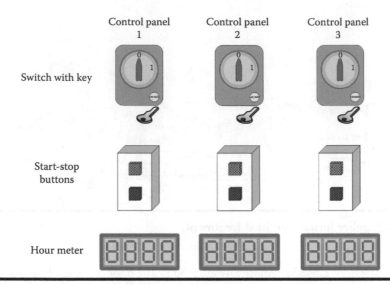

Figure 10.19   Working time monitoring of a machine.

10.20 Feeding an Assembly Machine with Components in Bulk

Figure 10.20 shows an arrangement for supplying an assembly machine with metallic components of Π schema that exist in bulk in a bucket. The rotating disk D has four curved blades, and each one can accept a single component which, due to blade rotation, is thrown on the pickup rail and simultaneously detected by the proximity switch PS. The component collection by any of the blades is a completely random event that may or may not happen one or more times. The assembly machine might need 4, 8, or 12 parts depending on the assembled device, which is selected by the operator by pressing the corresponding push-buttons b_1, b_2, and b_3.

The specifications of operation are the following:

1. The system is set ready for operation with the switch RS_{0-1}, otherwise nothing can work.
2. Subsequently, the operator must choose the desired number of parts, which is marked on the control panel.
3. With the condition that the number of parts has been selected, and by pressing momentary the START button, the disk rotation begins and continues to feed the selected number of components, and then stops.

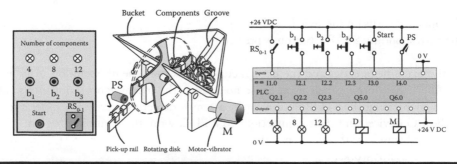

Figure 10.20   Feeder system of an assembly machine.

4. With a new START command, Step 3 is repeated, and so on.
5. If during the disk rotation it is spent without feeding a component for 2 min, then the vibrator motor M is automatically switched on for 10 s (without stopping the rotation of the disc), which shakes the bucket and the components, and thus facilitates their collection by a blade.
6. If during a feeding procedure (after START) a different number of components is selected, the system does not respond to this change. Only in the non-feeding phase (the disk does not rotate) can the desired number of components be changed.

If at any time the switch RS_{0-1} is opened, the system stops and all are reset.

Write the required PLC program in Boolean language.

10.21 A Roller Conveyor System for Wrapping Plastic Membrane

Figure 10.21 shows part of a roller conveyor for transferring a plastic membrane from a production machine to a wrapping machine. In this section of the conveyor, a floating folding is created deliberately, in order to have the potential difference between the rate of production and the rate of wrapping absorbed. The length of folding is measured with a laser-type photocell (PC) for distance measurement and should be kept between 40 cm and 50 cm limits through the regulation of the speed of the motor M_1, which performs the wrapping. The analog output 0–10 V of the distance meter is stored in the memory location MW200 of a PLC. With the START and STOP buttons, both motors M_1 and M_2 of the roller conveyor start and stop, respectively. The operating speed of the M_1 is regulated so that when:

Folding < 40 cm ⇨ Low speed
Folding > 50 cm ⇨ High speed
40 cm ≤ Folding ≤ 50 cm ⇨ Normal speed

In addition to the main programming task, tasks also include an additive digital output (not shown in the PLC) to be activated when the desired limits are violated, and write the overall PLC program.

10.22 Color-Based Separation of Plastic Balls

Figure 10.22 shows a mechanism that separates plastic balls depending on their color and packs them inside boxes in quantities of 6 or 12 pieces, depending on the position of the selector switch SS. The color sensor S_C detects the color of a ball (yellow or green) by measuring the light rays emitted by the motionless ball. The recognition time of a color from the S_C is 1 s. The activation of coil 1 happens after the S_C signal and allows the ball to drop. After the last pass from the proximity switch S_P, coil 1 is deactivated in order to stop the next ball in front of the sensor S_C. Coil 2 sets where each ball will fall according to its color, and the movable bearing has the position shown in the figure when coil 2 is deactivated.

When the chosen quantity is completed in every box, the corresponding indication is activated, and remains on until the next ball of the same color falls. The mechanism comes into operation (or pause) via the switch RS_{0-1}. The removal of the filled boxes is automatically handled by another machine and is not part of this application. Write the required PLC program in Boolean language.

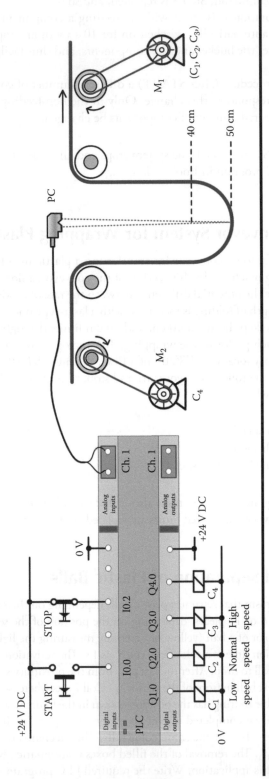

Figure 10.21   Roller conveyor for wrapping plastic membrane.

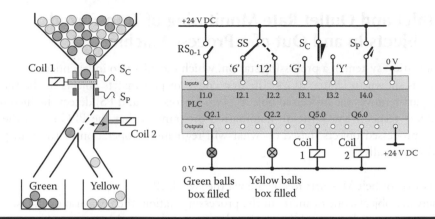

Figure 10.22   Color-based separation of plastic balls.

10.23 The Shearing Machine of an Unfolded Aluminum Sheet

A machine M unfolds and launches aluminum sheet in a laminate cutting table in order to be cut by a shearing machine SHM into pieces of a specified length. The machine M can drive the metal sheet at two speeds; at a low speed by the activation of the relay C_1, and at a high speed by the activation of C_2. The length of the metal sheet is measured via a pulse encoder ENC which is rotating during the sheet motion. The correspondence between pulses and sheet length is 1 pulse = 1 mm. It is desired for the machine to cut two possible lengths: 50 cm and 100 cm (Figure 10.23).

Operating specifications:

1. The length selection is done via the selector switch SS_{1-2}.
2. The pieces of small (large) length are cut at a low (high) speed, respectively.
3. When the desired length is achieved, the machine M stops and the shearing machine operates for 3 s. After 3 s passes, the cutting procedure continues by starting again the machine M.
4. With the START signal the operation of the whole machine begins, while with the STOP signal it stops. If during the operation of the machine at any cutting length, someone changes the position of the selector switch SS_{1-2}, there should be no change.

Write the required PLC program in any desired language.

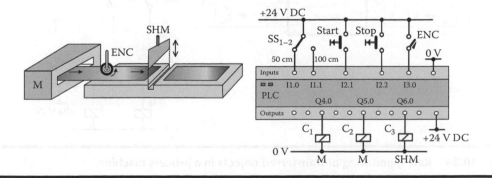

Figure 10.23   Shearing machine of an unfolded aluminum sheet.

10.24 Inlet and Outlet Rate Monitoring of Transferred Objects In and Out of a Process Machine

The transportation system of a processing station, which includes two independent conveyor belts M_1 and M_2, is shown in Figure 10.24. The first conveyor part inserts objects in the station and the second part removes the processed objects. Two sensors PS_1 and PS_2 detect the incoming and outgoing items respectively. Write an automation program so that the PLC stores the current counting of processed objects, based on what will regulate the operation of the transportation system and specifically:

1. The conveyor belt M_1 starts to operate via the switch RS_{0-1}.
2. If there are objects, one or more, in the processing station, then M_2 also operates.
3. If there are more than five objects inside the station, then the M_1 stops, and starts automatically again when objects become less than five.
4. If there are no objects in the station, then M_2 stops.
5. If the conveyor M_1 has stopped for a period exceeding 30 s, then the current counting is zeroed, considering that in the existing station the objects have been destroyed.
6. With the switch RS_{0-1} open, both conveyors stop.

10.25 A Metal Plate Rolling Mill Machine and Control of Their Thickness

Figure 10.25 shows an arrangement for rolling metal plates. The hot metal is pushed and squeezed by two driving rollers A and B through which a metal sheet with thickness d = 5 mm (under ideal conditions) is produced. The compression roller A may apply (simultaneously with the rotation) three different forces F_{-1A}, F_{0A}, and F_{+1A} via the activation of relays C_1, C_2, and C_3 respectively, and for which $F_{-1A} < F_{0A} < F_{+1A}$ is valid. The compression roller B may apply two different forces, F_{0B} and F_{+1B}, via the activation of the relays C_4 and C_5, respectively, and for which $F_{0B} < F_{+1B}$ is valid. The thickness of the metal sheet is measured with a suitable analog sensor, the measurement of

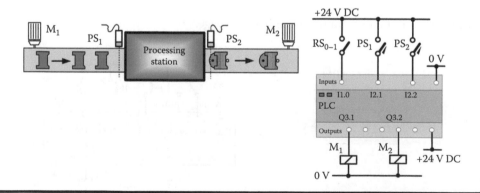

Figure 10.24 Rate monitoring of transferred objects in a process machine.

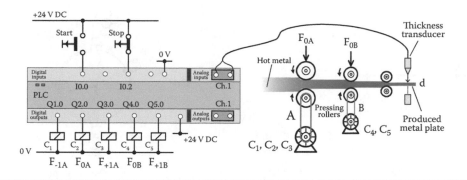

Figure 10.25 Metal plate rolling mill machine.

which is stored in the word memory MW100 of the PLC. Write the required PLC program in Boolean language that will adjust the thickness of the produced sheet (the conditions are never optimal) in accordance with the following specifications:

1. When the measured thickness is d = 5 mm, then the nominal forces F_{0A} and F_{0B} are applied.
2. When $5 < d \leq 5.1$ mm, the applied force is increased in roller B.
3. When d > 5.1 mm, the applied force is increased in both rollers.
4. When d < 5 mm, the applied force is decreased in roller A, while roller B applies the nominal force.
5. The operation starts by pressing the START button, and initially the nominal forces F_{0A} and F_{0B} are applied. By pressing the STOP button, both rollers cease to operate.

10.26 An Object Painting and Transporting System

Figure 10.26 shows a painting layout of objects that are transported on the conveyor belt M_1. For the color-dye injection system to function properly, the air pressure inside the tank should be between the limit values of 3 and 4 bar. The air pressure inside the tank is measured with the help of the pressure transducer P (the measured value is stored in MW 200–analog input 0–10 V) and is regulated by a simple ON-OFF control of compressor M_2. This means that:

■ If P ≥ 4 bar, then M_2 = OFF.
■ If P < 3 bar, then M_2 = ON and remains ON until P = 4 bar.

1. The whole layout is put into operation (or pause) via the switch RS_{0-1}.
2. The conveyor belt M_1 operates continuously (if RS_{0-1} = ON) until an object is moved underneath the painting nozzle and is detected by the photocell PC, then it stops.
3. On the condition that the pressure is 3–4 bar and the compressor M_2 is not operating, the valve V is energized for 3 s, which results in the spraying of color liquid on the object, i.e., painting the object.
4. After the elapse of 3 s, the valve V closes and the conveyor belt M_1 starts to operate again, until the next object arrives and the same process is repeated.

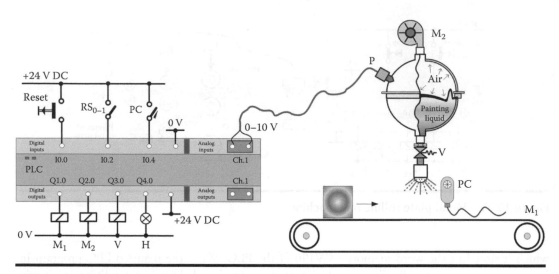

Figure 10.26 Object pointing and transporting system.

5. When 100 objects have been painted, the indicator H lights turn on and the conveyor belt stops. The pressure inside the tank continues to be controlled. By pressing the reset button, the conveyor belt starts again for the painting of the second 100 objects and so on.

The painting layout is controlled by a PLC, and for this operation, the overall automation should be programmed in a preferable language.

10.27 A Multiple Bottle Packing Station

A bottle packing station can accept bottles from three different bottling lines (M_1–M_3) that produce different products, as shown in Figure 10.27. The station packs bottles each time from one bottling line only, and they are forwarded toward it through the central conveyor belt M_0. Slat chain conveyors M_1–M_3 of bottling lines operate continuously and transfer bottles regardless of whether these are packaged or not. The movement of bottles that are not packaged is blocked from the corresponding movable barriers, whose open position is detected via the limit switches b_1–b_3 and is caused by the activation of the solenoids So_1–So_3. In order to avoid an error in the packaging procedure with products of different type on the same crate, the conveyor M_0 operates automatically (without a START-STOP command) if only one barrier is open. If for any reason, two or three barriers are found opened at the same time, the conveyor M_0 should immediately be stopped and the relative indicator H (meaning wrong packaging) should be activated. In such a case, the continuation of the packaging process is achieved by the intervention of an operator, which will remove the wrong bottles and manually will close the barrier that caused the problem.

When a crate has been fed with 12 bottles, then the conveyor belt M_4 must be enabled for 4 s in order for the filled crate to be removed, and a new empty crate to be placed from another machine, the operation of which does not interest us. During the time interval of 4 s, the conveyor belt M_0 remains stopped.

The packaging process starts with the closing of the switch RS_{0-1}. The opening of the latter results in an interruption of operation only if the current packaging order has been completed.

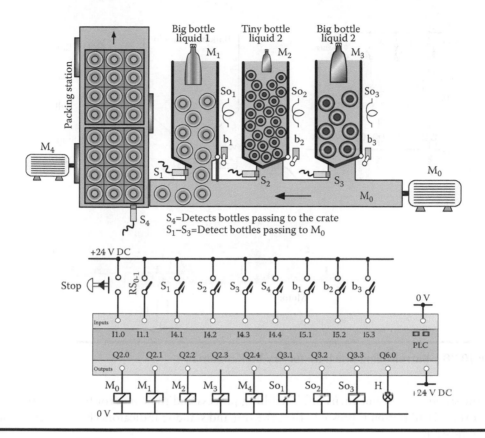

Figure 10.27   Multiple bottle packing station.

If any operating accident happens, then the pressing of the pushbutton emergency stop causes a general discontinuation, regardless of the point at which the packaging process is running.

The packaging process refers to the complete execution of a packaging order including a triple of crates from all the produced products (e.g., n_1, n_2, and n_3), depending on each customer's desire. This means that after closing the switch RS_{0-1}, the station will have to pack n_1 crates with products from M_1, n_2 crates with products from M_2, and n_3 crates with products from M_3.

A PLC program for this industrial application should be parametric to n_1, n_2, and n_3. As a first approach to the problem however, you are prompted to write the required program that implements a specific order such as (2, 1, and 3).

10.28 A Barrel-Filling System for Dry Bulk Material

Figure 10.28 shows a dry bulk material barrel-filling system. The filling process involves the transportation of barrels through the conveyor belt, the automatic positioning of the barrels, and their filling based on a level detection. The sequence of actions to fill a barrel is as follows:

1. Initially, the lights in the standby indicator (after putting PLC in run mode) turn on.
2. The filling process is activated by pressing the START button. Then the conveyor belt starts to operate (motor) by bringing empty barrels and the run indicator turns on, while

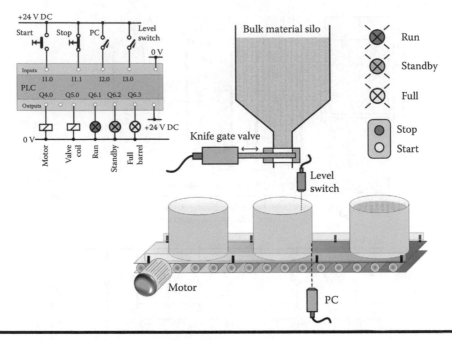

Figure 10.28   Barrel-filling machine for dry bulk material.

the standby indicator turns off. The filling process and the conveyor belt cease to operate if the STOP button is pressed. Then the run and standby indicators change their status of indication.

3. When a barrel reaches the proper position, which is detected by the photocell PC, the conveyor belt stops. With the barrel in the correct position, and the conveyor belt stopped, the knife gate valve is energized, and the dry bulk material falls into the barrel.

4. The knife gate valve will be disabled and the filling of the barrel will stop when the level switch is energized. Then, the indicator full turns on and will remain lit until the filled barrel is removed from the PC.

5. Once a barrel is filled (level sensor signal), the conveyor belt operates again in order to remove the filled barrel and to bring the next empty barrel, and so on. The next barrel arrival is a timely random event.

This automation problem can be further extended by adding the need for the kind of product and filling recipe selection in the PLC program.

10.29 An Electro-Pneumatic System for Pickup and Lay Down of Plastic Containers

Figure 10.29 shows a pickup and lay down mechanism of plastic cheese storage containers. The mechanism includes a special swivel bracket M_1 that initially is in a horizontal position, a pneumatic valve V for creating suction (vacuum), two proximity switches that detect the horizontal

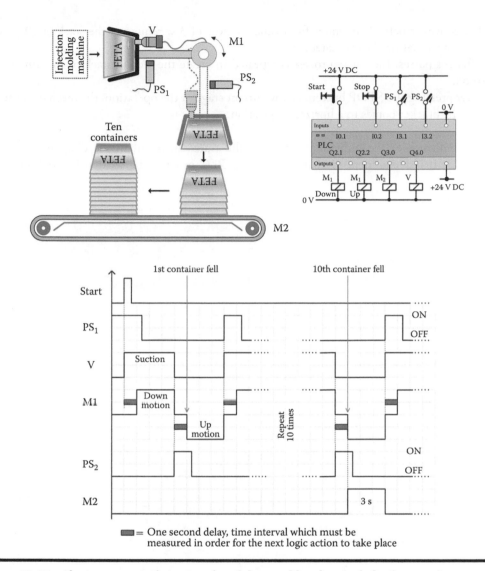

Figure 10.29 **Electro-pneumatic system for pickup and lay down of plastic containers.**

and vertical position of the bracket M_1, and the conveyor belt M_2 for deposition and removal of container stacks. The sequence of evolved operations is as follows:

■ With the bracket in horizontal position and by pressing the START button, a vacuum is created (V=On).

■ The plastic container is sucked from the injection molding machine. Following a delay of 1 s, the bracket (M_1) moves downward, and once detected by PS_2, the vacuum stops (V=Off).

■ Following a delay of 1 s, the bracket moves upwards.

■ Once detected by the PS_1 the vacuum is activated again (V=On), and the same process is repeated 10 times.

- The conveyor belt M_2 operates for a time window of 3 s, removing the stack of 10 plastic containers that has been created.
- After 3 s passes, the entire process is repeated to create the second stack of containers and so on.
- After pressing the STOP button, at any current stage of the operation, the mechanism stops and the bracket comes to a horizontal position.

The exact sequence of the individual actions is illustrated in detail in the diagram shown in Figure 10.29. It is recommended that the required PLC program be written in Boolean or IL language.

Appendix A: Arithmetic Systems

A.1 Introduction to Arithmetic Systems

Although knowledge related to arithmetic systems is the subject of books with mathematical or computer content, some basic issues of arithmetic systems are included in this book because it is absolutely necessary for an understanding of the operation and programming of PLCs. On the other hand, the reader's knowledge of arithmetic systems is usually satisfactory from basic computer courses, and it is much more necessary in using PLCs than personal computers. Even the most numerate user of a PC does not often come into contact with the concept of a "bit" since there is no such need. In contrast, in the world of PLCs, the programmer of an automation system starts with the utilization of bits, continued by integers, real numbers, or hexadecimal numbers, and ends again with the utilization of bits. The basic reason for this is the fact that the majority of devices that a PLC controls are of a binary status (binary digits or bits, such as 0 or 1, ON or OFF). Moreover, there are indication devices operating in a BCD code, where they are used, among others, in the octal system for the input/output address and corresponding program instructions for bit to bit operations, etc. Therefore, someone who knows the arithmetic systems and codes better will have an easier time programming and making use of even the most specialized instructions.

An arithmetic system is a code of symbols which are assigned a quantity. From the moment the code has been defined and memorized, it can be used to measure any quantity. The decimal arithmetic system uses ten basic (primordial) symbols or digits, where each digit represents a certain quantity. When several digits are grouped together, then larger quantities can be measured. The base of an arithmetic system is called the number of the primordial symbols or digits of the system. All arithmetic systems use position weighting to express the significance of each digit in a group.

A.2 Decimal Arithmetic System

In a decimal arithmetic system, ten unique digits or numbers are used, from 0 to 9, and hence the base is 10. The value of a decimal number depends on the amount of digits and the position weight of each digit. The first position to the immediate left of the decimal point corresponds to units, which follows that of decades, hundreds, etc., as shown in Figure A.1. Each position value can be expressed as a power of 10, starting from 10^0 to the left and 10^{-1} to the right of the decimal point.

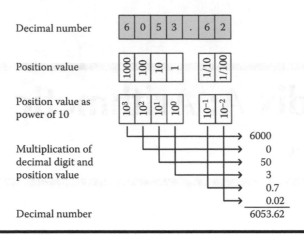

Figure A.1 Decimal number system.

A.3 Binary Arithmetic System

The binary arithmetic system uses the two digits 0 and 1, so its base is 2. Any number can be expressed as a combination of digits 0 and 1. In order for an electronic circuit to process numbers, which is called a "digital circuit", it is necessary to join the numbers with corresponding electric voltage signals. Therefore, if a digital device, such as the PLC, were to process decimal numbers, then ten different voltages would be required, one voltage level for each digit. Consequently, each decimal number will be expressed by a combination of these ten voltage levels. The implementation of a digital circuit for processing decimal numbers, due to many electric elements needed for creating the ten different voltages, will lead to a very complex circuitry. For this reason, to avoid complex electronic circuits, the binary arithmetic system is the most suitable for developing digital circuits for processing numbers, since only two voltage levels are required. Assuming that 5 V and 0 V represent the digits 1 and 0 correspondingly, a four-digit binary number (such as 1011) will be represented electronically by the combination of voltages 5 V, 0 V, 5 V, and 5 V, which can be recognized by a processor. It is widely known that a combination of eight binary digits (bits) is called a "byte" and a combination of 16 binary digits is called a "word".

The position weighting that is assigned to each digit of the binary number, due to the base 2, will be doubled for each step to the left of the number, with the first right position to correspond to the value 1 (2^0). As in the case of the decimal arithmetic system, all the position values of the binary arithmetic system can be expressed as powers of two, as shown in Figure A.2. The digit of

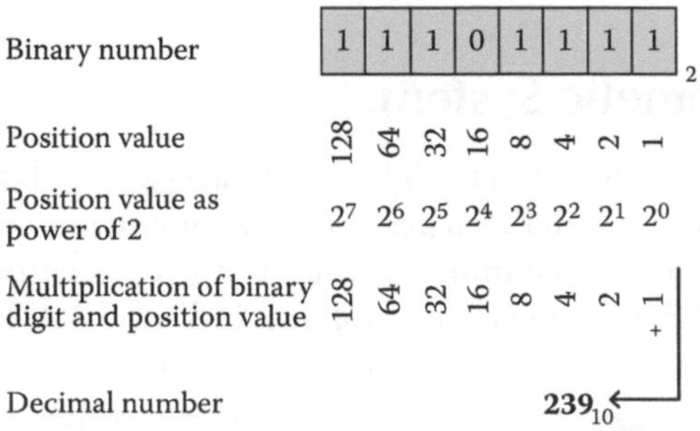

Binary number	1	1	1	0	1	1	1	1	$_2$
Position value	128	64	32	16	8	4	2	1	
Position value as power of 2	2^7	2^6	2^5	2^4	2^3	2^2	2^1	2^0	
Multiplication of binary digit and position value	128	64	32	16	8	4	2	1	+
Decimal number					239_{10}				

Figure A.2 Binary system of numbers and their conversion to the decimal system.

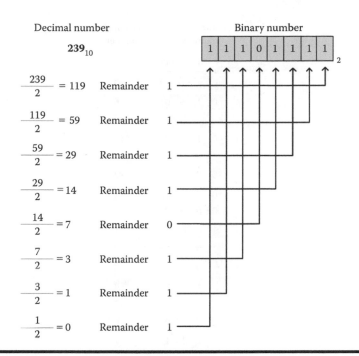

Figure A.3   Conversion of a decimal number to binary.

a binary number with the lowest position value is called the "least significant bit" or LSB, while the digit with the highest position value is called the "most significant bit" or MSB. In order to convert a decimal number to a binary one, the procedure illustrated in Figure A.3 is followed. This means dividing successively the decimal number by 2 and denoting the remainders 0 or 1 as digits of the binary number.

A.3.1 Negative Numbers

The CPU of various PLCs process the analog values in a binary form only. Analog input modules convert the analog process signal into digital form, and analog output modules convert the digital output value into an analog signal. The representation of a digitized analog value depends on the number of used bits (resolution), which has to include both positive and negative numbers. In the decimal arithmetic system, negative or positive numbers are denoted with a minus or plus sign in front of them, which means that each number has two parts: the sign that is a symbol and the magnitude. Therefore, this method of digitized value representation is not applicable in a digital device like the PLC.

The simplest method to represent a negative number as a binary number is to use an extra digit, called a "sign bit", which usually is the fifteenth bit (MSB) for a 16-bit resolution. If the sign bit is 0 (1), this means that the number is positive (negative), such as the signed binary numbers shown in Figure A.4.

A second method for the representation of negative numbers in a digital device is to use the complement of a binary number, and particularly the one's complement or two's complement. When a binary number, has the value 1 the complement is 0 and vice versa, that is called the one's complement. For example, the one's complement of the binary number 0110 is 1001.

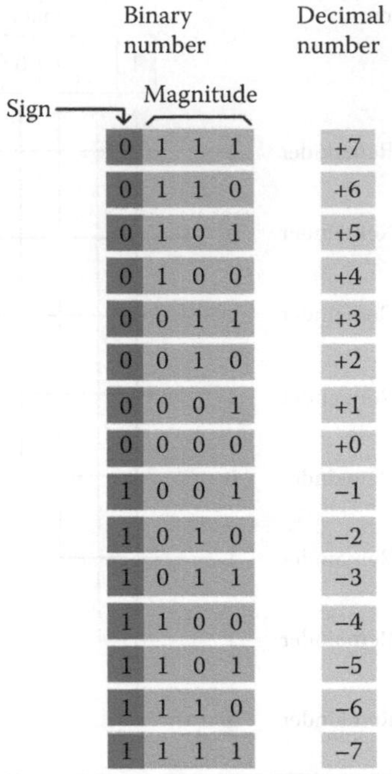

	Binary number	Decimal number
Sign → Magnitude	0 1 1 1	+7
	0 1 1 0	+6
	0 1 0 1	+5
	0 1 0 0	+4
	0 0 1 1	+3
	0 0 1 0	+2
	0 0 0 1	+1
	0 0 0 0	+0
	1 0 0 1	−1
	1 0 1 0	−2
	1 0 1 1	−3
	1 1 0 0	−4
	1 1 0 1	−5
	1 1 1 0	−6
	1 1 1 1	−7

Figure A.4   Signed binary numbers.

The magnitude of a negative number represented with regard to one's complement is the one's complement of the corresponding positive number magnitude. Some examples of negative numbers represented in one's complement, are shown in Figure A.5.

The two's complement of a binary number is the sum of one's complement plus 1, as expressed by Equation A.1,

$$2\text{'s complement} = 1\text{'s complement} + 1 \tag{A.1}$$

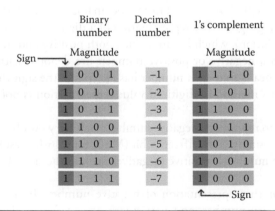

Figure A.5   Representation of negative numbers using the complement of a binary number (one's complement).

PLCs use the two's complement to perform subtraction of two binary numbers, which means that the subtraction is replaced by an additive operation, as clarified by Equation A.2,

$$(7)_{10} - (5)_{10} = (7)_{10} + (-5)_{10} = (0111)_2 + (1011)_2 = (0010)_2 = (2)_{10} \qquad \text{(A.2)}$$

2's complement

The subtraction operation using the two's complement facilitates its implementation in a PLC, since the same digital circuit will perform both addition and subtraction.

A.4 Binary-Coded Decimal Arithmetic System (BCD)

In a binary-coded decimal arithmetic system, each digit of a decimal number is coded via a four-digit binary number. This definition means that each digit of a decimal number is simply substituted by the corresponding binary number, as in the following example,

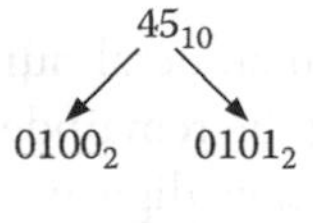

$$45_{10}$$
$$0100_2 \qquad 0101_2$$

In the past, computer systems had specific units for the execution of the arithmetic operations between BCD numbers in order to overcome memory storage problems. In some PLCs today, the BCD arithmetic system is even supported since it is compatible with the operation of some output devices as an indication display device, or input devices as an arithmetic switch. Figure A.6 shows an example of converting a BCD number to an equivalent decimal number.

A.5 Octal Arithmetic System

The octal arithmetic system is used by some manufacturers of PLCs, such as Allen-Bradley and Siemens, for addressing their digital inputs and outputs. The octal system uses the numbers 0 through 7 and hence has a base 8. The position weighting values of the octal system are derived in a similar way as that applied to the binary and decimal systems. Starting from the first right position with value the null power of 8 (8^0), the value at each next x position increases by the x

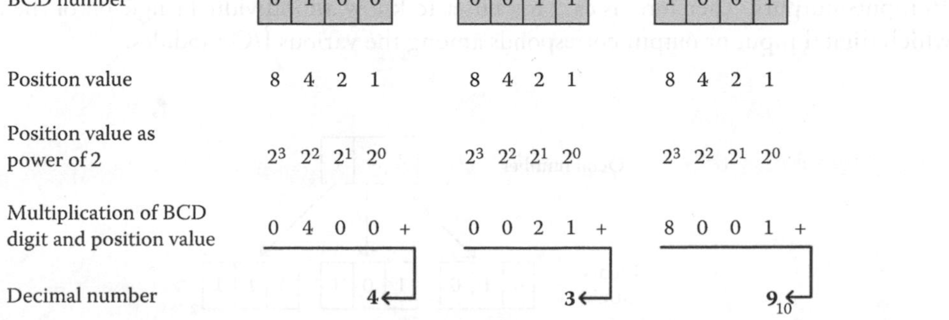

BCD number	0 1 0 0	0 0 1 1	1 0 0 1
Position value	8 4 2 1	8 4 2 1	8 4 2 1
Position value as power of 2	2^3 2^2 2^1 2^0	2^3 2^2 2^1 2^0	2^3 2^2 2^1 2^0
Multiplication of BCD digit and position value	0 4 0 0 +	0 0 2 1 +	8 0 0 1 +
Decimal number	4 ←	3 ←	9_{10} ←

Figure A.6   Conversion of a BCD number to a decimal.

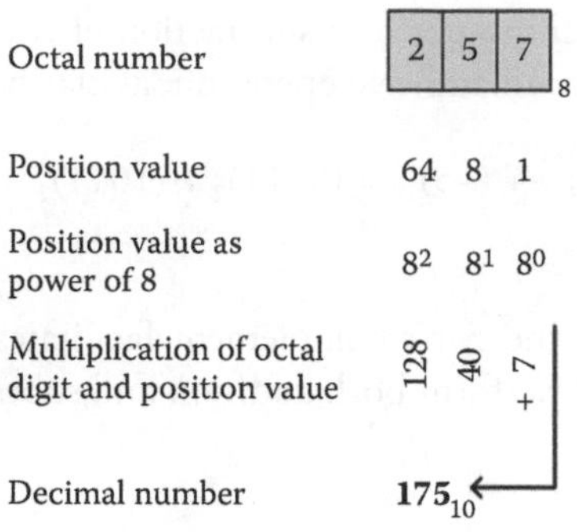

Octal number	$\boxed{2\quad 5\quad 7}_8$
Position value	64 8 1
Position value as power of 8	$8^2 \quad 8^1 \quad 8^0$
Multiplication of octal digit and position value	128 40 7 +
Decimal number	175_{10}

Figure A.7 Conversion of an octal number to a decimal.

power of 8. Figure A.7 shows an example of converting an octal number to an equivalent decimal number and the three first-position values.

The conversion of a decimal number to an octal number is performed by successive divisions of the decimal number by 8 and denoting the remainders as digits of the octal number. The first remainder corresponds to the least-significant digit of the octal number. An example of such a conversion procedure is the following:

215/8	= 26 with remainder 7
26/8	= 3 with remainder 2
3/8	= 0 with remainder 3, therefore the octal number is 327_8.

Since the largest one-digit number of the octal system is 7, three binary bits are enough to represent each of the eight digits of the octal system. The conversion of an octal number to an equivalent binary number is performed by replacing each octal digit by its 3-bit binary equivalent, as shown in Figure A.8.

The octal arithmetic system is the most user-friendly, since it is harmonized to the 8-bit general digital structure, independently of the 8-bit or 16-bit or 64-bit structure of memories and microprocessors. The manufacturers of large industrial PLCs use the octal arithmetic system to address words and bit locations in their memory as also digital I/O modules usually in groups of 4, 8, and 16 inputs/outputs. Therefore, is easy for a user to know an individual single bit of the memory to which digital input or output corresponds among the various I/O modules.

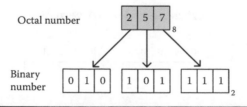

Figure A.8 Conversion of an octal number to binary.

A.6 Hexadecimal Arithmetic System

Another arithmetic system that is widely used in PLCs and PCs is the hexadecimal arithmetic system, also often referred to as HEX. This system has a base 16, which means that it uses 16-digit symbols that take decimal values from 0 to 15. But because it will create confusion if a hexadecimal digit consisted of two numbers, the digits from 10 to 15 are represented by the letters of the alphabet A to F, as shown in Table A.1. For example, the hypothetical number 15_{16} is not clear what expresses, because may be the hexadecimal 15 (equivalent decimal value 15) or two discrete digits, the hexadecimal 1 and the hexadecimal 5, which have an equivalent decimal value of 21. Although the use of letters in the hexadecimal system seems complicated at first sight, the hexadecimal code is the most appropriate for the digital representation of numbers. The reason is that it is possible with two hexadecimal digits to represent all the decimal numbers from 0 to 255 (0 to FF), while in a binary arithmetic system 8 binary bits or 1 byte are required.

The position weighting values of the hexadecimal system are derived in a similar way to that applied to binary and decimal systems. Starting from the first right position with the null power value of 16 (16^0), the value at each position of a hexadecimal digit rises to the left of the LSB by a power of 16. Converting a decimal number to a hexadecimal one becomes with subsequent

Table A.1   The Hexadecimal Digits and Their Equivalents in Both Binary and Decimal Systems

Hexadecimal Number	Binary Number				Decimal Number
0	0	0	0	0	0
1	0	0	0	1	1
2	0	0	1	0	2
3	0	0	1	1	3
4	0	1	0	0	4
5	0	1	0	1	5
6	0	1	1	0	6
7	0	1	1	1	7
8	1	0	0	0	8
9	1	0	0	1	9
A	1	0	1	0	10
B	1	0	1	1	11
C	1	1	0	0	12
D	1	1	0	1	13
E	1	1	1	0	14
F	1	1	1	1	15

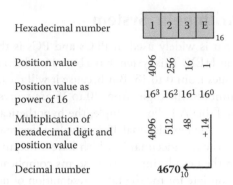

Figure A.9   Conversion of a hexadecimal number to a decimal.

divisions of the decimal number by 16, and by denoting the remainders of divisions as the digits of the hexadecimal number, as in the following example for decimal 4670:

4670/16	= 291 with remainder 14 (=E)
291/16	= 18 with remainder 3
18/16	= 1 with remainder 2
1/16	= 0 with remainder 1, therefore hexadecimal number is $123E_{16}$

The conversion of a hexadecimal number to a decimal number is performed by a procedure similar to this, which was applied in previous arithmetic systems, such as the one shown in Figure A.9. Finally, the conversion of a hexadecimal number to a binary number is done with a simple replacement of each hexadecimal digit with its equivalent binary form of 4-bits, according to Table A.1.

A.6.1 Parity Checking

Arithmetic digital data, independently of the utilized arithmetic system and the way that they are represented, do not stay static at some memory location of a digital device, but instead they are transported from one point to another. Digital data are transferred continuously, especially in PLCs, mainly due to the real-time control that PLCs perform (such as from a PLC to a peripheral device, from a PLC to another PLC via a communication network, etc.). It is very likely for a single bit of binary data to change value from 1 to 0, due to electromagnetic noise or a transitional phenomenon, or any other imponderable reason. For the proper detection of a possible error during the transmission or storage of binary data, the parity checking technique is applied which consists of adding an extra bit, called a parity bit, in the transferred data word.

Even Parity. The parity bit is added to the transferred data is such way that the total number of units (1) are even. For example, if the byte 01010100 is transmitted, then the parity bit 1 has to be added and becomes 010101001.

Odd Parity. The parity bit is added to the transferred data in such a way that the total number of units (1) are odd. For example, if the byte 01010100 is transmitted, then the parity bit 0 has to be added and becomes 010101000.

When receiving or reading the transmitted data word, the total number of units (1) are checked if they are even or odd, and thus it is determined whether there was an error in data transmission. When establishing a communication connection between two PLCs, the user is usually invited to select an even or odd parity in the corresponding software environment.

A.7 ASCII Code

A digital device, such as a computer or a PLC, should have the ability to process anything, except for numbers and letters, since this functionality is needed for enabling the programming of the PLC with text commands (e.g., an instruction AND). As in the case of numbers being represented based on an arithmetic system in a binary form that can be processed by a PLC, letters should also be represented in binary format. For this purpose, the coding of the letters of the alphabet and other symbols used to write text has been provided by the introduction of the ASCII code (American Standard Code for Information Language). Based on the ASCII code, 7 bits were initially used to represent the various letters and symbols. Since the number of letters and symbols surpassed the maximum 128 (2^7), which could cover the 7 bits, an eighth bit was added. A table with ASCII codes is not listed here, because it is a well-known issue and they appear in many books for someone who would like more details.

A.8 Gray Code

Gray code is a binary encoding method that does not use the position weighting of a digit like the other arithmetic systems do. Gray code defines only the transition from one number to the next where only one bit changes its status, and for this reason is not applicable in any types of arithmetical operations. However, it does have some applications in analog to digital converters and in some input/output devices as the encoders. In binary coding, two or more contiguous bits change their status to express a decimal number incremented by one; for instance, when going from 7 to 8 (0111 to 1000) there are four bits changing their state. In Gray coding, only one bit changes its status to express the same increment. For this reason, Gray code is ideal for use in PLCs and computers. In principle, it is the code that shows the minimum possible error because when only one bit changes from one state to another, the probability of error is drastically reduced. For the same reason, the transmission speed of Gray code is comparatively higher than others such as the BCD code. In Table A.2, Gray 4-bit codes are shown in relation to the equivalent binary codes for comparison purpose.

In the industrial world, automation technology, robotics, and especially in PLCs, Gray code is encountered often because the position encoders that base their operation on it may be input devices of a digital controller or PLC simultaneously. The position encoders (rotary and linear, absolute and incremental) have all been presented in Section 2.3.7. In general, position encoders connected to a PLC apply a pulse in an input module, which follows Gray code, i.e., only one bit changes at each step of a shaft rotation (rotary encoders) or linear movement of a machine carriage (linear encoders).

This appendix will end with a brief reference to the digital representation form of the numbers in PLCs. Generally in computers, the numbers are represented either as fixed-point or

Table A.2 Gray Code and Binary Equivalents

Gray Code				Binary Number			
0	0	0	0	0	0	0	0
0	0	0	1	0	0	0	1
0	0	1	1	0	0	1	0
0	0	1	0	0	0	1	1
0	1	1	0	0	1	0	0
0	1	1	1	0	1	0	1
0	1	0	1	0	1	1	0
0	1	0	0	0	1	1	1
1	1	0	0	1	0	0	0
1	1	0	1	1	0	0	1
1	1	1	1	1	0	1	0
1	1	1	0	1	0	1	1
1	0	1	0	1	1	0	0
1	0	1	1	1	1	0	1
1	0	0	1	1	1	1	0
1	0	0	0	1	1	1	1

floating-point numbers. In PLCs, both arithmetic possibilities are offered to a user. Specifically, most medium or large PLCs support the processing of:

- Single-precision integers (16-bit numbers with a range of values from -32768 to 32767)
- Double-precision integers (32-bit numbers with a range of values from -2147438648 to 2147438647)
- Floating point real numbers of single precision (32-bit numbers with a range of values from $-3,402823E+38$ to $3,402824E+38$)

It should be noted that the alteration of an arithmetic operation from another one (e.g., for single-precision integers from the corresponding double-precision integers) is performed using different programing instructions, such as the instructions +I and +D for this example, respectively. In Chapter 7, instructions have not been included for all kinds of numeric representation for the same reasons that have been explained regarding the advisability of the instructions included in Table 2.1.

Appendix B: Analog I/O Values Scaling

B.1 PLC Analog I/O Values Scaling

During the programming task of an automation application, including the processing of analog values, it is necessary to address some secondary issues, such as the violation of limits of an analog input, the scaling-conversion of analog I/O range into units of physical magnitude (also called engineering units), the overflow or underflow of digital arithmetic limits (>32768 or <–32768), etc. The correct or incorrect operation of the PLC program depends on proper knowledge of these issues. In the operation manuals with the technical characteristics of the PLC's analog I/O modules, one can find the required information concerning the digital representation of analog values for all the standard analog ranges that an analog module can accept, the measurement or overload limits of the A/D or D/A converter, and also the manner of address. Here, reference will be provided only to the analog value scaling that is illustrated in Figure B.1, and expresses the need for a human operator to use engineering units.

Let's assume that a temperature transducer has an analog output 0–10 V DC that is connected to an analog input of a PLC. The output voltage of the transducer is proportional to the temperature range 100 °C to 500 °C. The transducer measures the temperature of a thermal process, which should fluctuate from 250 °C to 300 °C by applying the required control. Figure B.2 illustrates the whole scaling procedure, particularly the linear relation between the input (voltage) and the output (arithmetic value that expresses °C). In the diagram, the maximum voltage 10 V corresponds to the arithmetic value 32767, but in some PLCs, this may be 27648 or another arithmetic limit, depending on the existence or absence of an overflow or over-range detection. The input/output linear relation is expressed by the following equations:

$$\text{Scaled Output Value} = (\text{Input Voltage} \times \text{Slope}) + \text{Offset} \tag{B.1}$$

$$\text{Slope} = \frac{\text{Max Scaled Value} - \text{Min Scaled Value}}{\text{Max Input Value} - \text{Min Scaled Value}} \tag{B.2}$$

$$\text{Offset} = \text{Min Scaled Value} - (\text{Min Input} \times \text{Slope}) \tag{B.3}$$

What is needed in the sequence, is to determine which voltage values correspond to the temperature limits (250 °C and 300 °C) that are controlled in order not to be violated. The calculation is based on Equation B.1, which is solved for "Input Voltage",

$$\text{Input Voltage} = (\text{Scaled Output Value} - \text{Offset})/\text{Slope} \tag{B.4}$$

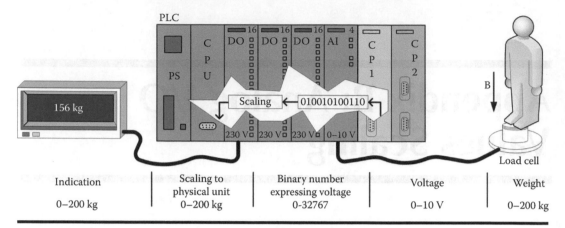

Figure B.1 **The scaling of an analog value is a necessary procedure to convert numeric data into meaningful units of the corresponding physical variable.**

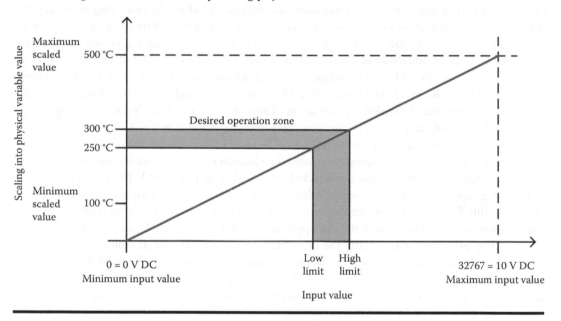

Figure B.2 **Diagram for derivation of the scaling formula.**

Applying Equation B.4 for the two temperature limit values, it is derived that,

$$\text{Low Limit} = (250 - 100)/(400/32767) = 12287$$

and

$$\text{High Limit} = (300 - 100)/(400/32767) = 16393$$

In some PLCs, system function blocks perform the required scaling operation. Otherwise, based on the above scaling equations, one has to program either the control of the temperature or any other simpler application as the indication of temperature in a digital panel, and then incorporate the required instructions to the whole PLC program of the automation.

Further Reading

Chapter 1

Ebel, F., S. Idler, G. Prede, and D. Scholz. 2008. *Fundamentals of Automation Technology*. Technical book, Festo Didactic GmbH & Co.

Levine, W.S. 1999. *Control System Fundamentals*. CRC Press.

Sysala, T., P. Dostál, and M. Adámek. 2006. Monitoring, Measuring, and Control Systems for Real Equipment Controlled by a PLC in Education. *XVIII IMEKO World Congress, Metrology for a Sustainable Development*, Rio de Janeiro, Brazil.

Chapter 2

Bishop, R.H. 2002. *The Mechatronics Handbook*. CRC Press

Considine, D. 1993. *Process/Industrial Instruments and Controls Handbook*. McGraw-Hill.

Herman, S. and W. Alerich. 1993. *Industrial Motor Control*. Delmar Publishers.

Kenjo, T. and A. Sugavara. 1994. *Stepping Motors and Their Microprocessor Controls*. Clarendon Press.

Kuo, B. 1979. *Incremental Motion Control: Step Motors and Control Systems*. SRL Publishing Company.

Sysala, T., P. Dostál, and M. Adámek. 2006. Monitoring, Measuring, and Control Systems for Real Equipment Controlled by a PLC in Education. *XVIII IMEKO World Congress, Metrology for a Sustainable Development*, Rio de Janeiro, Brazil.

Chapter 3

(No author). 2006. *Electrical Installation Handbook: Protection and Control Devices*. Vol.1. ABB SACE S.p.A.

(No author). 2006. *Electrical Installation Handbook: Electrical Devices*. Vol.2. ABB SACE S.p.A.

(No author). 2010. *Wiring Diagram Book*. Square D./Group Schneider.

Heumann, W., T. Kracht, B. Petrick, H. Riege, and R. Wiegand. 2011. *Command, Signaling, Automation, Motor Applications, and Power Management: Wiring Manual*. Eaton Industries GmbH. (http://www.moeller.net/binary/schabu/wiring_man_ en.pdf).

Riege, H. 2006. *Automation and Power Distribution: Wiring Manual*. Moeller GmbH.

Schmelcher, T. 1984. *Low Voltage Handbook: Technical Reference for Switchgear, Controlgear and Distribution systems*. Siemens-Aktiengesellschaft.

Chapter 4

Gupta, A.K. and S.K. Arora. 2013. *Industrial Automation and Robotics*. University Science Press.
Lin, W.C. 2000. *Handbook of Digital System Design*. CRC Press.
Pessen, D. 1989. *Industrial Automation: Circuit Design and Components*. John Wiley & Sons.
Sarkar, S.K., A. Kumar De, and S. Sarkar. 2014. *Foundation of Digital Electronics and Logic Design*. CRC Press.
Stanczyk, U., K. Cyran, and B. Pochopien. 2007a. *Theory of Logic Circuit. Vol. 1. Fundamental Issues.* Publishers of the Silesian University of Technology, Gliwice.
Stanczyk, U., K. Cyran, and B. Pochopien. 2007b. *Theory of Logic Circuit. Vol. 2. Circuit design and analysis.* Publishers of the Silesian University of Technology, Gliwice.
Vingron, S.P. 2004. *Switching Theory: Insight through Predicate Logic*. Springer Berlin.
Yanushkevich, S.N. and V.P. Shmerko. 2008. *Introduction to Logic Design*. CRC Press.

Chapter 5

Ayadi, A., S. Hajji, and M. Smaoui. 2014. Modeling and Sliding Mode Control of an Electro-pneumatic System. *22nd Mediterranean Conference of Control and Automation (MED)*.
Ebel, F., S. Idler, G. Prede, and D. Scholz. 2008. *Fundamentals of Automation Technology*. Technical book, Festo Didactic GmbH & Co.
Krivts, I.L. and G.V. Krejnin. 2006. *Pneumatic Actuating Systems for Automatic Equipment: Structure and Design*. CRC Press.
Prede, G. and D. Scholz. 2002. *Electro-pneumatics*. Festo Didactic GmbH & Co.

Chapter 6

Babb, M. 1995. MicroPLCs: A Most Favored Species. *Control Engineering*.
Batten, G. 1988. *Programmable Controllers: Hardware, Software, and Applications*. TAB Professional and Reference Books.
Bishop, R.H. 2002. *The Mechatronics Handbook*. CRC Press.
Cheded, L. and A. Fan. 1998. Controlling a Large Process with a Small PLC: A Senior Project Experience. *Int. J. Elect. Engin. Educ.*
Collins, D. and E. Lane. 1995. *Programmable Controllers: A Practical Guide*. McGraw-Hill.
Cox, R. 2007. *Guide to Programmable Logic Controllers*. Bookmark Inc.
Den Otter, J. 1988. *Programmable Logic Controllers: Operation, Interfacing, and Programming*. Prentice Hall.
Kissel, T. 1986. *Understanding and Using Programmable Controllers*. Prentice Hall.
McKeag, D., J.J. Blakley, and N.J. Hanson. 2009. University Industry Collaboration in the Development of PLC Training Material for Use in the Design and Development of Quarry Plant. *International Conference on Engineering and Product Design Education*. University of Brighton, UK.
Merte, P. and G. Ruhl. 1992. *The Electromagnetic Compatibility (EMC) of Automation Systems*. Klockner-Moeller.
Parr, E.A. 2003. *Programmable Controllers: An Engineer's Guide*. Newnes Publishing.
Pratumsuwan, P. and W. Pongaen. 2011. An Embedded PLC Development for Teaching in Mechatronics Education. *6th IEEE Conference on Industrial Electronics and Applications (ICIEA)*.
Sanders, S.A.C. 1988. The Role of Programmable Logic Controllers in Electrical Engineering Education. In *IEEE Conf. on Electrical Engineering Education,* University of Hull, (IEE New York).
Simpson, C. 1994. *Programmable Logic Controllers*. Prentice Hall.
Uzam, M. 2013. *Building a Programmable Logic Controller with a PIC16F648A Microcontroller*. CRC Press.
Wilhelm, R.E. 1985. *Programmable Controller Handbook*. Hayden Book Co.

Chapter 7

(No author). 2006. *Programming with STEP 7.* Manual, Edition 03, Siemens AG.

Baresi, L., S. Carmeli, A. Monti, and M. Pezze. 1998. PLC Programming Languages: A Formal Approach. *National Conference, Automation 98, ANIPLA.*

Batten, G. 1988. *Programmable Controllers: Hardware, Software, and Applications.* TAB Professional and Reference Books.

Crispin, A. 1997. *Programmable Logic Controllers and their Engineering Applications.* McGraw-Hill.

Jully, U. 2002. Programming Safety-Related PES with Standard IEC 61131-3, Application Burner Control. *PLC Symposium, Cologne.*

Lewis, R. 1998. *Programming Industrial Control Systems Using IEC 1131-3.* Institution of Electrical Engineers.

Plaza, I., C. Medrano, and A. Blesa. 2006. Analysis and Implementation of the IEC 61131-3 Software Model under POSIX Real-Time Operating Systems. *Microprocessors and Microsystems.*

Rinaldi, J. 2007. *The Fast Guide to Industrial Automation IEC 61131-3 Open Control Software.* Real Time Automation.

Tisserant, E., L. Bessard, and M. De Sousa. 2007. An Open Source IEC 61131-3 Integrated Development Environment. *5th IEEE International Conference.*

Webb, J.W. and R.A. Reis. 2003. *Programmable Logic Controllers, Principles and Applications.* Prentice Hall.

Wright, J. 1999. The Debate Over Which PLC Programming Language is State of the Art. *Journal of Industrial Technology.*

Chapter 8

Galloway, B. and G.P. Hancke. 2013. Introduction to Industrial Control Networks. *IEEE Communications Surveys & Tutorials* 15:860–80.

Gupta, R.A. and M.Y. Chow. 2010. Network Control System: Overview and Research Trends. *IEEE Transactions on Industrial Electronics* 57:2527–535.

Lin, Z. and S. Pearson. 2017. *An Inside Look at Industrial Ethernet Communication Protocols.* Texas Instruments Inc.

Liptak, B. 2002. *Process Software and Digital Networks, Instrument Engineers' Handbook.* CRC Press.

PROFIBUS International. 2010. PROFIBUS system description. (http://www.profibus.com/nc/downloads /downloads/profibus-technology-and-application-system-description/display.html.)

Sauter, T. 2010. Three Generations of Field-Level Networks—Evolution and Compatibility Issues. *IEEE Transactions on Industrial Electronics* 57:3585–95.

Thomesse, J.P. 2005. Fieldbus Technology in Industrial Automation. *Proceedings of the IEEE* 93:1073–101.

White Paper. 2016. *Industrial Communication Protocols.* Honeywell International Inc.

Wilamowski, B.M. and J.D. Irwin. 2011. *Industrial Communication Systems. The Industrial Electronics Handbook.* CRC Press.

Zurawski, R. 2005. *The Industrial Communication Technology Handbook.* Taylor & Francis.

Chapter 9

Åström, K.J. and T. Hägglund. 1995. *PID Controllers: Theory, Design, and Tuning.* Instrument Society of America, Research Triangle Park, North Carolina.

Åström, K.J. and T. Hägglund. 2005. *Advanced PID Control.* ISA—The International Society of Automation, Research Triangle Park, North Carolina.

Berner, J., T. Hägglund, and K.J. Åström. 2016. Asymmetric Relay Autotuning—Practical Features for Industrial Use. *Control Engineering Practice.*

Chaudhuri, U.R., and U.R. Chaudhuri. 2012. *Fundamentals of Automatic Process Control*. CRC Press.

Vandoren, V.J. 1996. Examining the Fundamentals of PID Control. *Control Engineering*.

Chapter 10

McKeag, D., J.J. Blakley, and N.J. Hanson. 2009. University Industry Collaboration in the Development of PLC Training Material for Use in the Design and Development of Quarry Plant. *International Conference on Engineering and Product Design Education*. University of Brighton, UK.

Index

Page numbers followed by f and t indicate figures and tables, respectively.